Ecological Studies

Analysis and Synthesis

Edited by

W.D. Billings, Durham (USA) F. Golley, Athens (USA)
O.L. Lange, Würzburg (FRG) J.S. Olson, Oak Ridge (USA)
H. Remmert, Marburg (FRG)

Volume 41

Forest Island Dynamics in Man-Dominated Landscapes

Edited by
Robert L. Burgess and David M. Sharpe

Contributors
M.C. Bruner   R.L. Burgess   D. Bystrak   D.L. DeAngelis
A.R. Ek   D.E. Fields   L.M. Hoehne   W.C. Johnson
M.K. Klimkiewicz   J.B. Levenson   T.E. Lovejoy   J.F. Lynch
P.E. Matthiae   R.M. May   R.J. Olson   D.C. Oren   J.W. Ranney
C.S. Robbins   V.A. Rudis   M.J. Scanlan   D.M. Sharpe
F.W. Stearns   A.L. Sullivan   B.L. Whitcomb   R.F. Whitcomb

With 61 Figures

Springer-Verlag
New York   Heidelberg   Berlin

ROBERT L. BURGESS
Environmental Sciences Division
Oak Ridge National Laboratory
Oak Ridge, Tennessee 37830
U.S.A.

DAVID M. SHARPE
Department of Geography
Southern Illinois University
Carbondale, Illinois 62901
U.S.A.

The illustration on the cover represents changes in wooded area of
Cadiz Township, Green County, Wisconsin, from 1831–1950. See
Fig. 1-1 for further information. (Reprinted from Curtis 1956, by
permission of the University of Chicago Press.)

Library of Congress Cataloging in Publication Data
Main entry under title:

Forest island dynamics in man-dominated landscapes.

(Ecological studies; v. 41)
Bibliography: p.
Includes index.
1. Forest ecology.   2. Forest ecology—United States.
I. Burgess, Robert Lewis, 1931–   II. Sharpe,
David M. III. Series.
QH541.5.F6F67   581.5′2642   81-2248
                                       AACR2

Printed in the United States of America.
9 8 7 6 5 4 3 2 1

ISBN 0-387-90584-7   Springer-Verlag   New York   Heidelberg   Berlin
ISBN 3-540-90584-7   Springer-Verlag   Berlin   Heidelberg   New York

Foreword: Pattern and Rates of Change

Discoveries of pattern in nature have provided both incentive and reward throughout the histories of science and art (1). Natural history is rich in the variety of biological and geographic information that is required in order to understand the landscape patterns that exist on earth. Natural philosophy, epitomized by the abstractions of physics, draws from all nature's observed complexity a few formal relationships expressing more or less general insight. Both approaches are woven together in this volume, and in the ecology of landscapes in general.

Historically, the natural history of species, of communities, and of biogeographic regions began to be expressed in a few numbers of birth and immigration rates and death and emigration rates, as population ecology became an abstraction of statistics and then of difference or differential equations. The net rate of change, equal to the difference between income and loss rates, was a formal expression not only for numbers of a given species, but for nutrients in the soil, or in various compartments of trophic levels or whole ecosystems (1, 2). My acquaintance with Robert MacArthur while he was a student at Yale was punctuated by a typically Hutchinsonian illumination. His island biogeography paradigm, referred to in many chapters of the present book, accounted for some variations in the number of species on "real" islands very parsimoniously: in terms of factors that could plausibly enhance rates of "income" by immigration and of "loss" by extinction.

Despite some limitations in stretching this analogy to forest "islands" in a terrestrial landscape, discussed throughout this book, the paradigm has already served its heuristic purpose in stimulating several kinds of analysis of landscape pattern explained by the authors. This beach-head is only one of several which have recently been made in landscape ecology, striving to invade and occupy a fairly new territory on the map of science.

This volume's editors and collaborators made another landing in analysis of space–time patterns of forest islands. Their contribution to the First International Conference on Landscape Ecology (3) and some related analyses (4, 5, 6, 7, 8) expressed the amount of area of a given landscape type as a function of rates of income minus rates of loss in simulation models for land use and cover change. Such models of landscape change as a Markov process complement others of ecological succession for replacement of one species by another (9, 10, 11, 12), and for competition in the growth and survival of individuals while competing for limited resources on a plot "island" in a "sea" of mixed landscape terrain (9, 13).

Further analysis of the meaning of terrain and the geologic and soil boundary conditions which constrain ecosystem equations is provided by George Bowen's recent thesis (14) analyzing forest island pattern in Ohio. Percent cover of forest and a density parameter (number of islands per unit area) or else a dissection index for the glaciated and (rougher) unglaciated terrain embodied much of the pattern information that was expressed more abstractly in a factor analysis.

A synthesis still awaiting consummation will link tangible patterns in the landscapes which grow our food and furnish our homes and enjoyment with the paradigms of soil forming factors (1) and of landscape geochemistry (15) that give more scientific meaning to why regional ecosystems are so different from one another. If that marriage is fruitful, biogeography and the scientific basis of geography in general will have been further enriched.

References

1. Olson, J. S. 1980. Foreword: Hans Jenny and Fertile Soil. In (2), pp. vii–xiii.
2. Jenny, Hans. 1980. The Soil Resource. Origin and Behavior. Ecological Studies 37, Springer-Verlag, New York; Heidelberg, Berlin.
3. Sharpe, D. M., F. W. Stearns, R. L. Burgess, and W. C. Johnson. 1981. Spatio-temporal patterns of forest ecosystems in man-dominated landscapes of the eastern United States. First International Conference on Landscape Ecology, Eindhoven, The Netherlands.
4. Johnson, W. C. and D. M. Sharpe. 1976. An analysis of forest dynamics in the Georgia Piedmont. Forest Science 22:307–322.
5. Hett, J. M. 1971. Land-use changes in east Tennessee and a simulation model which describes these changes for three counties. EDFB/IBP 71-8. Oak Ridge National Laboratory, Oak Ridge, TN. 56 pp.
6. Olson, J. S. and S. Cristofolini. 1965. Succession of Oak Ridge vegetation. In ORNL-3849. Oak Ridge National Laboratory, Oak Ridge, TN. pp. 76–77.
7. Olson, J. S. and G. Cristofolini. 1966. Model simulation of Oak Ridge vegetation succession. In ORNL-4007, Oak Ridge National Laboratory, Oak Ridge, TN. pp. 106–107.
8. Shugart, H. H., T. R. Crow, and J. M. Hett. 1973. Forest succession models: A rationale and methodology for modeling forest succession over large regions. Forest Science 19(3):203–212.
9. Weinstein, D. A. and H. H. Shugart. Ecological modeling of landscape dynamics. In H.

9. Mooney and M. Godron (eds.). Disturbance and ecosystems. Springer-Verlag, Berlin, Heidelberg, New York (in preparation).
10. Horn, H. S. 1976. Succession. In R. M. May (ed.). Theoretical ecology: Principles and applications. Saunders, Philadelphia and Toronto. pp. 187–204.
11. Olson, J. S. 1977. Climax community. In P. Lapedes (ed.). Encyclopedia of Science and Technology 3:199–202. McGraw-Hill, New York.
12. Shugart, H. H., D. C. West, and W. R. Emanuel. 1981. Patterns in the long-term dynamics of forests: An application of simulation models. In D. C. West, H. H. Shugart, and D. B. Botkin (eds.), Forest succession: Concepts and application. Springer-Verlag, New York, Heidelberg, Berlin. 496 pp.
13. Waggoner, P. E. and G. R. Stephens. 1970. Transition probabilities for a forest. Nature 225:1160–1161.
14. Bowen, G. W. and R. L. Burgess. 1981. A quantitative analysis of forest island pattern in selected Ohio landscapes. ORNL/TM 7759. Oak Ridge National Laboratory, Oak Ridge, TN. 111 pp.
15. Fortescue, John. 1980. Environmental geochemistry. A holistic approach. Ecological Studies 35. Springer-Verlag, New York, Heidelberg, Berlin.

Oak Ridge, Tennessee   Jerry Olson
July 1981

Preface

The International Biological Program (IBP), initiated in the mid-1960s, was a primary stimulus to many ecological studies over the past decade. As a cooperative effort among scientists of many nations, the IBP undertook to augment our meager knowledge of the biological processes that govern ecosystem function, and to analyze, synthesize, and interpret biological productivity and its relationship to human welfare.

The U.S. component of the IBP organized around a series of large, integrated research programs in Analysis of Ecosystems, essentially combining the international themes PT (Terrestrial Productivity) and PF (Freshwater Productivity). Five major "biome" programs evolved from this activity: Tundra, Grasslands, Desert, Coniferous Forest, and Eastern Deciduous Forest (Blair 1977). The latter, headquartered at Oak Ridge National Laboratory in eastern Tennessee, realized at an early date that site-specific investigations into ecological processes were unlikely to yield the kinds of information that would lead to significant advances in the understanding of ecosystem structure and function in a region as diverse as the eastern third of the United States.

Consequently, a subprogram in "Biome-wide Studies" was established, growing into "Biome and Regional Analysis" by 1972. Investigations of phenology, primary production, and forest succession led to documentation of "primary productivity profiles" in New York–New England, North Carolina–Tennessee, and southern Wisconsin. By 1974, strong indications were prevalent that certain processes normal to large areas of relatively undisturbed deciduous forest were not (necessarily) applicable to the present landscape in eastern North America. Previously contiguous forest had been fragmented through time by the collective and cumulative actions of European settlement, expanding population, industrialization, and changing land use pattern. A three-year project on "Forest Succession and Landscape Pattern" identified and brought together much of the preliminary work leading to this volume.

Forest patches (or "islands") exist today in most of the eastern United States embedded in a matrix of nonforest cover types or land use categories. The original forest has given way to continually increasing demands of agriculture, industry, and urbanization. Dam and reservoir construction, demands for energy production facilities, including pipeline and transmission corridors, a massive interstate highway system, and recent escalations in coal mining have all served to further subdivide and encroach upon existing forested lands.

MacArthur and Wilson's theory of island biogeography (1967) stimulated thinking and a great deal of research concerning the potential application to the dynamics of forest islands riding, as it were, in a "sea" of nonforest landscape. Questions of migration and extinction, succession rates and processes, species diversity, and optimal size of terrestrial nature preserves were beginning to be asked. Research was initiated to test hypotheses of island biogeographic theory, and evidence began to accumulate, particularly for plants and birds, that biologic processes operative within and between forest patches show both similarities and differences when compared to the dynamics of populations interacting on islands.

Consequently, a symposium was organized by the Ecological Society of America, meeting at Michigan State University in August 1977 with the American Institute of Biological Sciences (AIBS), to address this array of concerns. The present volume has its roots in that symposium. The chapters, however, are current and several additions have been made to broaden and deepen the concepts and examples portrayed in the text.

Many agencies and individuals have made substantial contributions to this volume, and we wish to acknowledge their efforts. The National Science Foundation, the U.S. Department of Energy, the Fish and Wildlife Service (USDI), and the Bureau of Plant Industry (USDA) have all supported portions of the work. Universities and state agencies also have contributed in many ways. The editors are thankful for the technical capabilities, editorial assistance, and graphic arts skills of Polly L. Henry, Natalie T. Millemann, Linda W. Littleton, the Word Processing Center of the Environmental Sciences Division, and the Graphic Arts Department at Oak Ridge National Laboratory.

We are also grateful to Springer-Verlag for including this work in the Ecological Studies series, and to Drs. W. Dwight Billings and Jerry S. Olson, members of the series Editorial Board, for their assistance in manuscript review.

July 1981   ROBERT L. BURGESS
                                                       DAVID M. SHARPE

Contents

Contributors

BRUNER, MARC C.

Aquatic Plant Management Laboratory, Science and Education Administration (USDA), 3205 S. W. 70th Avenue, Fort Lauderdale, Florida 33314

BURGESS, ROBERT L.

Department of Environmental and Forest Biology, SUNY College of Environmental Science and Forestry, Syracuse, New York 13210

BYSTRAK, D.

Migratory Bird and Habitat Research Laboratory, U.S. Fish and Wildlife Service, Laurel, Maryland 20811

DEANGELIS, DONALD L.

Environmental Sciences Division, Oak Ridge National Laboratory, Oak Ridge, Tennessee 37830

EK, ALAN R.

School of Forestry, University of Minnesota, St. Paul, Minnesota 55101

FIELDS, DAVID E.

Health and Safety Research Division, Oak Ridge National Laboratory, Oak Ridge, Tennessee 37830

HOEHNE, LINDA M.

2304 N. Grant Boulevard, Milwaukee, Wisconsin 53210

JOHNSON, W. CARTER   Department of Biology, Virginia Polytechnic
                             Institute and State University, Blacksburg,
                             Virginia 24060

KLIMKIEWICZ, M. K.   Office of Migratory Bird Management, U.S.
                             Fish and Wildlife Service, Laurel, Maryland
                             20811

LEVENSON, JAMES B.   Energy and Environmental Systems Division,
                             Argonne National Laboratory, Argonne,
                             Illinois 60439

LOVEJOY, THOMAS E.   World Wildlife Fund, 1601 Connecticut
                             Avenue, N. W., Washington, D. C. 20009

LYNCH, JAMES F.   Chesapeake Center for Environmental
                             Sciences, Smithsonian Institution, Edgewater,
                             Maryland 21037

MATTHIAE, PAUL E.   TERO, 13905 Grant Place, Elm Grove,
                             Wisconsin 53122

MAY, ROBERT M.   Department of Biology, Princeton University,
                             Princeton, New Jersey 08540

OLSON, RICHARD J.   Environmental Sciences Division, Oak Ridge
                             National Laboratory, Oak Ridge, Tennessee
                             37830

OREN, DAVID C.   Department of Biology, Harvard University,
                             Cambridge, Massachusetts 02138

RANNEY, JACK W.   Environmental Sciences Division, Oak Ridge
                             National Laboratory, Oak Ridge, Tennessee
                             37830

ROBBINS, CHANDLER S.   Migratory Bird and Habitat Research
                             Laboratory, U.S. Fish and Wildlife Service,
                             Laurel, Maryland 20811

RUDIS, VICTOR A.   Southern Forest Experiment Station,
                             USDA—Forest Service, 701 Loyola Avenue,
                             New Orleans, Louisiana 70113

SCANLAN, MICHAEL J.   Department of Biology, Virginia
                             Commonwealth University, Richmond,
                             Virginia 23284

SHARPE, DAVID M.   Department of Geography, Southern Illinois
                             University, Carbondale, Illinois 62901

STEARNS, FOREST W.   Department of Botany, University of
                             Wisconsin-Milwaukee, Milwaukee, Wisconsin
                             53201

SULLIVAN, ARTHUR L.   School of Design, North Carolina State
University, Raleigh, North Carolina 27607

WHITCOMB, B. L.   10271 Windstream Drive, Columbia, Maryland
21044

WHITCOMB, ROBERT F.   Plant Protection Institute, Beltsville
Agricultural Research Center (USDA),
Beltsville, Maryland 20705

1. Introduction

Robert L. Burgess
Environmental Sciences Division
Oak Ridge National Laboratory

David M. Sharpe
Department of Geography
Southern Illinois University

> Instead of an essentially continuous forest cover, . . . the landscape now presents the aspect of a savanna, with isolated trees, small clumps or clusters of trees, or small groves scattered in a matrix of artificial grassland of grains and pasture grasses . . .
>
> (Curtis 1956)

Such is the impact of man on the pattern of forest vegetation in the Wisconsin as Curtis knew it, which has more recently been documented by Galluser (1978). The pattern is repeated in much of the eastern United States (Burgess 1978), is impressed on the landscapes of Europe (Darby 1956, Falinski 1976, 1977, Olaczek and Sowa 1976), and is now being created in the tropics (Gomez-Pompa et al. 1972, Ranjitsinh 1979). Much former regional forest now exists as forest islands in a sea of agricultural, urban, and other land uses.

> Within remnant forest stands, a number of changes of possible importance may take place. The small size and increased isolation of the stands tend to prevent the easy exchange of members from one stand to another. Various accidental happenings in any given stand over a period of years may eliminate one or more species from the community. Such a local catastrophe under natural conditions would be quickly healed by migration of new individuals from adjacent unaffected areas . . . in the isolated stands, however, opportunities for inward migration are small or nonexistent. As a result, the stands gradually lose some of their species, and those remaining achieve unusual positions of relative abundance.
>
> (Curtis 1956)

These man-dominated landscapes are fundamentally under human control and their popular evaluations for agricultural, urban, and various other land uses. These regions are presumed today to differ significantly from the ecosystems of presettlement times, but in what ways and to what extent? Curtis expressed concern for changed balances of plant species and the possible extinction of certain examples. Gomez-Pompa et al. (1972) fear that the wholesale removal of forests will devastate the rich genetic pool of the humid tropics. The concern is thus worldwide and transcends species, biomes, and nations. There is much ecological insight to support these expressions of concern, but very little in the way of specific research to guide understanding and management. Yet, the ubiquity of such landscapes and the ecological values derived from them make their understanding and management an important goal. Clearly, it is timely to direct attention to the regional ecology of man-dominated landscapes.

This volume presents research on aspects of forest dynamics in several man-dominated landscapes in the eastern United States. These landscapes are laboratories in which ecological theories can be applied and tested. While restricted to one region, we view the landscape dynamics and their ecological consequences as analogies for many regions throughout the world. An overview of the evolution of these landscapes provides a glimpse of how diverse they are, and a backdrop for the issues dealt with in this volume.

Dissected Landscapes of the Deciduous Forest Biome

When Europeans colonized the eastern seaboard of North America, the vast region we now call the Deciduous Forest Biome was dominated by forest vegetation. Of course, patches of land were cultivated by Indians (Day 1953) and natural and other anthropogenic disturbance created openings and a general parklike aspect in some regions (e.g., Raup 1937). Edaphic, climatic, and historic conditions created sizable areas of nonforest, such as the prairie peninsula (Transeau 1935, Borchert 1950). The dynamics of the forested areas were paced by patterns of disturbance that spanned the spectrum from gap phase replacement (Watt 1947) to region-wide fire, insect infestation, and hurricane damage (Heinselman 1973, Swain 1980, Spurr 1956, Sprugel 1976), which created an ecological stability at the regional scale (Loucks 1970). Each region had a characteristic disturbance regime that resulted in an individualistic mosaic of regional ecosystems (Bormann and Likens 1979).

In the New World, European settlers continued the work of a millenium that had led to widespread removal of the forests of the British Isles and continental Europe (Darby 1956). Our knowledge of the landscape patterns that were created is most complete for New England and Wisconsin. Curtis (1956) graphically portrayed how deforestation in south-central Wisconsin had reduced the forest area by 70% by 1882, 90% by 1902, 95% by 1935, and more than 96% by 1950 (Fig. 1-1). Auclair (1976) and Gallusser (1978) have shown a similar history for Wisconsin's Driftless Area.

But such aggregate values fail to capture the essence of forest island landscapes—the small size, exposure, and isolation of the remnant forest patches, exemplified by some of the variables shown in Table 1-1. As portions of the forest were first cleared, rather large forest patches were left standing as woodlots. These patches have since been reduced in size, and some have been removed. Those that remain are the remnants of the forest fragments of 1882. Their average size has been drastically reduced, and the amount of edge per unit forest area has tripled. The level of isolation of the remnant forest is suggested by the average distance between forest islands. The connectivity

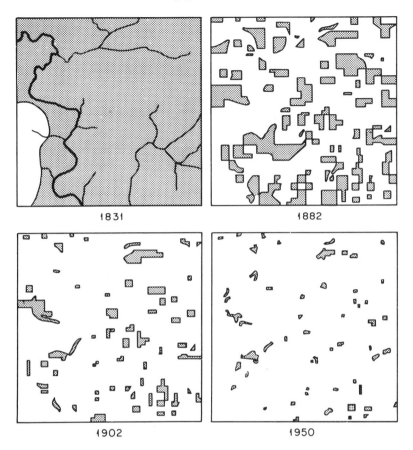

Fig. 1-1. Changes in wooded area of Cadiz Township, Green County, Wisconsin, during the period of European settlement. The shaded areas represent the land remaining in, or reverting to, forest in 1882, 1902, and 1950. (From Curtis 1956, by permission of the University of Chicago Press.)

Table 1-1. Landscape Pattern Variables for Cadiz Township, Green County, Wisconsin[a]

	Variables	1831	1882	1902	1935	1950
1	Total forest area (ha)	8724	2583	841	419	318
2	Number of forest islands	1	70	61	57	55
3	Average island size (ha)	8724	36.9	13.8	7.4	5.8
4	Total perimeter (km)	—	156.9	97.0	74.8	63.1
5	Average minimum dimension (m)[b]	9340	607	371	272	241
6	Average interisland distance (m)	—	153	332	336	339
7	Length of edge (m/ha of forest)	—	60.8	115	179	208
8	Connectivity index	—	32	6	3	0

[a] Variables 1-4 and 7 based on Curtis (1956).
[b] Islands are assumed to be square; deviation from this shape reduces the minimum average island dimension.

index is the number of common corners shared by discrete forest islands which would provide for a forest habitat corridor between them; fencerows and streamside vegetation provide linkages between some noncontiguous forest islands.

Deforestation of some regions has had to await a new technology. The Lower Mississippi Valley was still 48% forested in 1950, but forest cover had declined to 31% by 1969 (Dill and Otte 1971). Rapid deforestation has responded to worldwide demand for soybeans, and is projected to continue until only about 20% of the original forested area remains (Sternitzke 1976).

While this sequence is typical of much of the Midwest, other regions have had contrasting histories. Raup's (1966) chronicle of land use history of Harvard Forest, Massachusetts, notes that the area was 90% forested in 1771, when settlement was getting under way. The area in forest declined to 10% by 1850, just before wholesale farm abandonment reversed the trend. The remaining forest existed as woodlots, often in inaccessible areas, creating a pattern much like that in Wisconsin. By 1956, the region was again 85% forested. A similar history can be seen in other regions in the East. The Georgia Piedmont is now 65% forested, but Bond and Spiller (1935) estimated that 80% of the land had been cleared at one time or another. Old cotton fields, corn fields, and other cropland are clearly evident in the patchwork of old-field pines, which now occupy about 50% of the region (Johnson and Sharpe 1976, Boyce and McClure 1975). In both New England and in the Piedmont and Coastal Plain, the current spatial pattern, species composition, and size class distributions of the forests are directly attributable to this history of clearing and reversion to forest.

The location of remnant forest is far from stable, even in regions not undergoing wholesale deforestation or reforestation. For example, the total forest area in the Upper Piedmont of Georgia declined by 35,000 ha between 1961 and 1972 (Knight 1974); but in this period 116,000 ha formerly classed as nonforest reverted to forest, while 151,000 ha of forest were cleared. Throughout the Biome, the pattern of natural vegetation is unstable to varying degrees. The sequence of transient perturbations that creates a stable regional ecosystem envisioned for a natural landscape by Loucks (1970) has been superseded by a disturbance pattern tied to changing land use. The mix of perturbation agents, their frequency, intensity, and areal extent, now modify the characteristics of natural landscapes by a combination of factors tied to human land use, land ownership patterns, and the social forces that bring about landscape change.

Problems for This Volume

The analogy between islands in seascapes and remnant forest in agricultural and urban landscapes has suggested the extension of equilibrium island biogeographic concepts to the new circumstance. The hopes of transferring equilibrium island biogeography concepts to such landscapes is frustrated by the structural complexity of their pattern, their transience, and the residual impact of previous landscape states. Unlike islands in seascapes, habitat islands in man-dominated landscapes are not surrounded by a matrix that is de facto alien to their terrestrial biota. This interplay between islands and their matrix, and between islands themselves as mitigated by the intervening matrix adds new dimensions to their dynamics. These complexities have called into question how closely the relationship between species richness and island size and isolation applies to habitat islands in man-dominated landscapes.

Yet, there is a paucity of time series data both on landscape patterns, and on the species composition of forest tracts during the course of forest fragmentation, and reforestation in some regions, so that the response of the biota is difficult to assess. Explanations of the few extant observations are frustrated by the complexity of ecologic phenomena within islands and of interisland linkages.

Clearly, the challenge to formulate researchable hypotheses, to derive answers, and then to inject new understandings into the ongoing stream of landscape planning and management decisions is formidable.

The following chapters address these issues. Several chapters deal with forest islands in southeastern Wisconsin. Overstory and shrub composition (Levenson, Chapter 3), herbs (Hoehne, Chapter 4), and mammal populations (Matthiae and Stearns, Chapter 5), are considered. The enhanced importance of forest edge communities in such highly fragmented landscapes is also presented by Ranney, Bruner, and Levenson (Chapter 6). The impact of decreased seed dispersal between forest islands on invasion rates is studied using stand level simulations by Johnson et al. (Chapter 10). Levenson (Chapter 3) overviews land use and vegetation history of southeastern Wisconsin.

West-central Minnesota has scattered woodlots that were planted by European settlers in the mid-1800s, as well as remnant patches of natural forest and former savanna that has reverted to closed forest with fire protection. Scanlan (Chapter 7) has studied the floral richness of these as a function of their size and distance from blocks of extensive forest.

Eastern Maryland has a forest island landscape much like that of Wisconsin. Here, removal of a large proportion of the regional forest has altered the habitats of neotropical migratory birds, and the avian species composition of these forest habitats has changed accordingly, as documented by Whitcomb et al. (Chapter 8) and modeled by May (Chapter 9).

A second series of chapters stresses the management of forest island landscapes in the eastern United States. Lovejoy and Oren (Chapter 2) explore the minimum size of habitat islands that will meet specific ecological goals. Sullivan (Chapter 12) makes the point that planning needs to look toward specific ecological values of such landscapes. A modeling strategy to optimize the location and sizes of forest islands to meet specific objectives is discussed by Rudis and Ek (Chapter 11).

The special session of the ESA-AIBS meetings that preceded this volume was exploratory, as is this volume. We view it as a catalyst. The ecological dynamics of regions may prove to be an important new direction for ecosystem research, and these dynamics are inseparably linked to the patterns of landscape components, both in space and in time. The empirical and theoretical studies presented in the following chapters should provide a springboard for new studies, hypotheses, and emerging theory in an exciting area of ecology. If they do, then this volume will have served its purpose.

Acknowledgments. Research supported in part by the Eastern Deciduous Forest Biome, US-IBP, funded by the National Science Foundation under Interagency Agreement AG-199, BMS69-01147 A09 with the U.S. Department of Energy under contract W-7405-eng-26 with Union Carbide Corporation, and in part by NSF grant DEB 78-11338 to Southern Illinois University. Contribution No. 348, Eastern Deciduous Forest Biome, US-IBP. Publication No. 1647, Environmental Sciences Division.

2. The Minimum Critical Size of Ecosystems

THOMAS E. LOVEJOY
WORLD WILDLIFE FUND
WASHINGTON, D.C.

DAVID C. OREN
DEPARTMENT OF BIOLOGY
HARVARD UNIVERSITY

The surface of the planet is being developed at a quickening tempo. Vast segments of natural areas are disappearing while much of the remainder is being fragmented. As a consequence of the attendant increasing need to protect our biological patrimony, the effect of the fragmentation itself has become a matter of concern. If protection efforts are to be successful and significant, not only must some of the fragments be protected, but also they must be able to maintain their biological integrity over reasonable periods of time—or if not, at least no illusion to the contrary should be operating.

Island biogeographers in particular have, in recent years, drawn attention not only to the biological impoverishment that occurs when a natural area is reduced in size, but also to that which takes place *after* reduction in size (Diamond 1975a, Diamond and May 1976, Terborgh 1974, 1975, Wilson and Willis 1975). Both forms of reduction in species number follow from the basic theory of island biogeography (MacArthur and Wilson 1963, 1967). According to that theory, all other things being equal, larger islands hold more species than smaller ones, and so it is argued that reducing a forest area in size, essentially shrinking a large island to a small one, inevitably leads to a reduced number of species. The theory holds as well that species number for an island of given size is a function of both immigration rate and extinction rate, resulting in a predictable number of species at an equilibrium of the two rates. The rates, however, are also a function of island size, with immigration decreasing and extinction increasing with smaller area, so there will be a further species loss after reduction in area, as the two new rates approach a new and lower equilibrium.

In efforts to illuminate this problem, the island biogeographers, among others, have compared land bridge islands (natural equivalents of natural areas reduced in size by man) to oceanic islands (Diamond 1972, 1973, Terborgh 1974, 1975). While land bridge islands involve reductions of natural areas at a considerably more sedate pace than the often close to instantaneous reductions caused by man, they do present some interesting opportunities to investigate the biology of species loss, especially when it is possible to date age of isolation, as in the islands in the Sea of Cortez (Wilcox 1978).

There are some problems, however, in comparisons of oceanic and land bridge islands. The dynamics of loss from islands supersaturated with species because of reduction in area are not well understood, nor are they necessarily the mirror image of the processes by which species number is built up by colonization of a new island. The decline to a new and lower species number may well be a slower affair than rising to an equilibrium number in a similar, but colonizing situation. In the latter case, good dispersers, often *r*-selected species, are favored. In contrast, on a land bridge island declining in species number, species with good persistence but often low rates of reproduction (*K*-selected forms) will be favored. Diamond's analysis (1972) of islands in the vicinity of New Guinea tend to support this view.

It may not always be valid to consider an oceanic island as representing the end point a comparable land bridge will reach. Smaller ($<$ 250 km^2) oceanic land bridge islands do seem to have similar numbers of species (Diamond 1972, Terborgh 1974). Yet the equilibrium species number may not be the same for all islands after both declining and rising situations, for it may well be greater in some declining situations as a consequence of preceding on-island evolution.

Indeed, the spread between equilibrium number for oceanic and land bridge islands may well increase with area. The smaller oceanic and land bridge islands are, the more similar they may be not only in species number, but also in species composition (Diamond 1972), while larger ones may differ more in both aspects. Since larger land bridge islands may not have reached equilibrium and their end points are not completely predictable, it is difficult to base calculations on data from them with any precision.

The unintentional experiment of Barro Colorado Island (BCI), created when Gatun Lake was formed in 1914 for the Panama Canal, has provided tantalizing data. The number of bird species lost between then and 1971 (Willis 1974) is almost exactly that predicted by island biogeography (Terborgh 1974). Unfortunately, the effect of isolation and size reduction will remain forever partly confused with the effect of habitat changes of successional nature.

Experimental reintroductions of bird species lost from BCI provide interesting commentary on our primitive understanding of the process and that particular situation. One, the white-breasted wood wren (*Henicorhina leucosticta*) established itself in one of the few available areas of second growth and in tree fall areas (Morton 1978). This species (at last report) had not established itself as a reproducing species, but this was probably because of an unfavorable sex ratio. It would seem that this species was one lost because of successional vegetation changes rather than because of reduced area. Yet it is easy to imagine a situation where a patch of forest is too small to have a sufficient number of tree falls over long periods of time to sustain characteristic bird species although generally such species have high dispersal characteristics and may be likely to recolonize a forest patch.

The second reintroduced bird species, the song wren (*Cyphorhinus arada*), did reproduce successfully (Morton 1978). The birds weighed more on the average than their counterparts in forests of mainland Panama; competition for food, therefore, did not seem strong even with considerably higher populations than on the mainland of possibly competing antwren species. Nonetheless, the reintroduction eventually failed. Predation appears to be the cause, perhaps because of abnormally high populations of coatimundi (*Nasua narica*), which may well be a consequence of isolation and area and attendant lack of large predators. The story is not completely clear because the predation may well have related to the song wrens' building nests in conspicuous sites near BCI trails in the absence of less conspicuous sites along streams which were not available on BCI but which they would normally use. Reintroduction is, however, an important approach in studying the process of species loss.

The conclusion that species will be lost, both during and after reduction in area, can be derived without reference to island biogeography, and can lead to the derivation of the concept of a minimum critical size for ecosystems. The definition of an ecosystem for a terrestrial community has always tended to be somewhat diffuse because of the general absence of relatively sharp boundaries such as those possessed, for example, by a pond. We consider an ecosystem as a piece of presently or once continuous habitat at least large enough in extent to contain and maintain its characteristic species diversity and species composition. Of course, an ecosystem which has fewer species because of fragmentation or subsequent loss is still an ecosystem in the sense of being a functioning ecological system. It is, however, an impoverished version of the natural, undisturbed, unfragmented ecosystem, and thus is less than ideal for a conservation area.

An increase in species number with area has been familiar to biologists, particularly botanists, for some time (Cain and Castro 1959) and leads to the obvious conclusion that the more reduced an area is in size, the fewer species can be found in it. In studying this species/area relationship, phytosociologists were interested in more than an ever rising species number, an approach that would have essentially ignored the existence of ecosystems. The shape of a species/area curve is considered to be characteristic of the plant association, and is in large part a function of the basic diversity of the community involved, and thus different and more rapidly rising for a rain forest than for a meadow.

In this context, the concept of minimum area was defined as essentially the minimum area necessary to define a given association, i.e., to contain at least one of each of the characteristic plant species (Cain and Castro 1959). The minimum area concept was not intended to include a self-sufficient functioning ecological unit, at least not in terms of its dynamics were all the rest of the natural vegetation surrounding it to be stripped away. Perhaps there was little to suggest such drastic manipulation in those days.

The species/area curve continues to rise with increasing area beyond that minimum, (1) because new kinds of environmental patches are being included, (2) because of the substitution of ecological equivalents from the local species pool (e.g., as in diatoms, see Patrick 1967), and ultimately, (3) because of the replacement of one geographical form by another. The number of species from each of these sources will usually be greater with larger area, and the species/area curve can be thought of as the sum of these curves plus that describing minimum area as defined by the phytosociologists. To tease apart these different curves and demonstrate their separate contributions in nature, i.e., to derive an operational definition, is certainly difficult. It is probably more easily approached in biomes such as tundra and boreal forest where natural systems are less diverse and complex. Aquatic communities on shallow substrates may well follow principles similar to those followed by terrestrial ones, but pelagic communities are not easily described in these terms.

When it is possible to make the necessary separations and define the basic ecosystem curve, it should approach an asymptote, and indeed species/area curves, such as some described for plant associations in Perthshire, Scotland (Poore 1955) have been shown to do so. Nonetheless, the point at which that curve can be cut and the ecosystem expected to maintain its integrity over a reasonable time period is not clear. Certainly, were the minimum size needed taken to be at the point where the asymptote is reached (and the phytosociologists' minimum area attained), species loss would inevitably follow, because some species would consist of single individuals or tiny populations incapable of reproduction, whereas other low population species would be lost just from random fluctuations in their numbers. On this basis alone, minimum critical size of an ecosystem is clearly always greater than the minimum area of phytosociology.

The decay process of an ecosystem below minimum critical size stems from species loss also caused by other processes. One of these may be the genetic bottleneck problems encountered by small populations. As there are additional genetic problems just from continuing small population size (Franklin 1980), and as bottleneck problems are probably greater for K-selected than for r-selected species (Lovejoy 1979), genetic problems may occur with greater frequency in decay situations than in colonizing ones. There are also interlocking biologies such as those of obligate pollinators or seed dispersers, where the loss of one species inevitably leads to loss of all or many of the species in the coevolved system. There is no question that we need to know a great deal more about the various processes that contribute both to setting the minimum critical size of an ecosystem as well as to ecosystem decay. In this context it is vital that both natural experiments be studied and manipulative ones be undertaken.

We simply do not know at which point a species/area curve can be said to have reached the minimum critical size for an ecosystem. This is not surprising if the various sources of increase in species number with area have not been sorted out, for in cutting the species/area curve beyond the point of the phytosociologists' minimum area, there will inevitably be species losses from the other three sources of species number increase with area, and it is obviously important that species loss be confined to those sorts rather than to those integral to the ecosystem. Probably the minimum critical size will always have to be an approximation of some sort, and at least from a conservation point of view, this should not be considered too troubling. It must be realized, too, that the dynamics of intact ecosystems as well as the effects of external factors such as climate will inevitably render it impossible to maintain an ecosystem at 100% integrity even with elaborate management and a generous increment of area added as a safety factor.

One of the major difficulties will always be environmental patchiness, which will probably be encountered at any level or scale at which environment is considered. In Malaysia's Jengka Forest Reserve (Poore 1968), for example, patchiness may be so extreme that minimum critical size would include the entire forest (Poore, personal communication). In many situations, patchiness is probably a manageable factor, even in Amazonian forest, although minimum critical size is most likely substantial, on the order of 10^4 km^2.

An examination of the decay process of ecosystems below minimum critical size is needed to determine whether species loss is (I) random in sum and sequence, (II) predictable sequentially, resulting in impoverished ecosystems of similar species composition for similar sized ecosystem fragments, or (III) predictable sequentially resulting in impoverished ecosystems of different species composition depending on which species are contained in the ecosystem fragment at the time of isolation. There are several indications that hypothesis II—that leading to impoverished ecosystems of predictable and similar species composition—is, at least in part, being followed in nature. The work of Moore and Hooper (1975) on birds of British woodlands, that of Whitcomb and others on forest remnants near Washington, D.C. (Chapter 8), and that of Forman and colleagues (Forman et al. 1976, Galli et al. 1976) on birds of New Jersey piedmont oak forest patches suggest this may be the case, although the latter suggest that predictability decreases somewhat with smaller patches. In contrast, Diamond's analysis (1972) found more similarity of species numbers of birds among the smaller islands.

In part, this may be a function of the diversity of the fauna involved which would affect the probability of expectation of more abundant species occurring in a forest fragment of particular size. In forest patches near Belem, Brazil, which are large enough to have anything other than second growth bird species, the wedgebilled woodcreeper (*Glyphorynchus spirurus*) would almost always be among the species occurring initially because it is generally the most abundant bird species in those forests (Lovejoy 1975). For rarer bird species of those forest communities there would be less likelihood of any particular one being left in a given forest fragment, particularly the smaller the fragment.

It should be borne in mind that different taxa may tend to follow different hypotheses (Simberloff 1976a) and such differences may also exist between different ecosystems. Perhaps Simberloff's mangrove island experiments (see Simberloff and Abele 1976a) should be considered in this light?

Which hypothesis is correct or more general is obviously critical to the controversy which erupted in the pages of *Science* as to whether one large preserve will protect more species than a series of small reserves of equivalent total area or the reverse (Simberloff and Abele 1976a, 1976b, Diamond 1976a, Terborgh 1976, Whitcomb et al. 1976). If hypothesis II is correct, then a large reserve is the best option in terms of numbers of species conserved. The focus on numbers of species in that controversy tended to obscure, although not intentially, the very fundamental point that it is important to save functioning ecosystems (Simberloff and Abele 1976a, Terborgh 1976) and their processes, not just collections of species. The ecosystem concerned should be dictating the size of reserves, not just species number. This is apparent from the work on edge effects (Ranney et al., Chapter 6), which shows great species diversity in small forest patches with a high proportion of edge.

The great trap in all of this is to treat all species as being equal, which the simplicity of island biogeographic models tends to encourage. A similar kind of trap exists in the notion of a minimum critical size of ecosystems, for if attention is fixed on the species area curve it would be easy to ignore the last species added, perhaps even to consider it unimportant. This could be a grave error. In tundra, for example, the first caribou would probably be added long after the several square meters (< 20) needed for bryophytes and seed plants.

From this point of view, the contribution of Diamond (1978) on critical area for maintaining viable populations of species (which incidentally tends to support hypothesis II) is useful. Since little is known of the biological requirements defining a minimum viable population for a species, it is helpful to get some idea of minimum area from real island or environment remnant situations, provided it is possible to determine whether the isolated ecosystems under consideration are at equilibrium or still in decay. Minimum area determined in this fashion for large, high-trophic-level organisms, might approach the minimum critical size for the ecosystem in question. When it is less, not only would the ecosystem protected be of less than minimum critical size, but also the ensuing decay process might even erode the ecological support system of the particular species used to define the protected area. As Diamond points out, the dispersal biology of the species in question is an essential consideration in this approach and in conservation planning in general.

Even without considering the additional problem of turnover (Whitcomb et al., Chapter 8), it is apparent that there are many important questions relating to the minimum critical size of ecosystems to be answered. Answered they must be if the function and structure of ecosystems are to be understood, and if twenty-first century biology is to consist of more than paleontology, the biology of weedy species, laboratory and zoo biology and what Scott McVay calls the science of pickled parts.

Acknowledgments. The authors gratefully acknowledge the help and counsel of R. O. Bierregaard, Jr., Jared M. Diamond, Charles S. Elton, G. Evelyn Hutchinson, Frances C. James, Robert E. Jenkins, Scott McVay, Susan McGrath, Larry Marshall, Eugene S. Morton, Sarah Najafi-Kopourchali, Ruth Patrick, Ghillean T. Prance, M. E. Duncan Poore, John Terborgh, and Bruce A. Wilcox. This paper is dedicated to G. Evelyn Hutchinson for his concern for the kindly fruits of the earth and his insights into their workings.

3. Woodlots as Biogeographic Islands in Southeastern Wisconsin

JAMES B. LEVENSON
DEPARTMENT OF BOTANY
UNIVERSITY OF WISCONSIN-MILWAUKEE

The concepts of island biogeography have been broadly tested with data on inverte-brates (Wilson and Simberloff 1969, Simberloff and Wilson 1969, 1970), mammals (Crowell 1973), and birds (Diamond 1969, 1973, Terborgh and Faaborg 1973). Plant communities on islands have also been examined in Puerto Rico (Heatwole and Levins 1973), in Lake George, New York (Slack et al. 1975), and in Lake Mockeln, Sweden (Nilsson 1978). For terrestrial "habitat islands," i.e., small patches of habitat in a regional landscape, Vuilleumier (1970), Brown (1971), and Picton (1979) related species richness to "island" size and isolation after examining the bird and mammal populations of mountaintops. Culver (1970a) utilized the concepts to consider the biota of caves, while Janzen (1968) compared individual host plants to islands in time and space for the individual insects that feed on them.

Kolata (1974), Diamond (1975a), Sullivan and Shaffer (1975), and Goeden (1979) all have suggested using island biogeographic concepts as the basis for establishing a net-work of wildlife preserves. On a smaller scale, the same basic concepts have been applied to county parks and rural woodlots (Forman and Elfstrom 1975, Galli et al. 1976, Whitcomb 1977, Gottfried 1979). Extension to woodlots and parks of urban systems seems to be an appropriate application of the habitat island concept (Davis and Glick 1978).

In the metropolitan Milwaukee, Wisconsin area, major alterations of the landscape pattern such as the addition or deletion of islands, resulting in changes in interisland distances and matrix qualities, are caused by human activity. In the process, the rates of colonization and extinction are continually altered. In natural communities, succession is a mechanism for such change. However, successional patterns resulting from urban influences may differ from those expected under natural conditions.

In this chapter, we consider the parks and woodlots in metropolitan Milwaukee as biogeographic islands in order to examine the applicability of the concepts of island biogeography to isolated woodlots in an agro-urban matrix. Specific objectives were (1) to survey and assess the forest plant communities that are embedded and isolated from each other in nonforest matrix; (2) to test the applicability of the concepts of island biogeography that relate island size to species richness of the forested "habitat island"; and (3) to provide concepts useful for regional landscape planning and resource management.

Description of the Study Area

The study area was metropolitan Milwaukee, an area of about 525 km^2, including Milwaukee County, southern Ozaukee County, and the eastern portions of Waukesha and Washington Counties (Fig. 3-1).

Climate

The region has a continental climate, modified by Lake Michigan. Average monthly temperatures range from –6.2°C in January to 21.5°C in July. Average annual precipitation is about 76.2 cm with nearly two-thirds falling during the frost-free season.

Fig. 3-1. Map showing the distribution of civil townships in Milwaukee and Ozaukee counties, Wisconsin.

Areas adjacent to the lake are generally cooler in the summer and warmer in the winter than more westerly locations. The average length of the growing season near Lake Michigan in Milwaukee County is 180 days, but drops to less than 160 days within 50 km to the northwest (Fig. 3-2).

Physiography and Soils

Glacial features dominate the landscape. Two to three undulating moraines with intervening troughs parallel the Lake Michigan shoreline (Fig. 3-3). Drainage, poorly integrated and controlled by the moraines, is largely confined to the troughs that harbor the Milwaukee, Root, and Menomonee Rivers (Fenneman 1938, Thornbury 1965). Generally, local relief is slight, but the northwestern portion of the study area is characterized by kettle and kame topography.

The pattern of soil associations in southeastern Wisconsin (Fig. 3-3) is closely correlated with glacial features, primarily the end moraines and adjacent ground moraines. Ozaukee-Morley-Mequon soils occupy morainic uplands, accounting for about 80% of the study area. The Ozaukee and Morley soils (Typic Hapludalfs) are well drained to moderately well drained and are characterized by moderately slow permeability and moderate fertility. The Mequon series (Udollic Ochraqualf) consists of somewhat poorly

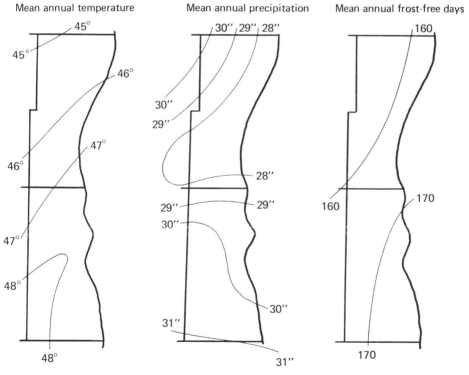

Fig. 3-2. Climatic summary of the metropolitan Milwaukee region (redrawn from SEWRPC 1963).

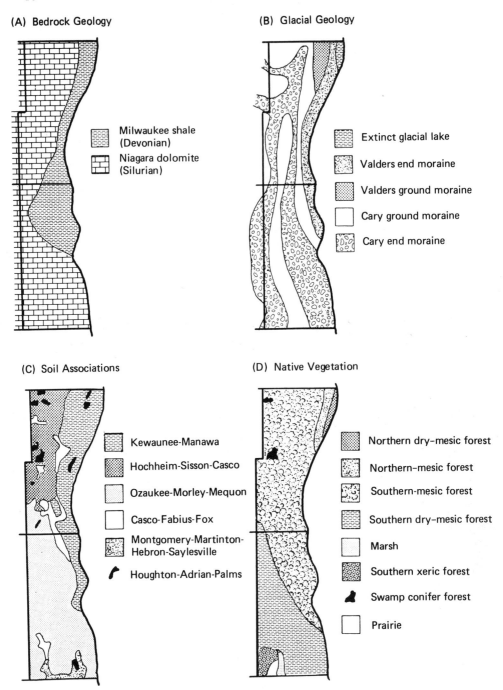

(A) Bedrock Geology

Milwaukee shale
(Devonian)

Niagara dolomite
(Silurian)

(B) Glacial Geology

Extinct glacial lake

Valders end moraine

Valders ground moraine

Cary ground moraine

Cary end moraine

(C) Soil Associations

Kewaunee-Manawa

Hochheim-Sisson-Casco

Ozaukee-Morley-Mequon

Casco-Fabius-Fox

Montgomery-Martinton-
Hebron-Saylesville

Houghton-Adrian-Palms

(D) Native Vegetation

Northern dry-mesic forest

Northern–mesic forest

Southern-mesic forest

Southern dry-mesic forest

Marsh

Southern xeric forest

Swamp conifer forest

Prairie

Fig. 3-3. Bedrock geology (A) and glacial geology (B) of the metropolitan Milwaukee region (redrawn from SEWRPC 1963), with maps of major soil associations (C) (redrawn from USDA-SCS 1970, 1971), and the native (presettlement) vegetation (D) (redrawn from Chamberlin 1973).

drained, silty soils typically found near waterways, foot slopes and depressions (USDA 1971). These soils are typical of most of Milwaukee County and southwestern Ozaukee County.

The Hochheim-Casco series (Typic Argiudoll and Typic Hapludalf) occur in western Ozaukee County. These well drained soils are formed in thin loess and loamy glacial material on ground moraines and outwash plains, and are characterized by moderate fertility and permeability (USDA 1970).

Vegetation

The pre- and postsettlement vegetation of metropolitan Milwaukee has been described by several authors (Chamberlin 1877, Shinners 1940, Whitford and Salamun 1954, Ward 1956, Curtis 1959). Three major vegetation types (prairie, northern forest, and the southern forest—after Curtis 1959) were originally present (Fig. 3-3).

The general vegetational pattern in the study area is oriented on a southwest-northeast axis. Prairie entered the extreme southwestern corner of Milwaukee County. Scattered oak openings were present in association with the prairie. These stands graded into the southern xeric (oak-hickory) forest. Southern dry-mesic (maple-red oak) forest covered most of Milwaukee County (Shinners 1940). The southern mesic (beech-maple) forest was discontinuous, primarily in the northeastern half of Milwaukee County and most of Ozaukee County. Patches of the southern xeric and southern low-land forest were found interspersed on xeric and depressional sites, respectively. Northern forest occurred in extreme eastern Ozaukee County and extended into northeastern Milwaukee County (Fig. 3-3). The northern mesic forest (northern hard-wood or hemlock-hardwood forest), originally mapped as a narrow strip, entered Ozaukee County from the north, stretching southward along the lake shore to Port Washington (Chamberlin 1873). Patches of northern lowland, or swamp conifer forest (the wet phase of the northern forest) were also scattered throughout the study area (Hansen 1933).

Only a small portion of the two-county area was covered by prairie or northern forest. The greatest area was vegetated by southern forest. Three major upland southern forest communities were distinguished by Chamberlin (1877), Whitford (1951), and Curtis (1959) (Table 3-1). Chamberlin (1877) was probably the first to note that "no abrupt line of demarcation existed between" the communities.

At the moist end of the continuum, the southern lowland forest was once wide-spread throughout the study area. This type occurred along rivers and streamways, and in poorly drained upland depressions. There is little quantitative information describing the original lowland forest of this region, but it is probable that it was common before alteration of the water tables through deforestation and tiling of agricultural lands. The southern lowland forest has a great variety of tree species but exhibits wide-spread regional uniformity (Stearns 1965). Species composition normally includes silver maple (*Acer saccharinum*), black willow (*Salix nigra*), American elm (*Ulmus americana*), and swamp white oak (*Quercus bicolor*). Transition communities between the lowland and mesic forest were present in the upland depressions with green and black ash (*Fraxinus pennsylvanica* and *F. nigra*), red maple (*Acer rubrum*) and bass-wood (*Tilia americana*) gaining dominance (Ware 1955, Curtis 1959).

Table 3-1. Sources, Nomenclature, and Dominant Species in Various Treatments of the Upland Forests of Southern Wisconsin

Sources			
Chamberlin (1877)	Whitford (1951)	Curtis (1959)	Dominant tree species
Oak group	Oak-hickory	Southern xeric	*Quercus alba* *Q. velutina* *Q. macrocarpa* *Q. borealis*
Oak-maple group	Intermediate	Southern dry-mesic	*Quercus borealis* *Q. alba* *Tilia americana* *Acer saccharum*
Maple group	Maple-basswood	Southern mesic	*Acer saccharum* *Tilia americana* *Fagus grandifolia* *Ulmus rubra* *Quercus borealis* *Ostrya virginiana*

Land Use History

The present landscape pattern reflects its historical development. Before European settlement, the resident and transient Indian population had little impact. The Public Land Survey of exterior lines for the area (township and range) was completed by 1836 (SEWRPC 1970).

Early settlement was primarily by farmers. The process of clearing farms from the virgin forest was the "labor of a generation" (Schafer 1927). Settlers first came to the heavily forested region of Milwaukee and Ozaukee Counties in the early 1830s. The movement of settlers into the region was well under way by the spring of 1836. A territorial census taken in July 1836 assigned 2893 inhabitants to the original, much larger, Milwaukee County, which was established in 1834 (Schafer 1927).

The process of land selection by the settlers provides insight into land use patterning. During the great land sale of February and March 1839, nearly 500,000 acres were sold at $1.25 per acre. Since all the public lands were offered at a uniform price, the usual criteria for homestead selection included proximity to a spring or brook, proximity to a market, or to a market road, and/or proximity to the lake ports for sale of fuel wood and forest products. Another consideration was the general belief in the superiority of forested land over prairie land. The soils of the region were generally suitable except for some lowland areas and the narrow belt of heavy clay loam east of the Milwaukee River. In addition, it was a distinct advantage to have and maintain a woodlot for fuel and building materials. High, well-drained morainic ridges were used for roadways, a practice still evident in the region today. Homesteads were located near the ridges.

By 1860, over one-third of the land in Milwaukee County had been converted to plowfields, with another third also in farm ownership (Whitford and Whitford 1972). Wheat was the main cash crop from the time farming first started until about 1880.

The wheat monoculture soon depleted the upland soils, resulting in reduced yields and declining returns (USDA 1971). After the Civil War, the malting industry began to include barley and hops. After 1890, the popularity of these crops declined and gradually agriculture shifted to dairying, the major agricultural pursuit by 1920 (SEWRPC 1963). Grazing of woodlots was common.

The trends in farming were related to industrial and urban growth. The census of 1850 showed 31,077 persons in Milwaukee County. Between 1850 and 1860, the population more than doubled. Slow growth occurred between 1860-1870, but with increasing industrialization, a fivefold increase occurred between 1860 and 1900, doubling again between 1900 and 1930 (SEWRPC 1972). By 1970, Milwaukee County reported 1,054,249 residents.

For over 100 years, 1840 to 1950, urban growth and development occurred in a general outward expansion from the earlier established urban centers. But

> From 1950 to 1970 . . . a dramatic change occurred in this pattern of urban development in that large, scattered tracts of rural lands were subdivided for urban use, resulting in a highly dispersed, discontinuous, low-density development pattern, a pattern which has become known as "urban sprawl" (SEWRPC 1971).

The rapid population growth coupled with urban sprawl accounted for the conversion of more than 16,194 ha (40,000 a and 920 farms) of farmland to urban use in Milwaukee County between the late 1940s and 1959 (USDA 1971). In 1964, only 10,393 ha (25,670 a) of farmland remained in Milwaukee County, a decrease of 22% in five years (USBC 1964). The trend accelerated between 1964 and 1969 with only 7049 ha (17,412 a) remaining in farmland, a decrease of another 32%. The overall change during the decade, 1959 to 1969, was a 47% decrease in farm acreage and a 56% reduction in the number of farms (USBC 1969). Woodland, including woodland pasture, averaged only 8% of the county's farmland, and showed an overall decrease of 15% in area between 1964 and 1969.

In Ozaukee County, the total land in farms was 43,808 ha (108,205 a) in 1964. About 9% was wooded and 43% of the woodlands were actively grazed (USDA 1970). Ozaukee County experienced a 10% decrease in farmland between 1959 and 1969. The number of farms decreased by nearly 23% over the same period.

In Ozaukee County, the woodlands remaining for study were generally privately owned. In Milwaukee County, most were part of the Milwaukee County Park System (one of the finest in the country). The County Park Commission began an active program of land acquisition in 1910. Urban expansion in the county has eliminated many potential sites, and those remaining have accrued increased value for their open-space character in the heavily urbanized landscape (SEWRPC 1965).

Methods

Site Selection

In the metropolitan Milwaukee region, topography, soils, and climate are relatively uniform. Widely dispersed woodlots within an urban and agricultural matrix include southern mesic and southern dry-mesic forest types (Curtis 1959), dominant upland vegetation types now, as they were in presettlement time.

Forest island selection was based on several criteria. Each forest island ideally should:

1. be isolated from other islands, surrounded by a matrix of urban or agricultural land.
2. contain sugar maple in combinations representative of the southern mesic forest type (Curtis 1959).
3. be a remnant of original upland vegetation, not a newly established stand.
4. include all structural vegetation layers.
5. have no evidence of recent major disturbance.
6. have existed as a discrete unit long enough to have developed a mature forest edge.

Few stands fulfilled all of the criteria. Land use history of each stand was obtained by questioning owners, referring to earlier reports, and examining aerial photographs taken in 1937.

Forty-three woodlots of the southern upland mesic type were sampled in the metropolitan Milwaukee region between May and October, 1975 (Fig. 3-4). By design, islands were chosen to include a variety of sizes ranging from 0.03 to 40 ha.

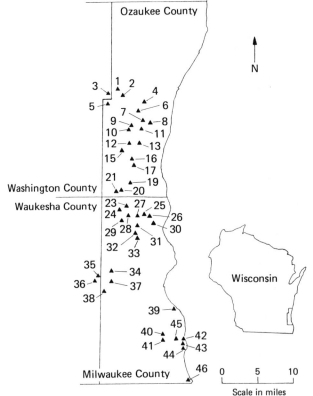

Fig. 3-4. Map showing the locations of the 43 forest islands studied in the metropolitan Milwaukee region.

Field Sampling Methods

Only the area bounded by the "edge trees" was sampled. Edge trees were defined as those trees exhibiting an asymmetrical bole, with a considerable clear-length on the interior side (forest-grown side) and heavy branching to the outside (open-grown side). These edge trees were not always present at the physical edge of the woodlot, but may have been some distance toward the island's interior. Woodlots adjacent to intensively managed land parcels develop mature edges and edge trees over time. If the adjacent land is later abandoned, or less intensively managed, succession occurs to the outside of the formerly stabilized edge (Wales 1972). However, the original forest area will remain delimited by its original edge trees. In some cases, the forest core was identical to current size and shape, but often the original core was considerably smaller and embedded in an expanding, second growth community. Vegetation outside of the edge trees was not sampled.

The vegetation of each island was sampled using the stratified-random line-strip method (modified from Lindsey 1955). Shrub, understory, and canopy strata were sampled contemporaneously using a series of nested rectangular plots (Fig. 3-5). In each stand, line-strips, separated by 25 m, were located systematically throughout the entire area. Species lists for each woodlot were compiled from the sample data plus extensive reconnaissance between plots and transects.

Canopy and understory strata were sampled in 10 X 25 m plots. All stems greater than 5 m tall and 2.5 cm diameter breast height (dbh) were recorded in three size categories: 2.5-5 cm (1-2 in.) dbh; 5-10.1 cm (2-4 in.) dhb; or, if over 10.2 cm (4 in.) dbh, by the exact diameter. Dead stems, recent windthrows, and stumps were also recorded. In most stands, 20 such plots were sampled, collectively totaling 0.5 ha. For stands > 10 ha, a larger sample was obtained but line-strips were separated by a distance of 50 m. This modification insured representative coverage. Stands of 1 ha or less were full-tallied for canopy and understory strata.

The shrub layer was sampled in plots 2 X 12.5 m beginning at the midpoint of each 10 X 25 m plot. All woody species between 0.5 and 5 m tall were recorded. The num-

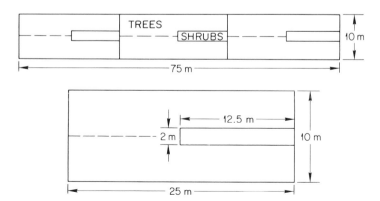

Fig. 3-5. Example of a line-strip, three plots in length, showing the arrangement of the nested plots. In the enlarged single plot (below), trees were sampled in the 10 X 25 m plot and shrubs in the 2 X 12.5 m plot.

ber of stems of each species present was recorded for each plot. Twenty plots were sampled in most stands, collectively comprising an area of 0.05 ha. In the smaller stands full-tallied for the canopy and understory, the shrub stratum was sampled using the line-strip technique, but with only 15 m separating the transects. Nomenclature follows Gleason (1952).

Data Analysis

Tables of species attributes were calculated for each structural stratum of each site (Levenson 1976). Species density was recorded as the number of stems per hectare. Frequency was defined as the percentage of plots in each stand in which the species occurred. Basal area (m^2/ha) was recorded for each species in each island. Relative density, relative frequency, and relative dominance were calculated and averaged to obtain the importance value expressed as a percentage for each species (Lindsey 1956). The importance value for species of the shrub stratum is the average of relative density and relative frequency only, since cover values were not measured.

Interisland comparisons were made using a hierarchical cluster analysis (Ward 1963, Veldman 1967). The technique produces clusters of mutually excluded stands, whose composition is more similar to others in the cluster than to stands in other clusters. The procedure reduces the number of clusters from n to $n-1$ to minimize information loss. Without modifying clusters already formed, the procedure was repeated until the number of clusters was systematically reduced from n to 1 (Ward 1963). With each grouping, an error index was provided, determined from the sum of within-group sums of squares. The analysis was based on the importance value of species present and on the "no value" information contributed by absent species.

Nonlinear regression analysis was performed to test the various models proposed by MacArthur and Wilson (1967), when examining the relationships between island size and degree of isolation with species richness. Likewise, the colonization and extinction models were nonlinear.

Results and Discussion

Overstory Structure and Composition

The forest woodlots ranged from dense, successional forests dominated by basswood (*Tilia americana*) with low basal areas to old-growth stands dominated by sugar maple (*Acer saccharum*) with a relatively high basal area. Stem density ($\geqslant$ 10.2 cm) ranged from 240 to 929 stems/ha with an average of 442 (179 trees/acre). The average basal area (stems $\geqslant$ 10.2 cm) for the 42 woodlots was 29.8 m^2/ha (130 ft^2/acre), but ranged from 21.5 to 43.3 m^2/ha (Table 3-2). This compares favorably with the 29.9 m^2/ha reported by Goff and Zedler (1968) for 125 xerophytic stands dominated by oaks (*Quercus* spp.) in southern-central Wisconsin. Such broad ranges of stem density and basal area are indicative of the structural variability found within the forest islands in metropolitan Milwaukee.

Of the 33 species in the canopy, nine species had a constancy value over 50% (Table 3-3). These nine species, in different proportions, dominated the 43 forest stands (Fig.

Table 3-2. Island Size and Woody Species Richness for Canopy and Shrub Strata of 43 Forested Islands in the Metropolitan Milwaukee Region, Wisconsin[a]

Site no.	Size (ha)	Canopy stratum			Shrub stratum		Total no. of woody species
		Total no. species	Density (no./ha)	Basal area (m²/ha)	Total no. species	Density (no./ha)	
31	0.03	2	— —	— —	15	23,332	17
28	0.36	12	929	28.86	19	11,300	24
11	0.57	8	337	21.56	19	14,340	24
46	0.59	13	285	33.23	16	10,560	22
15	0.61	9	486	28.46	16	10,200	18
02	0.65	11	847	26.37	13	14,865	19
29	0.73	10	269	28.54	14	14,440	22
10	1.21	14	476	24.27	24	15,223	31
32	1.40	16	331	22.78	20	13,872	27
13	1.50	9	448	30.52	24	9,480	28
08	1.54	11	445	32.36	13	13,700	19
20	1.58	10	494	22.81	25	20,075	27
09	1.62	11	265	21.51	17	18,720	24
26	1.70	8	465	30.63	22	30,050	24
12	1.98	13	360	28.30	17	8,984	27
05	2.06	10	495	31.25	24	16,700	28
21	2.19	17	542	29.62	35	14,060	42
03	2.23	9	516	32.76	22	10,900	25
23	2.35	11	447	31.00	24	14,175	27
35	2.39	14	540	37.44	27	12,820	35
24	2.43	12	446	29.99	22	29,900	27
07	2.43	18	576	33.65	16	21,900	31
25	2.47	15	506	29.49	25	18,200	33
27	2.51	16	684	29.52	23	23,360	32
43	2.83	10	326	33.79	14	13,420	20
19	2.91	14	478	30.13	23	15,600	31
33	3.12	17	358	25.68	26	28,540	34
42	3.24	13	304	28.42	13	6,260	25
44	3.97	15	436	43.30	8	21,080	21
30	4.09	13	366	33.49	18	23,520	28
41	4.13	12	358	32.03	11	10,058	20
17	4.25	8	464	31.10	21	18,960	23
37	4.49	11	330	31.44	20	18,800	28
34	5.47	11	240	22.49	19	23,360	28
06	6.48	7	356	27.22	18	16,180	19
16	7.21	12	550	32.00	13	21,420	20
40	7.81	11	416	36.12	20	16,040	24
45	7.93	12	276	31.04	17	12,050	21
39	11.46	15	474	33.02	16	14,380	24
04	14.53	7	398	31.94	19	12,317	21
38	18.34	11	366	31.07	16	12,200	25
36	21.05	12	348	29.80	26	21,100	27
01	39.96	13	524	24.30	18	18,057	21
	Mean	12	442	29.84	19	16,616	25
	SD	± 3	±141	± 4.36	± 5	± 5,641	± 5

[a] Canopy stratum includes all species more than 5 m tall and 2.5 cm dbh, but density and basal area were calculated only for stems greater than 10 cm dbh. The shrub stratum includes all woody species from 0.5 to 5 m tall.

Table 3-3. Summary Values for All Canopy Species (Ranked by Importance Values), Occurring in 43 Forest Islands in Southeastern Wisconsin[a]

Species	Constancy (%)	Density (no./ha)	Basal area (m²/ha)	Mean importance value
Acer saccharum Marsh.	100.0	118	7.76	26.24
Fagus grandifolia Ehrh.	76.7	98	5.76	15.32
Tilia americana L.	88.4	81	4.84	14.59
Quercus borealis Michx.	83.7	35	5.92	10.16
Fraxinus americana L.	95.3	38	2.95	9.34
Quercus alba L.	60.5	27	3.33	5.56
Ostrya virginiana (Mill.) K. Koch	88.4	33	0.55	5.21
Prunus serotina Ehrh.	79.1	28	0.84	3.83
Ulmus rubra Muhl.	72.1	15	0.76	2.78
Acer rubrum L.	37.2	17	1.23	1.41
Carya cordiformis (Wang.) K. Koch	44.2	6	0.22	0.70
Crataegus succulenta Link.	30.2	10	0.15	0.66
Juglans cinerea L.	30.2	4	0.49	0.53
Carya ovata (Mill.) K. Koch	14.0	10	0.54	0.48
Betula papyrifera March.	18.6	10	0.62	0.43
Ulmus americana L.	41.9	4	0.13	0.39
Fraxinus pennsylvanica March.	23.3	8	0.37	0.38
Acer saccharinum L.	4.7	36	3.72	0.35
Quercus macrocarpa Michx.	9.3	6	1.32	0.33
Juglans nigra L.	7.0	11	1.87	0.26
Fraxinus nigra Marsh.	11.6	9	0.20	0.21
Populus grandidentata Michx.	2.3	47	1.61	0.13
Quercus bicolor Willd.	9.3	4	0.35	0.11
Amelanchier laevis Wieg.	11.6	3	0.05	0.10
Populus tremuloides Michx.	9.3	4	0.14	0.10
Catalpa speciosa Warder.	2.3	12	0.80	0.08
Acer negundo L.	4.7	6	0.11	0.07
Carpinus caroliniana Walt.	9.3	3	0.04	0.06
Betula lutea Michx.	4.7	3	0.28	0.04
Celtis occidentalis L.	2.3	4	0.50	0.03
Pyrus malus L.	4.7	3	0.05	0.03
Crataegus punctata Jacq.	2.3	2	0.03	0.02
Cornus alternifolia L.	2.3	1	0.02	0.02
Totals		442	29.84	99.95

[a] Values for the mean density and mean basal area are based on only the stands in which the species was present. Mean Importance Value is the average of the sum of the relative values of frequency, density, and dominance (maximum = 100).

3-6). Forty of the habitat islands were dominated by nearly every possible combination of only six species: sugar maple, American beech, white ash, basswood, red oak, and white oak. The final three clusters of the dendrogram (Fig. 3-6) were sufficiently different to be recognized as major groups: Maple and Beech Group, Maple Group, and Basswood Group. The separation of the Maple Group from the Maple and Beech Group

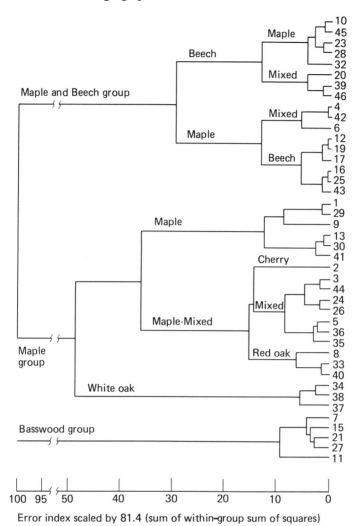

Error index scaled by 81.4 (sum of within-group sum of squares)

Fig. 3-6. Dendrogram of the results of the cluster analysis of the canopy stratum data for 42 forest islands based on species composition and importance values. Site 31 was excluded from the analysis because of insufficient sample size (see Table 3-2). The leading dominant(s) of each major group is listed.

was the result of the dominant role played by beech. Interestingly, this separation is the same as that recognized by Chamberlin (1877) over a century ago. The Basswood Group was sufficiently different in composition and structure to remain separated from the other groups. The five stands in this group appeared to be successional in nature. Three were solely dominated by basswood (Sites 7, 11, and 15). In the other two, basswood, red oak and white ash shared dominance.

Shrub Stratum Composition and Structure

Sixty-six different woody species were found in the shrub stratum of the 43 forest habitat islands (Table 3-4). Thirty-six species were considered true shrubs, species whose maximum potential height restricted them to the subcanopy. The remaining 30 species had the potential of entering the canopy.

Stem density in the shrub layer varied greatly from stand to stand. Stem densities for the 43 forest islands averaged nearly 16,700 stems/ha (Table 3-2). Choke cherry and sugar maple were the dominant species of the shrub layer (Table 3-4). Choke cherry was sampled in all islands, and was the dominant or codominant shrub in 95% of the stands. Sugar maple, the second dominant species, followed by white ash were each absent from only one stand. These three species accounted for 67% of all stems recorded. Dogberry (*Ribes cynosbati*) shared dominance in 24% of the stands and the hybrid honeysuckle (*Lonicera bella*) in 14%.

Only nine species in the shrub stratum were classified as exotics, i.e., not native to eastern North America. Of the nine, only three occurred in more than two islands. Nightshade (*Solanum dulcamara*) and common buckthorn (*Rhamnus catharticus*) were present in 42% and 30% of the stands, respectively. Bird dispersal was suggested for both species as they produce multiseeded berries and drupes. Both species appeared most successful in disturbed sites with wet-mesic conditions. Only the hybrid honeysuckle (*Lonicera bella*) was widely distributed, occurring in 58% of the stands, often with locally high densities. Honeysuckle is an opportunistic species, entering the forest in an opening or edge, prospering, and finally declining as the canopy closes. None of the exotic shrub species appeared able to survive under the closed forest canopy of the southern mesic forest.

In general, the shrub flora of the islands reflected the disturbance history with varying degrees of species richness attributable to the balance of elements of both the pioneer and terminal communities, as well as exotics. Stands with a high species richness in the shrub layer were characterized either by mixed communities without dominants or by only weakly dominant species, i.e., low relative densities of choke cherry and/or sugar maple.

Species-Area Relationship

The species-area relationship of Arrhenius (1921), Preston (1960) and MacArthur and Wilson (1967) is depicted for all woody species greater than 0.5 m tall (Fig. 3-7). The function

$$S = kA^z$$

where S is the number of species in an area A, and k and z are constants, was fitted to the points. The best fit was the near-linear function approximating the average number of species for the 43 islands but relatively independent of island size, indicating little correlation between island size and species richness ($R^2 = 0.01$). It is evident that, as attractive as the ability to predict plant species richness by island size appears, additional variables must be considered. Although many researchers have shown an

Table 3-4. Summary Values for All Species Occurring in the Shrub Stratum of 43 Forest Islands in Southeastern Wisconsin[a]

Species	Constancy (%)	Density (no./ha)	Importance value
Prunus virginiana L.	100.0	6863	30.28
Acer saccharum Marsh.	97.7	3012	16.64
Fraxinus americana L.	97.7	1391	10.18
Prunus serotina Ehrh.	90.7	615	6.35
Ribes cynosbati L.	74.4	1026	4.67
Lonicera bella Zabel.[b]	58.1	1273	4.19
Fagus grandifolia Ehrh.	69.8	472	3.47
Tilia americana L.	81.4	431	3.18
Ribes americanum Mill.	72.1	601	2.82
Ostrya virginiana (Mill.) K. Koch	65.1	197	1.51
Vitis riparia Michx.	58.1	214	1.42
Carya cordiformis (Wang.) K. Koch	62.8	127	1.30
Cornus stolonifera Michx.	25.6	1513	1.27
Viburnum lentago L.	53.5	315	1.22
Ulmus rubra Muhl.	58.1	138	1.04
Cornus racemosa Lam.	37.2	384	0.99
Populus grandidentata Michx.	2.3	300	0.99
Hamamelis virginiana L.	25.6	399	0.76
Rubus occidentalis L.	44.2	173	0.76
Viburnum opulus L.	20.9	708	0.72
Solanum dulcamara L.[b]	41.9	244	0.70
Zanthoxylum americanum Mill.	41.9	115	0.70
Viburnum acerifolium L.	23.3	406	0.62
Crataegus succulenta Link.	32.6	106	0.48
Viburnum rafinesquianum Schult.	20.9	347	0.47
Cornus alternifolia L. f.	25.6	177	0.45
Menispermum canadense L.	25.6	359	0.42
Dirca palustris L.	23.3	136	0.42
Sambucus canadensis L.	20.9	76	0.42
Fraxinus pennsylvanica Marsh.	9.3	167	0.34
Quercus borealis Michx. f.	32.6	70	0.34
Rhamnus catharticus L.[b]	30.2	69	0.34
Crataegus punctata Jacq.	32.6	45	0.29
Acer negundo L.	20.9	61	0.22
Carpinus caroliniana Walt.	18.6	35	0.17
Lonicera prolifera (Kirchner) Rehder	14.0	132	0.16
Amelanchier sp. L.	16.3	55	0.14
Acer rubrum L.	18.6	38	0.14
Sambucus pubens Michx.	7.0	247	0.12
Parthenocissus quinquefolia (L.) Planch.	14.0	55	0.12
Viburnum lantana L.[b]	4.7	163	0.11
Fraxinus nigra Marsh.	7.0	87	0.11
Euonymus atropurpureus Jacq.	4.7	293	0.10
Rhamnus frangula L.[b]	4.7	173	0.08
Ulmus americana L.	11.6	25	0.07
Quercus macrocarpa Michx.	4.7	80	0.06
Smilax hispida Muhl.	2.3	280	0.05
Carya ovata (Mill.) K. Koch	4.7	95	0.05
Juglans nigra L.	4.7	30	0.04
Betula papyrifera Marsh.	2.3	100	0.03

Table 3-4 (continued)

Species	Constancy (%)	Density (no./ha)	Importance value
Acer saccharinum L.	2.3	120	0.03
Juglans cinerea L.	2.3	63	0.03
Forsythia suspensa (Thunb.) Vahl.[b]	2.3	120	0.02
Ptelea trifoliata L.	2.3	40	0.02
Celtis occidentalis L.	4.7	44	0.02
Rosa sp. L.	2.3	100	0.02
Rhus radicans L.	4.7	20	0.02
Berberis thunbergii D. C.[b]	2.3	150	0.02
Euonymus alatus (Thunb.) Sieb.[b]	2.3	20	0.01
Populus tremuloides Michx.	2.3	20	0.01
Quercus alba L.	2.3	20	0.01
Thuja occidentalis L.	2.3	25	0.01
Rhus typhina L.	2.3	15	0.01
Celastrus scandens L.	2.3	72	0.01
Rubus hispidus L.	2.3	15	0.01
Acer ginnala Maxim.b	2.3	40	0.01
Totals		16,627	100.00

[a] Mean density is based on stands in which the species was present. Importance Value is the average of the sum of the relative values of density and frequency (maximum = 100).
[b] Introduced exotics.

increasing number of species with increasing island size, another major consideration is the environmental heterogeneity of the community. Niering's (1963) floral study of Kapingamarangi Atoll indicated that increasing the areas of sand did not add to the floral richness. The flora remained restricted to the few species adapted to the xeric habitat. An increase in area added more of the same habitat, but no additional new habitats. However, the establishment of a freshwater lens in islets of 1.4 ha and larger (Wiens 1962) created increased habitat variety which supported more species. Johnson et al. (1968) found island area to be the best single predictor of plant species diversity, but concluded that environmental richness was also significant. Qualitatively different environments result from increasing topographic, edaphic, and/or climatic diversity (Heatwole and Levins 1973).

The woody species composition of forest woodlots in southeastern Wisconsin was restricted by similar factors (Levenson 1980). An increase in area very often added more of the same habitat. It was not until variations in soil, relief, or disturbance were present that changes in species variety were realized. The southern mesic forest community of southeastern Wisconsin is composed of a relatively few, shade tolerant woody species adapted to the mesic forest's interior: sugar maple, beech, ironwood, and slippery elm (Curtis 1959). Vegetative reproduction in the form of root and stump sprouts increases the chances for these species to persist and compete under the full canopy. Additional species present in the forest islands such as white ash, red oak and black cherry are less shade tolerant. Their presence is often the result of disturbance and redevelopment through gap-phase reproduction (Watt 1947, Bray 1956).

Disturbance, whether natural or man-induced, is a major variable controlling woody species richness in forest woodlots. In the southern mesic forest "... it is clear

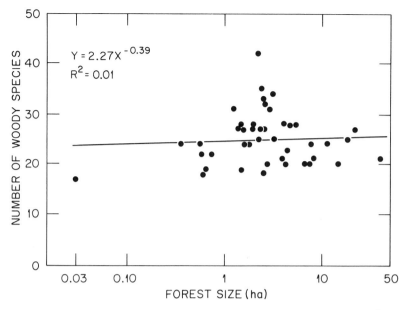

Fig. 3-7. Total number of woody species as a function of island size ($S = kA^{Z}$).

that the degree of success of the intolerant trees and hence the degree of species mixture to be found in the forest is proportional to the chances for disturbance" (Curtis
1959). In the agro-urban matrix of metropolitan Milwaukee, the forest islands owe
their existence, isolation and composition primarily to man-induced disturbances. In
northwestern Ohio, Tramer and Suhrweir (1975) concluded that human interference
was a major variable in affecting species richness of the elm-ash-maple woodlots. Slack
et al. (1975) also indicated that increased disturbance contributed to an increased
number of exotic species as well as to an increased overall diversity of vascular plants
and bryophytes on 54 islands in Lake George. Islands of the Puerto Rican Bank have
also received heavy human disturbance by littering, cutting of trees and shrubs, making
of paths, and abandonment of unwanted pets which leads both to local extinctions
and to establishment of previously absent species (Levins and Heatwole 1973). Many
of the consequences of human usage of the islands in Ohio, New York, and Puerto
Rico are similar to those in metropolitan Milwaukee. Auclair and Cottam (1971) indicated that, coupled with forest island isolation, heavy usage was the most critical
aspect of woodlot ecology.

Frequent disturbance results in a shift toward a more xeric habitat with greater
light, greater transpiration stress, and more variable temperature and moisture levels.
Disturbance destroys the mesic "stabilizing mechanisms" to favor species with pioneer
tendencies, in contrast to more mesic species (Curtis 1956, Auclair and Cottam 1971).

Disturbance in the isolated forest woodlots can be quantified indirectly by a number of variables. Stem densities, percentage canopy cover and size-class distributions
are all variables traditionally used to determine the amount of disturbance a woodlot
has undergone. In the southern mesic forest, canopy closures may be one of the major
variables. Gaps in the canopy reduce the mesic character of the forest interior. In the

old-growth, southern mesic forest, the gaps in the canopy are a result of the number of dead standing, cut or fallen stems. Following this logic, Table 3-5 presents a list of possible indicators of disturbance and values for each site.

Using a stepwise multiple linear regression program, three of these variables were found to be significant in determining the species richness, a function of the basal area of sugar maple, the proportion of dead stems and the amount of dead basal area in the stand. More mature stands in this region have greater dominance by sugar maple, so that species richness should be inverse to the density of large stems of sugar maple. Sugar maple forms a very dense canopy and is largely responsible for maintaining the mesic conditions of the forest interior. On the other hand, dead stems may indicate that the canopy has been opened and that intolerant, or successional, species were able to invade the resultant gap. Therefore, the proportion of dead stems is an indicator of the potential number of disturbance sites within the stand. In the mature forest, the size of the stems is also important. That is, the larger the dead trees, the larger will be the gap in the canopy, to a point. Trees of small diameter may not reach the canopy and may be understory components. However, a spurious correlation exists between the proportion of dead stems and their basal area (Table 3-6). Excluding the dead basal area, the remaining variables account for approximately 21% of the variance of species richness of the forested woodlot islands. The following equation is statistically signifi-cant at the $P < 0.01$ level:

$$\text{Species richness} = 24.8 - 0.2X_1 + 0.3X_2$$

where X_1, basal area of sugar maple/ha; X_2, proportion of dead stems.

The variables of woodlot size and disturbance level are not necessarily mutually exclusive and may be analogous for woodlots less than some threshold size. As islands become smaller, younger, or more disturbed, the entire island may become functionally edge. Figure 3-8 depicts several hypothetical woodlots of varying sizes, each surrounded by an edge, X units wide. The depth of the edge has been assumed to remain constant. As woodlot area decreases, edge and interior areas also decrease; but the edge/interior ratio increases. At some point, designated (a), the woodlot has become sufficiently small that the edges from all aspects merge at the woodlot's center. Woodlots of sizes smaller than (a) are not capable of maintaining mesic conditions and have become, functionally, entirely edge. Woodlot size (a) is theoretically the minimum size at which

Table 3-5. Potential Indicators of Disturbance in Southern Wis-consin Forest Islands

Number	Characteristic
1.	Human population density per census tract
2.	Living basal area per hectare
3.	Dead basal area per hectare
4.	Living stem density per hectare
5.	Dead stem density per hectare
6.	Sugar maple density per hectare
7.	Sugar maple basal area per hectare
8.	Sugar maple importance value
9.	Proportion of dead basal area [i.e., $3 \div (2+3)$]
10.	Proportion of dead stems [i.e., $5 \div (4+5)$]

Table 3-6. Matrix of Correlation Coefficients of Species Richness and Each Index of Disturbance (Table 3-5)

	Species richness	Forest island size	Log$_{10}$ human population	Living basal area	Dead basal area	Live stem density	Dead stem density	Sugar maple stem density	Sugar maple basal area	Sugar maple importance value	Proportion dead basal area
Forest island size	-0.08										
Log$_{10}$ human population	0.12	-0.03									
Living basal area	-0.26	0.04	0.08								
Dead basal area	-0.10	-0.13	-0.40	-0.33							
Live stem density	0.17	-0.20	-0.33	-0.06	0.40						
Dead stem density	0.31	-0.21	-0.33	-0.30	0.60	0.66					
Sugar maple stem density	-0.23	-0.10	-0.30	0.34	0.10	0.05	-0.19				
Sugar maple basal area	-0.38	0.01	-0.04	0.54	-0.07	-0.30	-0.38	0.65			
Sugar maple importance value	-0.35	-0.00	-0.14	0.34	0.04	-0.31	-0.38	0.83	0.90		
Proportion dead basal area	-0.04	-0.18	-0.37	-0.44	0.97	0.33	0.59	0.04	-0.10	0.04	
Proportion dead stem density	0.38	-0.11	-0.22	-0.43	0.44	0.29	0.88	-0.28	-0.36	-0.34	0.48

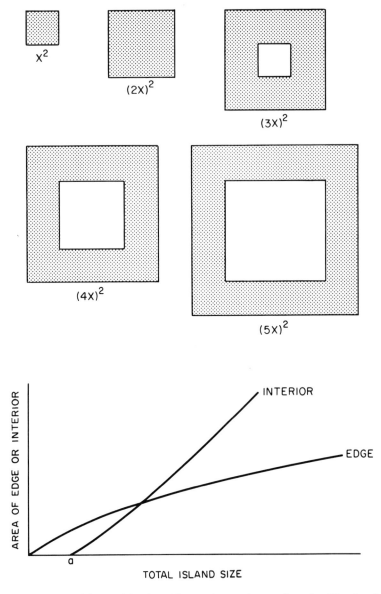

Fig. 3-8. Hypothetical forest islands with varying perimeter lengths. The depth of the edge community (stippled) has been held constant at X units. The area of edge increases with an increase in island size. At point "a" (lower graph), interior conditions begin to form. The area of interior increases at a steeper rate than the area of edge for nearly square or circular islands.

"interior" can be differentiated from "edge" and where mesophytic conditions might exist. The relationship between the area of edge, area of interior, and total area can be further demonstrated for forest remnants of a generally round or square shape (Fig. 3-8). Actual slopes of the edge and interior functions could be determined from field analysis.

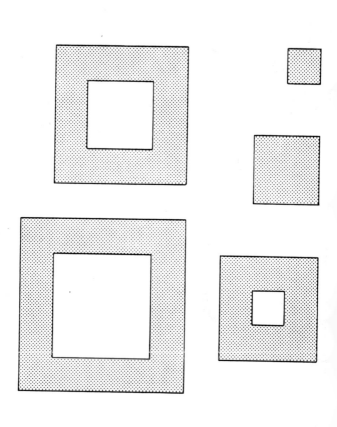

These assumptions were related to the species richness and community dynamics of the mesic forest woodlots of metropolitan Milwaukee (Levenson 1976). When the interior is of sufficient size to maintain the low light, mesophytic conditions, sugar maple, and beech (if present) are favored over the less tolerant species. With smaller interior areas, the mesic species are "rejected" in deference to those species which possess pioneer tendencies. The relationship between the area of edge (or disturbance) and the potential number of woody species can be demonstrated graphically (Fig. 3-9). Given enough area and variation in conditions, the number of woody species which could exist in an edge environment increases to become asymptotic to the number of woody species of the region, i.e., in southeastern Wisconsin. The function increases rapidly since the major constraint is adequate physical space to accommodate an individual of a new species. The locally high stem densities generate a steep curve of the general form of the species-area function. With the formation of interior conditions, theoretically at size (a), the relationship of woody species in the interior of a woodlot, as approximated by the area bounded by edge trees, follows the general form of a depletion curve with increasing island size (Fig. 3-9). The curve approaches the number of shade-tolerant, mesic species of the region. A general expression describing the woody species richness for the interior of southern mesic forest woodlots can be derived by combining the two previous functions (Fig. 3-9). The resultant function illustrates how species richness responds to the transition of woodlot character from one of all edge to one of largely mature forest. Oscillations occur in the number of species present in the mesophytic environment, whose amplitude and frequency are a function of the intensity and duration of perturbations.

Fitting the theoretical curve to sample data, total woody species richness of the sampled core increased with woodlot size to approximately 2.3 ha. Beyond this size, species richness declined. Levenson (1976) interpreted this as the theoretical point (a)—the minimum size at which the edge community could be differentiated from the interior community. This interpretation compares reasonably with the 2 ha island size at which tree species richness stabilized in the mature oak forests of New Jersey (Forman and Elfstrom 1975). Earlier, Vestal and Heermans (1945) established a similar value (1.6 ha) for "minimum stand" size (the minimum reference area to identify a stand as a particular association or variant) in central Illinois.

The species depletion curve leveled off near an island size of approximately 3.8 ha (9.4 acres), designated as point (b). Point (b) is significant as it probably represents the smallest size at which a mature, southern mesic forest can perpetuate its interior conditions while sustaining limited, random perturbations (Loucks 1970).

The species composition and structure of the remnant forests is more readily understood when examined in light of relatively recent historical events and land use patterns throughout the region. The reduction of the regional forest complex over the past century did not preserve the full variety of environments and associated species of the original forest. Most fertile upland sites were converted to agriculture (Curtis 1959) while the land remaining in forest was generally wet, stony, or rough and otherwise undesirable for agriculture. Thus, by the time urbanization was underway, the choicest upland forest sites had already disappeared. Similarly, Auclair and Cottam (1971) indicated that the majority of the remnant forests in south-central Wisconsin were located on steep slopes of low agricultural value. This study was restricted to upland sites, but the numerous lowland species present in the stands suggest that many

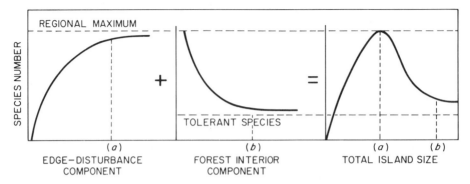

Fig. 3-9. The relationship between the amount of edge and potential woody species richness. Left: Given enough area and variety of conditions, the number of species that could exist becomes asymptotic to the number of woody species of the region. At some point (a), interior conditions begin to form. Center: Response of woody species to an interior environment characterized by low light levels and mesic conditions. The depletion curve levels off at the number of tolerant, mesophytic species in the region, point (b). Right: Curve combining the exponential and depletion functions that simulates the field data from southeastern Wisconsin forest islands.

of these remaining upland sites are actually imperfectly drained upland depressions. The presence of the somewhat poorly drained Mequon silt loam soil further supports this position (Levenson 1976).

Initial isolation and disturbance of the remnant stands occurred between 1850 and 1940. All woodlots of the region were probably selectively logged during the last century and some were clear-cut, while grazing of the woodlots in southern Wisconsin was a widespread practice (Auclair and Cottam 1971, Whitford and Whitford 1972). Owners and neighbors of isolated forests indicated that nearly all woodlots studied were grazed 40-50 years ago (1930-1940). Grazing was a constituent of a major regional ecosystem disturbance during the drought and depression years of the mid-1930s when forage crops were in limited supply and the woodlots were a source of inexpensive, emergency forage. Coupled with the drought, the impact of the grazing and soil compaction by the large livestock population was devastating. Although grazing was largely terminated by the 1950s, the forest structure today is, in large part, a result of natural recovery following that period.

The species turnover in mesic forest communities was viewed by Loucks (1970) as "repeating waveform phenomena triggered by random perturbations" at intervals of 20-200+ years. He suggested that peak species diversity in the mesic forest was achieved 100-200 years after secondary succession had been initiated, a period when members of both the pioneer and stable communities were present. As intolerant species drop from the community with time, species diversity declines. The self-perpetuating forests thus are less diverse than the later successional stages leading to them.

Species diversity of the canopy and understory strata of the 43 habitat islands appears to conform to the model described by Loucks (1970), but at a local, not a regional scale. His random perturbations were large or widespread, and in the framework of a regional forest. Perturbations in the isolated islands may be much smaller in magnitude, i.e., windthrow, death by old age, vandalism, etc., but occur more frequently, maintaining the stand in a subclimax condition. Instead of destruction of all species triggering widespread secondary succession as in Louck's (1970) model, the woodlots suffer chiefly localized disturbance with patches of successional species intermixed among the older, large climax species. More disturbance patches result in a higher species richness for the island. Suhrweir (1976) also interpreted this phenomenon as maintaining species richness near the peak of the waveform. Frequent disturbances coupled with isolation prevent some habitat islands from reaching the expected structure and composition of the terminal stages and result in the moderately high species richness.

Isolated forest woodlots are completely surrounded by a protective edge when the concept is expanded to include the canopy—the upper edge. When a large tree dies, or is windthrown, forest structure and composition are likely to change rapidly. Gaps within the forest function in much the same manner as the edge. The frequency of gaps can contribute greatly to making a large woodlot in reality no more than a highly convoluted edge.

Similarly, removing an edge or dividing a large woodlot with an access road or transmission line also may destroy the mesophytic environment of the system. A large woodlot is more capable of sustaining localized perturbations. However, as woodlot size declines, a perturbation in the edge-canopy becomes more catastrophic. A small woodlot (< 2.3 ha) dominated by sugar maple probably could not persist longer than the remaining canopy trees themselves. Conditions within a small woodlot, with its protective edges removed, shift to favor intolerant species and to reject the "more conservative mesic types" (Auclair and Cottam 1971). Recently isolated segments most commonly shift from a mesophytic to a more xerophytic community which is in dynamic equilibrium with the surrounding environment. Over time, with the formation of a complete canopy and dense edges, the woodlot may again provide conditions favorable to the development of a mesophytic community, if it is of sufficient size.

The inverse relationship between relative density and success of sugar maple and white ash seedlings in the urban and disturbed rural habitat islands is a manifestation of the dynamics. Sugar maple seedling densities were lower in the urban and disturbed sites while white ash appeared to reach somewhat higher densities (Levenson 1980). The situation was reversed in the less disturbed rural islands, even though sugar maple and white ash occupied similar positions in the canopy. The reduced success of sugar maple in the metropolitan woodlots had been observed earlier and was partially attributed to heavy rabbit browsing—an impact of reduced woodlot size and increased edge habitat. There was also some indication that suitable seedbeds for sugar maple were absent in the more xeric and compacted soils of the urban sites. Greller (1975) reported a general scarcity of seedlings of canopy species in the municipal forests of Queens County, New York, both in high disturbance areas as well as in the less-disturbed portions of the park. These problems require further investigation as they have a direct bearing on the successional dynamics and future composition of urban and other isolated forests.

The loss of American and slippery elm due to Dutch-elm disease is another perturbation that has created conditions favorable for white ash. The relative contribution of white ash in the upland forests of the Milwaukee region has increased in recent years.

Whitford and Salamun (1954) reported an average importance value for white ash of 4.3% compared to 9.3% in this study. The number of stands in which white ash was present are comparable (95% in each study), but the relative contribution to the importance value has more than doubled. In sites common to both studies (Sites 30, 33, 37, 38 and 40), an increase of white ash is evident while the elms showed a marked decline. Most of the other species have retained similar importance values over the past 24 years. Relatively high abundance of white ash in the smaller size classes accounts for its increased importance values (Levenson 1980). White ash reached a high average importance value in the shrub stratum just as it had in the canopy, averaging almost 1400 stems/ha (Table 3-4).

The replacement of elm by white ash is logical as the genera have similar ecological requirements. White ash has become a major component of the second growth forests of southeastern Wisconsin. Less frequent in old-growth forests, its presence is attributed to the ability to persist in dense shade until a canopy opening occurs (Cope 1948, Fowells 1965). White ash is shade tolerant when young, capable of surviving with little growth under a full canopy at less than 3% full sunlight (Cope 1948), but becomes less tolerant with increasing age. Canopy openings formed by the death of elms, or other disturbance, create areas of localized secondary succession within the stand. When exposed to full light, ash seedlings are capable of rapid growth (Guenther 1951).

The role of white ash in the mesic forest is similar to that described for black cherry (*Prunus serotina*) in the xeric forests of south-central Wisconsin (Auclair and Cottam 1971). The relative importance value of black cherry was reported to average over 50% in the sapling and small tree sizes, and in some stands it was the only species in those strata. To a lesser extent, I recorded relatively high densities of black cherry in the smaller size classes only in the dry-mesic sites dominated by white oak and character-ized by less dense canopies (Levenson 1980). White ash is more successful in the southern mesic forest, although both species act as opportunists responding to disturb-ances in the canopy, as suggested by higher stem densities in the shrub layer (Table 3-4).

For the community to approach a terminal equilibrium, a source of propagules for each potential component of the community must be close enough to insure the arrival of viable seeds in the stand. Once a species is eliminated from a woodlot, it may not be possible for it to reestablish. Auclair and Cottam (1971) report that species which are bird-dispersed (*Prunus* spp. and *Celtis* spp.) have a considerable advantage over the heavy-seeded mammal-dispersed or wind-dispersed species (*Quercus* spp., *Carya* spp., *Acer* spp.). They further report that the isolation of woodlots in south-central Wisconsin has localized the distribution of sugar maple and is the major obstacle to the further successional development of the oak forests (*Quercus* spp.) to sugar maple-basswood, the terminal forest for the region. American beech, another climax species in the forests of southeastern Wisconsin, offers a similar example. As beech is unable to produce a viable mast in this region (Ward 1956), its reproductive potential is limited to root sprouts. Thus, the elimination of beech from an existing island greatly diminishes the probability of natural reestablishment.

The majority of isolated forest habitat islands in metropolitan Milwaukee are bounded by agricultural, urban, or other intensive land uses. The retention of extensive tracts of original forest was proposed by Gomez-Pompa et al. (1972) as the only way to reconstruct tropical rain forests which are unable to develop toward maturity when the tracts are widely separated by agricultural land. Considerable distances exist between the few large islands remaining in the agro-urban landscape (Levenson 1976, Davis and Glick 1978). Various authors have suggested that islands be connected by strips of protected habitat, or green belts. However, land availability and cost almost prohibit such acquisition in the metropolitan matrix. In the rural sectors, fencerows, windbreaks, utility and transportation rights-of-way often provide a source of seeds, cover, breeding grounds, and avenues for seed dispersal. These habitat corridors contribute greatly to the overall texture of the landscape. However, most of the woody species of the fencerow corridors are the same as those of the forest edge, not its interior. Fencerows are not unlike edge communities in terms of stem and foliage density, and are often extensions of the forest edge community with the same wildlife-initiated origins. They are of particular value when they link otherwise isolated woodlots.

Aerial reconnaissance of the Milwaukee metropolitan area indicated an extensive interstitial forest (Detwyler 1972) composed of rapidly growing, traditional lowland species, e.g., American elm, silver maple, and various ashes. In recent years, rapid development and urban sprawl compounded by heavy elm mortality have left large areas devoid of interstitial forest cover. These circumstances provide an unusual opportunity to restore a portion of regional ecosystem variety. Encouragement of the use of ecotypically local, upland trees as street and yard plantings could provide seed sources, increase species richness, and overcome the problems associated with interisland distance. Landscaping with native shrubs and ground cover in residential and commercial areas could supplement the lesser forest strata while providing widespread habitat and food sources for birds and small mammals.

In addition to removing associated migration and fencerow corridors, urban systems tend to select for smaller woodlot island sizes. These changes in landscape structure can have serious consequences for the vegetation and associated fauna since they may be functionally edge. Galli et al. (1976) indicated that woods of 0.2 ha contained only bird species characteristic of forest edges. Birds characteristic of the forest interior began appearing in islands of about 0.8 ha. Foliage and stem density are greatest per unit area in the edges and canopy, and penetration of the edge into the forest can be approximated in these terms (Bruner 1977). Height of the edge foliage can be measured or estimated. Similarly, expanding the concept of edge to include the canopy, vertical thickness of the canopy layer can also be quantified. Mawson et al. (1976) developed a model using geometric shapes to quantify tree and shrub crown volumes by layers. The resultant edge-canopy volume represents the major source of cover and nesting for birds as well as the region of greatest primary production. This volume provides a meaningful indicator of avian habitat quality (MacArthur et al. 1962, MacArthur 1964, Sturman 1968) since the remainder of the mesic forest interior is essentially open space.

Conclusions

The richness of woody species in isolated woodlots is largely a function of disturbance, whether natural or human-induced. Heavy human usage of the urban islands maintains a continual state of disturbance resulting in an increased edge effect and a high species richness from colonization of less tolerant species. Human disturbance maintains species richness near the regional peak (Loucks 1970) and, in agreement with Suhrweir

(1976), retards the island's transition to more mesic species. In rural areas, where islands are under less human pressure and disturbances are more random, an island can redevelop successively resulting in more frequent extinction of intolerant species and a lower overall species richness in the mesic interior.

The classical species-area relationship was shown to be incapable of predicting the number of woody plant species as a function of island size for these southern mesic forests. Limits were placed on the function by (1) the finite number of woody plant species in the region, and (2) mortality of shade intolerant species under the low light levels and mesic conditions of undisturbed interiors of large woodlots. Lag in community response to earlier disturbance, size reduction, and isolation of islands obscured direct correlation.

Species richness for trees and woody plants generally increased with island size to approximately 2.3 ha. Islands smaller than 2.3 ha functioned essentially as edge communities composed of a mix of intolerant and residual tolerant species. Islands larger than 2.3 ha showed a general decline in species richness as interior, mesic conditions became established with only tolerant species persisting. Species richness ceased to decline at approximately 3.8 ha, suggesting that only the shade-tolerant, mesophytic species remained.

White ash is a vigorous, opportunistic competitor in the forest islands of metropolitan Milwaukee that responds to disturbance. It appears to be replacing the elms (*Ulmus* spp.) and may be considered a major community component of all structural strata of the southern mesic forest.

Sugar maple seedling densities were often lower in the urban and heavily disturbed sites while white ash reached somewhat higher densities. Locally heavy rabbit browsing and seedbeds unsuitable for sugar maple have been suggested to account for its low numbers. Further investigation of this relationship is needed as it has a direct bearing on the successional dynamics and future composition of the urban forest.

Community development, viewed as a series of extinctions and colonizations, is a function of the distance to the seed source of each component species. Encouraging the use of local ecotypes of upland trees and shrubs as street and yard plantings in the urban system could help to reduce interstand distance by providing seed sources. In rural areas, the retention of fencerows and other environmental corridors could provide the much-needed "stepping stones" for dispersal.

Management priorities must include protecting the large (> 4 ha), less disturbed woodlots as only they have the potential to provide self-perpetuating examples of the once regional southern mesic environment. The larger islands also function as refugia for the rarer species requiring the southern mesic habitat.

The notion of obtaining maximum species diversity in the landscape is much in vogue. The mesic forest is the terminal association, but is characterized by a relatively low species diversity. A management strategy that would preserve maximum species diversity would maintain a high proportion of edge through frequent disturbance. Diamond (1973) warns ". . . the smaller the tract, the more rapidly will forest species tend to disappear and be replaced by the widespread second-growth species that least need protection." A more rational approach would maintain forested islands large enough to support the region's stable natural community. Or, as Usher (1979) suggests, ". . . manage a few reserves that approximate to the normal species-area relation for that region since . . . they are more 'typical' than a selection of particularly species-rich sites."

Nevertheless, retention of the smaller islands and fencerows is desirable as they harbor a diverse mix of exotic, pioneer and terminal plant community components. In agreement with Forman et al. (1976), the smaller areas function as "stepping stones" from which species can be reintroduced when community equilibrium is disturbed (Sullivan and Shaffer 1975).

A top priority should include research programs and management strategies for American beech (*Fagus grandifolia*) and other species with specialized or low dispersal potential. Once eliminated from an island, natural reestablishment is not possible. The small populations of beech found in some of the Milwaukee County Parks are presently in danger of local extinction.

Acknowledgments. I express my appreciation to Dr. Forest Stearns for his advice and support throughout the study, and to Robert L. Burgess and David M. Sharpe for their suggestions and help in preparing the manuscript. Special thanks go to Professors John Ong, Alan Ek, and Richard Forman for stimulating discussions, and to Mr. David Dralle and Ms. Linda Hoehne for their aid in the field. Special acknowledgments are expressed to the Milwaukee County Park Commission and to the individual owners of the private tracts for their interest and kind permission to conduct the study on their grounds. Finally, I would like to thank my wife, Linda, for her patience and capacity to understand, help and encourage.

This research was supported by the Eastern Deciduous Forest Biome, US/IBP, funded by the National Science Foundation under Interagency Agreement AG-199, BMS76-00761 with the Department of Energy, Oak Ridge National Laboratory. Contribution No. 349 from the Eastern Deciduous Forest Biome, US/IBP.

4. The Groundlayer Vegetation of Forest Islands in an Urban–Suburban Matrix

LINDA M. HOEHNE
DEPARTMENT OF BOTANY
UNIVERSITY OF WISCONSIN-MILWAUKEE

Forest islands are areas of woodland surrounded by a matrix of cultivated land, residential development, or other nonforest land use. The vegetation of forest islands can be separated into four strata: a canopy, an understory, a shrub layer and a groundlayer. While the arboreal vegetation is more conspicuous, the structure and composition of the groundlayer may be affected discretely by island size, past history, degree and kind of current disturbance, and relative isolation of these forest patches.

Curtis (1959) described the vegetation of the southern mesic (beech-maple) forest in Wisconsin's southeastern counties, and Whitford and Salamun (1954) examined the vegetation of upland forests in the Milwaukee area. Levenson (Chapter 3) described the woody vegetation of 43 urban and suburban forest patches in Milwaukee and Ozaukee counties, including four previously studied by Whitford and Salamun (1954). Both studies dealt with the effect of urbanization on the native flora.

Objectives of this study were to survey the groundlayer component of the forest islands, determine the factors influencing groundlayer composition, and identify and account for changes in the species composition of selected stands in the 24 years since the Whitford and Salamun study.

Description of the Study Area

The forest islands studied are all in Milwaukee County, the southern half of Ozaukee County and the eastern edge of Waukesha and Washington counties (Fig. 4-1). All wooded islands in this study lie within the climatic influence of Lake Michigan, al-

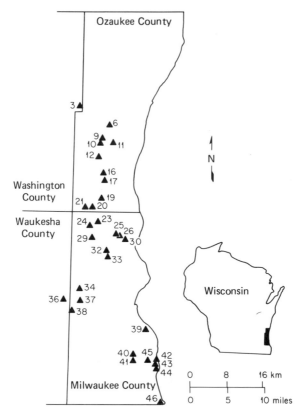

Fig. 4-1. Location of the 31 forested islands in the metropolitan Milwaukee region. Key: 3, Fechter's Woods; 6, Buckskin Bowmen; 9, R & R Excavating; 10, Cedarburg Woods; 11, Grafton Bank Woods; 12, Fox Farm Woods; 16, Mee Kwon Park; 17, Highland Woods; 19, Garvey's Woods; 20, Gengler's Woods; 21, Stauss Woods; 23, Convent Woods; 24, Bradley Woods; 25, Brown Deer Park; 26, Rangeline Woods; 29, Haskell Noyes Park; 30, Kletzsch Park; 32, U.S. Army Reserve; 33, McGovern Park; 34, Underwood Parkway; 36, Bishop's Woods; 37, Milwaukee County Zoo; 38, Greenfield Park; 39, St. Francis Seminary; 40, Cudahy Woods (north); 41, Cudahy Woods (south); 42, Grant Park (old growth); 43, Grant Park (north); 44, Grant Park (south); 45, Rawson Park; 46, Oak Creek Power Plant.

though the degree to which the influence is felt depends on the distance from the shoreline and on prevailing weather conditions. The effects of Lake Michigan are greatest in spring and early summer when the prevailing winds are northeasterly and least in the winter when the prevailing winds are westerly. Temperatures near the lake are lower in summer and higher in winter than farther inland.

Milwaukee and Ozaukee counties, both adjacent to the western shore of Lake Michigan, are characterized by morainic ridges parallel to the lake that control the drainage pattern (Fenneman 1938). Local relief is less than 30 m (100 ft) with general relief up to 115 m (380 ft) above Lake Michigan. The soils belong chiefly to the Ozaukee, Morley, Hochheim, and Theresa series. These are moderately drained to well drained silty soils with little surface runoff or erosion. The Mequon series, also represented, occurs in depressions with poor drainage and slow runoff. Levenson (Chapter 3) describes the region and the forest islands in detail.

Methods

Study Site Selection

Thirty-one patches of southern mesic and dry mesic forests (Curtis 1959) were sampled between June and September, 1975. The stands were located in two north-south transects matching the north-south climatic bands and topographic ridges found in the area. These transects traverse a continuum through urban, suburban and rural areas and permit the effects of urbanization to be investigated. Four forest islands were also located near the Milwaukee-Waukesha county boundary, the western limits of beech-maple forests in this area (Fig. 4-1).

Stand Selection

Stands were selected initially on the basis of tree composition. Sugar maple (*Acer saccharum*) was a required indicator species for the mesic stands (Curtis 1959). Associated species included beech (*Fagus grandifolia*), white ash (*Fraxinus americana*), red oak (*Quercus borealis*), white oak (*Q. alba*), and basswood (*Tilia americana*). All stands had fully developed ground, shrub, understory, and canopy layers. Stands chosen had fully developed edges, as indicated by trees exhibiting heavy branching on the open-grown side of the trunk and clear length on the forest grown side. There was no evidence of recent major disturbance, such as grazing, fire or timber harvesting, but disturbance from foot paths, bicycles, campfires, and minor vandalism was noted. All islands were discrete units with no interconnections such as fencerows or parkways. Island size varied from 0.59 ha (1.5 acres) to 21.05 ha (52.0 acres).

Sampling Methods and Materials

The groundlayer was sampled using the stratified random strip method (Lindsey 1955). A starting point was located at one corner of the island approximately 10 m from the closest edge tree and a series of east-west transects was established. Transects were 25 m apart in most islands, but 50 m separated the line strips in larger islands and a 15 m separation was used in smaller islands. Plots 1 m^2 were used to sample the groundlayer beginning with a quadrat at the starting point and plots thereafter every 12.5 m. Generally 40 plots, totaling 0.004 ha, were sampled in each island; however, in some smaller or larger islands, 30 to 50 plots were used. The 1 m^2 plots were nested within the shrub and canopy sampling plots (Fig. 4-2) used by Levenson (Chapter 3).

   Woody stems under 0.5 m high and all herbaceous stems were counted. Since many species reproduce vegetatively and occur in clones, each stem was counted separately. Each tuft of grasses and sedges was counted as a unit. This was also true for species such as *Geranium* where leaf tufts arose from a shortened stem. Species present in the stand but not occurring in the plots were listed. Nomenclature follows Gleason (1952).

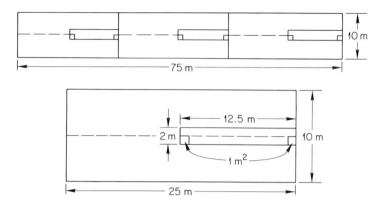

Fig. 4-2. Example of a line-strip, three plots in length, showing the arrangement of nested plots. In the enlarged view (below), trees were sampled in the 10 X 25 m plots, shrubs in the 2 X 12.5 m plots, and the groundlayer vegetation in the 1 m² plots.

The type and relative extent of disturbance was recorded for each wooded island. Examples of disturbance included foot and bicycle trails, campfire areas, tree girdling, rubbish disposal, and cutting of trees for firewood. Disturbance in each plot was noted.

Density (stems/unit area) and frequency (percentage occurrence in plots) were determined for each groundlayer species and were converted to realtve values. Relative density is the proportion of the individuals of one species expressed as a percentage of the total individuals of all species. Relative frequency is a measure of the frequency of one species compared to the frequency of all species expressed as a percentage (Curtis and Cottam 1962). Importance value was calculated as the average of relative density and relative frequency (Lindsey 1956).

Interisland comparisons, by means of hierarchical cluster analysis, were based on the importance values of those species present. This analysis indicated which stands were most similar in groundlayer composition.

A relative disturbance value was estimated for each stand taking into account visible evidence of trails, campfires, vandalism and other human activity, and the recorded and observed information on recent grazing and logging. Stands were rated from one to five ranging from minimum disturbance to stands that were severely disturbed.

The Shannon-Wiener Index (H') was used to calculate species diversity for each island (Odum 1971). This was expressed as

$$H' = - \Sigma \left(\frac{n_i}{N} \right) \ln \left(\frac{n_i}{N} \right) = - \Sigma P_i \ln P_i$$

where n_i was the number of individuals for each species, N was the total number of individuals of all species and P_i was the importance probability for each species $= n_i/N$. This index utilizes both the number of species and the numerical distribution of individuals among species, i.e., the equitability component (Odum 1971). The equitability component (J') was calculated as $J' = H'/H'_{max}$ where H' is the island's Shannon-

Wiener index and H'_{max} is the natural logarithm of the number of species (Pielou 1966). This index is independent of sample size. H' values near 0 indicate low diversity while higher values indicated greater diversity. Values of J' near 1.0 indicate greater evenness of distribution while lower values indicate less equitability among the species. Data were analyzed using a computer program developed by Zar (1968).

Results and Discussion

Nature of the Groundlayer

Disturbance of the groundlayer in the 31 islands varied from relatively light to heavy, depending primarily on the quantity and type of human use. Highland Woods (17) was little affected, while wide footpaths and bicycle trails were common in some Milwaukee County Park stands. Although the paths were bare, many nonforest species were found along the trails and in other disturbed sites.

The largest group of southern mesic forest species are spring blooming plants (Curtis 1959). Those spring ephemerals that retained their leaves during the summer and hence were included in this study were bloodroot (*Sanguinaria canadensis*), Jack-in-the-pulpit (*Arisaema triphyllum*), wild ginger (*Asarum canadense*), blue cohosh (*Caulophyllum thalictroides*), hepatica (*Hepatica acutiloba*), woods phlox (*Phlox divaricata*), mayapple (*Podophyllum peltatum*), and two species of trillium (*Trillium grandiflorum* and *T. gleasoni*).

Distribution patterns of herbaceous species varied from random to clumped. For example, mayapple was usually found in clumps, a result of vegetative reproduction. Snakeroot (*Sanicula gregaria*), reproducing chiefly by seed, was randomly distributed. Species such as clearweed (*Pilea pumila*) and yellow jewelweed (*Impatiens pallida*) often were associated with wet areas or with canopy openings, respectively. Beech-drops (*Epifagus virginiana*), a root parasite, was found in beech stands.

Nineteen species, including eight woody species, were found in 50 percent or more of the 31 stands (Table 4-1). Choke cherry (*Prunus virginiana*), sugar maple, enchanter's nightshade (*Circaea quadrisulcata*), white ash, and false Solomon's seal (*Smilacina racemosa*) were present in 90% or more of the islands.

White ash showed the highest average density (43,659 stems/ha) and frequency (57%). Choke cherry, false Solomon's seal, white ash, and sugar maple were similar in average density and were found in 33 to 46% of the plots. Other species reaching high average densities (over 8000 stems/ha in all stands) and found in more than 30% of the islands included Jack-in-the-pulpit, wild geranium (*Geranium maculatum*), early meadow rue (*Thalictrum dioicum*), Virginia waterleaf (*Hydrophyllum virginianum*), and broad-leaved goldenrod (*Solidago flexicaulis*).

Only four species (21% of the total), choke cherry, white ash, sugar maple and false Solomon's seal, had an average frequency per stand greater than 30% while 39% of the species had an average frequency of 3% or less (Fig. 4-3). Most species had low frequencies while few had medium to high frequencies, suggesting the absence of invaders, a characteristic of stable stands.

Table 4-1. Percent Presence, Average Density (stems/ha), Standard Deviation of Density, and Average Frequency per Stand for Groundlayer Species Present in 50% or More of the Stands

Species	Presence	Density	SD	Frequency (%)
Prunus virginiana	96	26,366	17,561	45
Fraxinus americana	93	43,659	72,837	57
Circaea quadrisulcata	93	24,911	22,469	18
Acer saccharum	93	23,766	32,909	46
Smilacina racemosa	90	26,365	39,292	33
Tilia americana	77	6,169	14,997	16
Arisaema triphyllum	74	16,247	31,675	22
Geranium maculatum	70	12,874	14,009	20
Allium tricoccum	70	3,068	4,371	9
Geum canadense	67	1,759	1,588	8
Podophyllum peltatum	64	3,776	3,943	8
Trillium grandiflorum	64	2,843	4,116	6
Fagus grandifolia	64	2,419	2,652	10
Vitis spp.	64	1,477	2,038	7
Hydrophyllum virginianum	61	8,371	8,945	8
Prunus serotina	61	1,628	1,662	7
Ribes cynosbati	54	5,779	8,380	13
Thalictrum dioicum	51	10,398	8,837	9
Actea alba	51	939	623	6

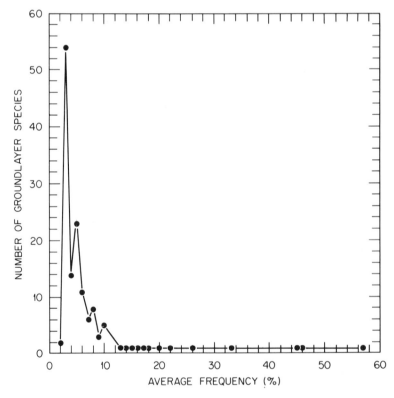

Fig. 4-3. Species frequency distribution for the groundlayer in 31 forested islands in southeastern Wisconsin.

Cluster Analysis

Cluster analysis (Ward 1963) suggested relationships among the 31 islands based on the ground vegetation (Fig. 4-4, Table 4-2). The dendrogram is based on importance values using the 100 species with the highest values. Leading dominants are designated in each cluster.

Six species dominated the groundlayer. These were sugar maple, choke cherry, white ash, false Solomon's seal, Jack-in-the-pulpit and enchanter's nightshade. No single species was dominant in all islands, but importance values of the six dominants were generally high. The islands clustered into six groups (Fig. 4-4).

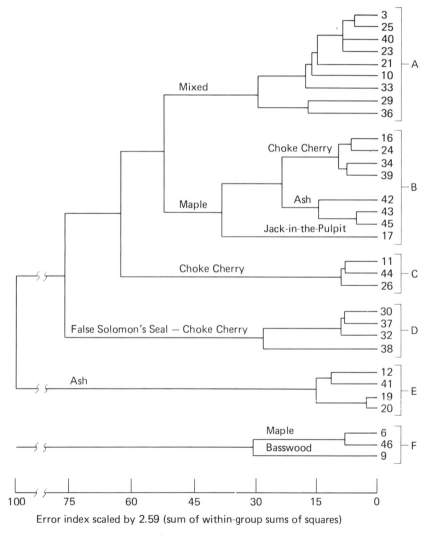

Fig. 4-4. Dendrogram of the cluster analysis for the groundlayer of 31 forested islands based on species composition and importance values. The leading dominant is listed for each major group.

Table 4-2. Characteristics of Stands in Six Cluster Groups of Forest Islands in Southeastern Wisconsin

Stand	Size (ha)	Canopy dominants	Basal area (m²/ha)	Soil types	H'	J'	Disturbance rating[a]
Group A							
3. Fechter's Woods	2.23	Maple-mixed	32.76	Hochheim	3.07	0.84	3
25. Brown Deer Park	2.47	Beech	29.49	Ozaukee, Mequon	2.85	0.76	4
40. Cudahy Woods (north)	7.81	Red oak-maple	36.12	Morley	2.46	0.71	1
23. Convent Woods	3.00	Maple	31.00	Ozaukee	2.71	0.72	3
21. Stauss Woods	2.80	Basswood	29.62	Ozaukee	2.81	0.74	4
10. Cedarburg Woods	1.21	Maple	24.27	Hochheim	2.73	0.72	4
33. McGovern Park	3.12	Red oak-maple	25.68	Mequon	2.51	0.75	4
29. Haskell Noyes Park	0.73	Maple	28.54	Ozaukee, Mequon	2.12	0.63	5
36. Bishops Woods	21.05	Maple-mixed	29.80	Ozaukee	2.26	0.63	3
Group B							
17. Highland Woods	4.25	Maple-beech	31.10	Hochheim, Ozaukee	1.74	0.58	1
16. Mee Kwon Park	7.21	Maple-beech	32.00	Ozaukee	2.30	0.68	2
24. Bradley Woods	2.43	Maple-mixed	29.99	Ozaukee, Mequon	2.84	0.78	2
34. Underwood Parkway	5.47	Beech-mixed	22.49	Ozaukee, Morley	2.69	0.74	4
39. St. Francis Seminary	11.46	Maple-mixed	33.02	Blount	2.69	0.74	4
42. Grant Park (old growth)	3.24	Maple-mixed	28.42	Morley	2.62	0.81	4
43. Grant Park (north)	2.83	Maple-beech	33.79	Morley	2.35	0.68	2
45. Rawson Park	7.93	Beech-maple	31.04	Morley	2.45	0.71	4
Group C							
11. Grafton Bank Woods	0.57	Basswood	21.56	Hochheim	2.23	0.69	5
44. Grant Park (south)	3.97	Maple-mixed	43.30	Morley	2.22	0.65	2
26. Rangeline Woods	1.70	Maple-mixed	30.63	Ozaukee	1.97	0.63	5

Group D							
30. Kletzsch Park	4.00	Maple	33.49	Fox	2.15	0.71	3
31. Milwaukee County Zoo	4.50	White oak	31.44	Ozaukee	2.05	0.74	3
32. U.S. Army Reserve	4.21	Beech-maple	22.78	Mequon	2.25	0.62	5
38. Greenfield Park	18.34	White oak	31.07	Ozaukee, Mequon	1.34	0.44	3
Group E							
12. Fox Farm Woods	1.98	Maple-beech	28.30	Hochheim, Theresa	1.19	0.34	2
41. Cudahy Woods (south)	4.13	Maple	32.03	Morley, Adrian	1.82	0.69	1
19. Garvey's Woods	2.91	Maple-beech	30.13	Ozaukee	2.21	0.66	2
20. Gengler's Woods	1.58	Beech-mixed	22.81	Kewaunee	2.09	0.59	3
Group F							
6. Buckskin Bowmen	6.48	Maple-mixed	27.22	Hochheim	1.30	0.43	2
46. Oak Creek	0.59	Beech-mixed	33.26	Morley	1.92	0.52	1
9. R & R Excavating	1.62	Maple	21.51	Hochheim	1.72	0.52	5

[a] Disturbance rating ranges from 1 (least disturbed) to 5 (very heavily disturbed).

Group A consists of nine islands that lack one or two clear-cut dominant species. No pattern was evident among the species with greatest importance values. Diversity (H') in these islands was usually greater than in other groups. Island size ranged from 0.73 to 21.05 ha and all of the major soil types were represented. These nine islands were located from Waukesha County on the west to Cudahy on the south and northward into Washington County.

In eight islands (Group B), sugar maple, choke cherry, Jack-in-the-pulpit, and white ash were high in importance value. Diversity (H') was usually average or above average. These islands ranged from 2.83 to 11.46 ha, and most of the major soil types were represented.

Choke cherry was the dominant species in three islands (Group C). Again, island size and soil type varied. Species diversity (H') was slightly below average.

False Solomon's seal and choke cherry were the leading dominants in four islands (Group D). Island size ranged from 4 to 18.3 ha and three different soil types were present. Diversity (H') was below average.

Four islands (Group E) were dominated by white ash. The soils varied between stands. Island size ranged from 1.98 to 4.13 ha while diversity (H') was below average. Three of the four islands were in Ozaukee County while the fourth was in southeastern Milwaukee County.

Two of the three islands in Group F had sugar maple as a single leading species while in the third, sugar maple and basswood were codominants. Size ranged from 0.59 to 6.48 ha and the location from far southeastern Milwaukee County to mid-Ozaukee County. The two islands in Ozaukee County had similar soil types. Diversity was below average.

Thus, when these 31 islands were clustered by species present or absent and by importance value, no relationship to size, soil type or location was evident. Similarly, there was no apparent relationship between cluster groups and canopy basal area (Table 4-2).

There is, however, a relationship between the number and kind of species present and the amount of groundlayer disturbance. Islands receiving heavy human pressure with wide footpaths, campfire areas, or open canopies showed greater diversity. For example, the woods at Brown Deer Park and at St. Francis Seminary both received heavy use and showed higher than average diversity. The types of plants varied, with more weedy and open field species found in disturbed areas. In general, diversity increased from the least disturbed to the heavily used stands but dropped to lower levels in stands that had been severely abused.

Presence and distribution of species are also affected by local environment. Local high densities of yellow jewelweed and wood nettle (*Laportea canadensis*) were associated with openings in the canopy (Curtis 1959). Clearweed was often found in moist areas.

Recent Changes in Vegetation

Curtis (1959) compared species presence in three regions of the southern mesic forest. He included species present in 80% or more of the stands in at least one of these regions. Washington and Ozaukee counties comprised one region. Since that region and the area of this study overlap and are similar in location, topography and soils, the vegetation data were compared using a χ^2 test. The species Curtis listed and their percent presence were used as the expected values and percent presence in this study as the actual values. The two sets of data were not similar ($\chi^2 = 971.23$).

Approximately 40% of the species listed by Curtis were spring flowering plants. Since the 1975 sampling was done in summer, a second χ^2 test was performed using only the summer flowering plants from both lists; this also indicated a significant difference between the species in the two studies ($\chi^2 = 515.11$).

Two ferns, maidenhair fern (*Adiantum pedatum*) and Virginia grape-fern (*Botrychium virginianum*) were not sampled in the 1975 study while Curtis listed presence values near 50%. There may be several possible reasons for the absence of these plants. Ferns are often collected for home gardens and some may be unable to tolerate disturbance, soil compaction or air pollution. The grass *Brachyelytrum erectum* also was not encountered in this study while Curtis listed it with a presence value of 43%.

Curtis also listed higher presence values for the bedstraws (*Galium aparine* and *G. concinnum*) and tick-trefoil (*Desmodium glutinosum*) than those recorded in 1975. Three other species not present as often in 1975 as in the earlier work were hairy Solomon's seal (*Polygonatum pubescens*), elm-leaved goldenrod (*Solidago ulmifolia*), and rattlesnake root (*Prenanthes alba*). The wild leek (*Allium tricoccum*) was found considerably more often in 1975 than earlier. Most remaining species had similar presence values in both years.

In 1951, the upland forests of the Milwaukee area were surveyed by Whitford and Salamun (1954); four of the same sites, Greenfield Park, the Milwaukee County Zoo, McGovern Park and Grant Park (old growth stand), were sampled again in 1975. Frequency and presence values from the 1951 and 1975 studies were compared for shrubs, summer flowering herbs, and spring flowering herbs that remained through the summer. Fifty-nine shrub and herbaceous species, three present chiefly in the spring, were encountered by Whitford and Salamun at Greenfield Park; only 22 herbs and shrubs were found there in 1975. Of those species present in both years, only two showed large changes in frequency; wild geranium dropped from 82.5% in 1951 to 12.5% in 1975 while choke cherry increased from a frequency of 5.0% in 1951 to 32.5% in 1975. These differences, including the loss of about 60% of the species, may be attributed to increased disturbance and soil compaction.

By 1975, the Milwaukee County Zoo woods had lost about 60% of the species recorded in 1951. Jack-in-the-pulpit, wild geranium, and enchanter's nightshade each decreased considerably in frequency while choke cherry greatly increased. Frequency of false Solomon's seal changed little. These declines in frequency and in number of species result from groundlayer disturbance, soil compaction and a 75% reduction in stand size. Choke cherry is clearly more successful under disturbed conditions.

McGovern Park showed a decrease of about 50% in species richness. Choke cherry and false Solomon's seal had similar frequencies in the two studies but frequency of wild geranium and strawberry (*Fragaria virginiana*) decreased drastically.

At Grant Park (old growth stand), 25% fewer species were present in 1975 than in 1954. Most species showed similar frequencies with the exception of Jack-in-the-pulpit and choke cherry which increased from 40 to 98% and 29 to 48%, respectively.

I also compared tree seedling data from 1951 (Whitford, unpublished data) with my 1975 data. White ash and basswood seedlings generally decreased in frequency in the four stands. This could signify either fewer seed trees in the canopy layer, or closure of the canopy or other environmental conditions unfavorable to seed germination and growth. Black cherry (*Prunus serotina*) also decreased in frequency in the four stands, while choke cherry increased.

In 24 years, these four islands have changed significantly in the number of species present and in the frequencies of those remaining. Two stands have been reduced in size and disturbance has increased in all four stands. Increases in choke cherry may have reduced the numbers of other species through competition and shading, especially if those species were intolerant of disturbance. Sampling dates, weather, and sampling methods may have influenced these results. However, Whitford and Salamun (1954) used approximately the same numbers and sizes of plots as those used in the present study.

Dispersal Strategies

Haase (1965) catalogued the seed dispersal methods of upland hardwood forest species in southeastern Wisconsin. Based on his work, the plants of the forest islands in this study were found to depend heavily on vegetative reproduction or on bird dissemination of seeds. The number of species depending on wind dispersal apparently has declined since 1951 (Table 4-3). However, the small sample size precludes a statement of statistical significance.

Some changes in dispersal method were noted for species of the four islands in common to the Whitford and Salamun (1954) work and this study (Table 4-3). Dispersal mechanisms did not show any significant change at McGovern Park. Greenfield Park showed an increase in species dependent on insect transport while Grant Park (old growth stand) showed an increase in species characteristically using vegetative reproduction. Data from the Milwaukee County Zoo and Greenfield Park indicate that species whose seeds adhere to animals declined in 24 years. All four islands showed a decrease in wind-dispersed species. Although the small number of species precludes a test of significance, these decreases in wind-dispersed plants and in those depending on animal adhesion may relate to increased isolation of these islands.

Conclusions

Disturbance, whether old or recent, natural or human related, appeared to be the most influential factor regulating composition of the groundlayer vegetation. Disturbance affects species presence, density, and frequency. Disturbance usually resulted in greater diversity, but in the most heavily used stands, groundlayer destruction caused by human use, such as footpaths, minibike trails, and campsites, sometimes reduced diversity. Species intolerant of trampling or compaction disappeared while others increased.

The number of species present in the Milwaukee County Zoo, Greenfield Park, and McGovern Park stands decreased 50-60% while the Grant Park old growth stand decreased 25% between 1951 and 1975. Species composition changed as native herbs such as wild geranium, Jack-in-the-pulpit, enchanter's nightshade, and strawberry decreased in frequency. Of the woody species, choke cherry increased while black cherry decreased in frequency. Both of these species are found in Wisconsin southern mesic forests (Curtis 1959) and often occur in disturbed areas (Gleason 1952). The reason for the contrasting response is not clear.

Table 4-3. Methods of Herb and Shrub Dispersal Characteristic of Species from Islands 33, 37, 38, and 42 in Southeastern Wisconsin, and the Number of Times (N) Each Strategy Was Tabulated for the 20 Groundlayer Species with the Highest Relative Frequency Values in Each Site[a]

Method of dispersal	McGovern Park (33) 1951		McGovern Park (33) 1975		Milw. Co. Zoo (37) 1951		Milw. Co. Zoo (37) 1975		Greenfield Park (38) 1951		Greenfield Park (38) 1975		Grant Park (42) 1951		Grant Park (42) 1975	
	N	T	N	T	N	T	N	T	N	T	N	T	N	T	N	T
Vegetative reproduction	12	13.5	9	40.9	6	23.0	3	27.2	7	25.9	5	33.3	6	24.0	6	40.0
Passed by birds	4	12.5	3	13.6	8	30.7	5	45.4	7	25.9	4	26.6	6	24.0	3	20.0
Water or rainwash	3	9.3	2	9.0	1	3.8	0	0.0	1	3.7	1	6.6	1	4.0	1	6.6
Ants or other insects	2	6.2	3	13.6	0	0.0	1	9.0	0	0.0	2	13.3	1	4.0	1	6.6
Animal transport	2	6.2	2	9.0	2	7.6	1	9.0	2	7.4	0	0.0	2	8.0	1	6.6
Animal adhesion	3	9.3	2	9.0	5	19.2	1	9.0	5	18.5	1	6.6	3	12.0	2	13.3
Plant discharge	2	6.2	1	4.5	1	3.8	0	0.0	1	3.7	1	6.6	1	4.0	0	0.0
Wind	4	12.5	0	0.0	3	11.5	0	0.0	4	14.8	1	6.6	5	20.0	1	6.6
Totals	32	100.0	22	100.0	26	100.0	11	100.0	27	100.0	15	100.0	25	100.0	15	100.0

[a] T is percent of total. Methods of dispersal are after Haase (1965).

Exotic or weedy species did not show a great increase in frequency over that found by Whitford and Salamun (1954) or recorded by Curtis (1959), but the presence of several species was noteworthy. Jerusalem artichoke (*Helianthus tuberosa*) and day-lily (*Hemerocallis fulva*), for example, are not native to the beech-maple forest. The former occurs in disturbed sites (Curtis 1959) and the latter has escaped from cultivation (Gleason 1952). Both appear to be successful invaders of disturbed stands. Species not encountered by Whitford and Salamun (1954) but found in my 1975 survey were privet (*Ligustrum vulgare*), dandelion (*Taraxacum officinale*) and common burdock (*Arctium minus*).

Orchids and ferns declined in presence and abundance since Curtis (1959) published his lists. Maiden-hair fern was not encountered in 1975 and Virginia grape-fern showed a large reduction in abundance. Bracted orchid (*Habenaria viridis*) was encountered only twice in 1975 but earlier had a presence value of 11% (Curtis 1959). In contrast, the European helleborine (*Epipactis latifolia*), appeared to have adapted as successfully to the Wisconsin southern mesic forest as it had in New England (Dowden 1975). Apparently a recent invader in this area, helleborine had a 1975 presence value of 16% while it was not mentioned by either Curtis (1959) or Whitford and Salamun (1954).

In summary, between 1951 and 1975, native herbaceous species have decreased in presence and frequency. Stands receiving moderately heavy human use have a greater diversity resulting from additional exotic and weedy species, than do lightly used stands. While the frequency of some native herbaceous species seems to be declining, they are not being replaced by species foreign to the maple-beech forest. The weeds and exotics increase diversity, but their numbers are small. Introduced species, those found in disturbed areas and those native species not characteristic of southern mesic forests, each comprised about 10% of the total number of species. Whether the native species will continue to decrease in frequency while the weeds and exotics increase remains to be seen. Tolerance to human disturbance will determine success or failure of individual species.

Presumably, theories of island biogeography could be applied to the management of urban islands as natural areas. More information will be needed on the life history and tolerance of disturbance of each species. It will be important to determine which native species can survive and reproduce under stress such as that caused by trampling, collecting for gardens, air pollution, and changes in drainage. Woodlots associated with parks normally receive heavier usage than relatively isolated and privately owned wooded islands. Criteria might be established to restrict use of these areas only to the point where the natural flora would not be injured beyond recovery. Park managers might reserve certain wooded islands as natural areas with limited use where the native flora might survive. Other areas could be established for heavier recreational use. In this way the two objectives, preservation and recreation, could both be satisfied.

5. Mammals in Forest Islands in Southeastern Wisconsin

PAUL E. MATTHIAE
DEPARTMENT OF BOTANY
UNIVERSITY OF WISCONSIN-MILWAUKEE

FOREST STEARNS
DEPARTMENT OF BOTANY
UNIVERSITY OF WISCONSIN-MILWAUKEE

The deciduous forest that once covered vast areas of the Midwest has been dissected and decimated until today only isolated patches remain. These islands of forest range from less than one to many hectares in size, but the great majority are smaller than 50 ha. Some are connected by wooded fencerows or drainageways, others are completely isolated in an agricultural or urban matrix.

A rich presettlement fauna included large herbivores and carnivores such as buffalo, moose, elk, white-tailed deer, cougar, lynx, bobcat, wolverine, and black bear (Jackson 1961) and a great variety of smaller mammals as well as the wetland fur-bearers—mink, otter, beaver, and muskrat. With the rapid spread of agriculture and gradual urbanization, many of the larger mammals were extirpated although some, such as the white-tailed deer and coyote, have returned as land use patterns have changed. At least one mammal, the opossum, is a recent immigrant and is now well established in the region. Knowledge of current populations is sketchy, and is largely limited to game animals with estimates based on hunter and highway-caused mortality and on direct observation.

The work presented in this chapter is a portion of a cooperative study of forest island habitats in southeastern Wisconsin, carried out concurrently with studies of the plant communities. Specific questions were, first, is there a relationship between species richness in a given island and the area of that island? Second, does the character of the surrounding landscape affect the species richness of a particular patch?

To answer these questions, the study sought to (1) determine the mammalian species present in each of a number of forest islands, (2) determine the relative abundance of the mammals present, (3) examine the relationship of presence and abundance to island size, and (4) consider the influence of the surrounding matrix on mammalian presence and success (Matthiae 1977).

Description of the Study Area

Sites chosen were in the northern half of Milwaukee County, the southern half of Ozaukee and extreme eastern Washington County, all within the Milwaukee metropolitan area (Fig. 5-1). The study area coincides with that portion of Ozaukee County most heavily urbanized, a complex mixture of agriculture and suburbia; the Milwaukee County portion is almost entirely converted to urban uses. Forest islands examined

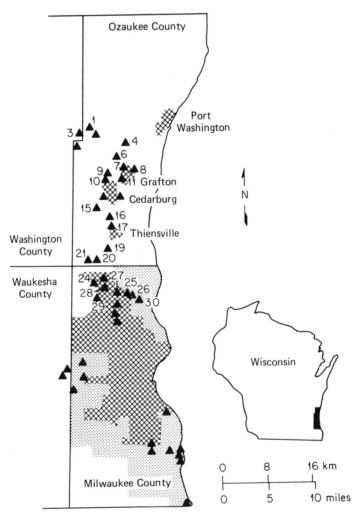

Fig. 5-1. The metropolitan Milwaukee region of southeastern Wisconsin showing the distribution and location of the 43 forest islands in the total project. Numbered islands in the northern portion are those used in the study of mammals. Urban areas are shown cross-hatched, while the developed suburban zone is indicated by stippling.

were restricted to those classified as southern mesic forest and characterized by the presence of sugar maple and (generally) beech (see Chapter 3). Islands ranged from 0.03 ha to 40 ha in size and consisted of woodlots isolated from others by agricultural or urban landscapes. Physiography, soils, and vegetation have been described by Levenson (Chapter 3).

Methods

Twenty-two forest islands were selected. The islands lie in a north-south corridor between McGovern Park on the north side of the city of Milwaukee and the University of Wisconsin-Milwaukee Field Station in the town of Saukville, Ozaukee County (Fig. 5-1). The corridor is bounded on the east by the Milwaukee River and on the west by the Milwaukee and Ozaukee County lines. The area includes urban, suburban, and rural areas, grading from south to north.

The vegetation of these forest islands was studied by Levenson (Chapter 3). Site criteria were (1) the islands should be isolated from others with no direct connections by fencerows, etc., (2) the island must consist primarily of southern mesic (maple-beech) forest, (3) the island must be a remnant of the original upland vegetation, (4) the forest must include all vegetational strata from the groundlayer through to the canopy, (5) there should be little or no evidence of major recent disturbance, and (6) the island should have existed as a discrete unit long enough to develop a mature edge. For the mammalian studies, two additional criteria were essential: (1) authorization from the owner for trapping, and (2) reasonable security from theft or tampering with traps.

Determination of mammal populations was based on intensive spring observation, summer and fall live-trapping, observation during the live-trapping, and a winter track census following the development of uniform snow cover. Visual observations noted tracks, scats, burrows, dens, nests, feeding stations, browsing evidence, and trails, as well as direct sighting. Two trapping transects were established in each woodlot, one across the long and another across the short axis of the woodlot, from the inside of the forest edge on one side to the inside on the other (Fig. 5-2). The total length of the trap lines was proportional to the area being sampled. Trapping inside the forest edge reduced the chances of unintentional capture of nonforest (including forest edge) species other than those which also frequent the forest interior.

Wire mesh (Tomahawk) and aluminum or sheet metal (Sherman and Longworth) live traps were used in sizes appropriate to the target animal. All traps were baited with peanut butter, and the large traps were also baited with apples, which did not appear to improve success. Traps were checked twice daily for five days, in the early morning after sunrise and in the evening at sunset. After recording species, age, and sex, animals were marked and released at the trap site.

Results and Discussion

Only those mammals which could be readily live-trapped or otherwise observed were included in the study, a total of 13 species (Table 5-1). No attempt was made to trap shrews, moles, weasels, and bats, although presumably these species were present in some stands. Shrews may be live-trapped using peanut butter as bait during periods of high abundance, but were not encountered in this study.

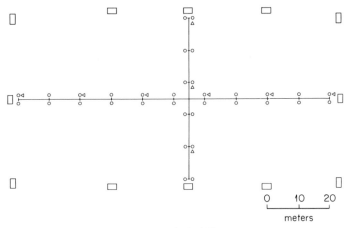

○ SMALL SHERMAN LIVE TRAP (MOUSE)
△ MEDIUM SHERMAN LIVE TRAP (CHIPMUNK)
☐ LARGE TOMAHAWK LIVE TRAP (MEDIUM SIZE MAMMALS)

Fig. 5-2. Trapline layout for study of small mammals in forest islands in southeastern Wisconsin. Small, medium, and large traps were arranged along short and long axes in each woodlot.

Of the 22 islands sampled, four could be grouped as primarily rural (Table 5-1). Of these, the 40 ha Cedar-Sauk Woods at the UWM Field Station supported all 13 species. A second rural and relatively large (14.5 ha) site, Kurtz Woods, supported 11 of the 13 species, while the smallest rural site supported only six species, an average of 9.75 species per rural site (Table 5-1). In contrast, each of the seven primarily urban sites averaged 6.57 species with a range from only six to eight species.

The remaining 11 islands are intermediate, influenced both by the agricultural matrix and by nonfarm rural residences and subdivisions or industrial areas. In this portion of the transect, most fencerows have been eliminated, and islands are effectively isolated from each other. Increased human pressure results in noise, habitat disturbance, hunting, fires, and pedestrian and motorbike traffic. These 11 intermediate sites averaged 5.8 species per island and both predators and omnivores were markedly fewer in the intermediate zone compared to either rural or urban sites.

Peromyscus leucopus (white-footed mouse) was the only mammal occurring in every island. *Sciurus carolinensis* (gray squirrel) was present in 20 sites (91% presence). Other mammals generally present, i.e., occurring in 15 to 17 sites, included *Canis familiaris* (dog), *Odocoileus virginianis* (white-tailed deer), *Procyon lotor* (raccoon), *Sylvilagus floridanus* (cottontail rabbit), and *Tamius striatus* (chipmunk). Save for the white-tailed deer, which was infrequent in the urban islands, these species were distributed over the entire transect.

Dogs, the free-roaming component of the residential and farm pet population, become predators on mice, rabbits, squirrels, and white-tailed deer. Dogs in the urban sites represent both domestic strays and feral animals and also play a predatory role (Cauley and Schinner 1973). *Didelphis marsupialis* (opossum) was found in 13.6% of the islands in both the rural and urban matrices but was absent in the transitional or

Table 5-1. Sites, Island Sizes, and Presence, Abundance, Frequency, and Percent Presence of Mammals in Forest Islands in Southeastern Wisconsin

SITE	AREA (ha)	Canis familiaris (dog)	Didelphis marsupialis (opossum)	Marmota monax (woodchuck)	Mephitis mephitis (skunk)	Odocoileus virginianus (white-tailed deer)	Peromyscus leucopus (white-footed mouse)	Procyon lotor (raccoon)	Sciurus carolinensis (grey squirrel)	Sciurus niger (fox squirrel)	Sylvilagus floridanus (cottontail rabbit)	Tamias striatus (chipmunk)	Tamiasciurus hudsonicus (red squirrel)	Vulpes vulpes (red fox)	TOTAL NO. SPECIES	AVE. NO. SPECIES PER MATRIX TYPE	MATRIX TYPE
1 CEDAR - SAUK FOREST	40.0	1	3	2	4	4	4	4	4	1	4	3	3	3	13	9.75	RURAL
3 FECHTER'S WOODS	2.2	0	0	0	2	2	2	2	1	0	0	0	0	0	6		
4 KURTZ WOODS	14.5	2	0	1	4	4	4	3	4	2	4	3	0	3	11		
6 BUCKSKIN BOWMAN	6.5	0	2	0	3	3	3	0	3	3	0	2	0	0	9		
7 WOODVIEW SCHOOL WOODS	2.4	0	0	3	0	0	4	0	4	0	2	3	0	0	6	5.80	AGRO - URBAN TRANSITIONAL
8 GRAFTON HIGH SCHOOL	1.5	0	2	0	0	4	4	2	4	0	0	2	0	1	7		
9 R & R EXCAVATING	1.6	0	0	0	2	2	4	0	4	0	2	0	0	0	5		
10 CEDARBURG WOODS	1.2	0	0	0	0	2	3	0	4	0	0	0	0	0	4		
11 GRAFTON BANK WOODS	0.6	0	0	0	0	2	2	0	4	0	3	0	0	0	4		
15 NIEMAN WOODS	0.6	1	0	1	0	3	3	0	2	0	1	0	0	1	4		
16 MEE KWON PARK	7.2	0	0	1	1	2	3	3	3	0	1	2	0	0	6		
17 HIGHLAND WOODS	4.2	2	3	3	3	1	3	3	3	0	2	3	3	3	8		
19 GARVEY'S WOODS	2.9	1	0	3	3	3	2	2	0	0	2	3	3	0	9		
20 GENGLER'S WOODS	1.6	0	2	0	2	2	2	0	0	0	3	1	0	0	5		
21 STAUSS WOODS	2.2	2	0	1	3	4	3	0	4	0	3	1	0	2	6	6.57	URBAN
24 BRADLEY WOODS	2.4	4	2	1	4	4	3	0	4	0	4	3	0	1	7		
25 BROWN DEER PARK	2.5	3	0	0	4	4	4	0	4	0	4	4	0	3	8		
26 RANGELINE WOODS	1.7	0	1	0	4	4	4	0	4	0	4	3	0	0	6		
27 TRIPOLI COUNTRY CLUB	2.5	1	3	1	4	4	4	0	4	0	4	3	0	0	6		
28 BRYNWOOD COUNTRY CLUB	0.4	3	0	0	1	1	3	3	1	0	3	3	0	0	6		
29 HASKELL NOYES PARK	0.7	1	0	0	3	3	3	0	4	0	3	3	0	0	7		
30 KLETZSCH PARK	4.1	3	0	0	0	0	3	2	4	0	4	3	0	2	6		
FREQUENCY OF OCCURRENCE		16	6	3	7	15	22	15	20	2	16	17	2	8			
PERCENT PRESENCE		72	27	14	31	68	100	68	91	9	72	77	9	36			

Table 5-1. Sites, Island Sizes, and Presence, Abundance, Frequency, and Percent Presence of Mammals in Forest Islands in Southeastern Wisconsin

intermediate islands. *Mephitis mephitis* (skunk) and *Vulpes vulpes* (red fox) were found in only 9% of the islands, primarily in the rural and transitional area, although both skunks and foxes have been observed in downtown Milwaukee.

Three other mammals, *Marmota monax* (woodchuck), *Sciurus niger* (fox squirrel), and *Tamiasciurus hudsonicus* (red squirrel), were less frequent. The woodchuck (an animal of forest edge, brushy woodlands, and fields) appeared to have taken advantage of the reduced overstory in rural sites affected by a recent severe ice storm. A third location for woodchuck was an urban park. Red and fox squirrels were seen in only two sites, each adjacent to a conifer stand, cedar-tamarack in one case and a pine plantation in the other.

The winter survey indicated some differences in seasonal use. Gray squirrels, red fox, and white-tailed deer were present in some wooded islands during the winter although their summer use may not have been observed.

Abundance

Relative abundance is a subjective but useful concept. For the smaller species (mice, chipmunks) an abundance rating can be based on actual numbers per unit area. For the larger mammals, absolute numbers are difficult to determine, and abundance values are chiefly derived from such indirect measures as presence of browsed stems and scat and track frequency, or from actual observation of the animals. The scale of relative abundance, compared to absolute numbers, differs for each species. It was developed and used here as follows. Category 0 indicates that no animals were present, sighted, or suspected to exist in the woodlot. Category 1 indicates that only an animal or two were observed or trapped sometime during the study. Category 2 suggests a limited population or occasional use as shown by trapping of several individuals, by occasional observations, or other indication (browse, trail, scats) of a visit. Abundance category 3 indicates a normal population, i.e., the species was observed on a number of visits or the number trapped approximates the density believed to be typical of the area and vegetation. Category 4 indicates an unusually high population, higher than normal when compared to long-term averages for the area or as determined from the literature. It may reflect a peak population, or one equal to or exceeding maximum carrying capacity, perhaps evidenced by the extent of browsing. Categories 0 and 1 are self-explanatory, while categories 2-4 are further defined by examples (Table 5-2). The population estimates used reflect other factors intrinsic to the dynamics of many populations, including expected seasonal variation in use, diurnal movement, population cycles, feeding behavior, time of annual peak population, and food availability.

The concept of carrying capacity does not apply uniformly to all mammals in forest islands, as carrying capacity is strongly tied to the surrounding matrix. Urban islands frequently shelter raccoon and gray squirrel far in excess of "normal" carrying capacities when the animals are able to forage freely in surrounding areas. Mammals of small islands often need supplemental food or shelter except for the limited period when seasonal seed crops mature. Dog abundance was difficult to determine although dogs were clearly more common in urban areas or in proximity to rural subdivisions than in the rural sites.

The omnivorous opossum was most abundant in the far rural sites that have greater habitat diversity. The opossum was also relatively abundant in the urban forest islands

Table 5-2. Description of Estimated Abundance Levels (Catgories 2-4) for Selected Species, Indicating Differences Resulting from Animal Size and Behavior

Species	Catgory 2	Category 3	Category 4
Mouse (no./ha)	5-15	20-40	45-100
Chipmunk (no./ha)	5-8. Burrows frequent. Frequently heard, occasionally seen.	14-20. Burrows common. Feeding stations frequent. High degree of visibility.	25-40. Burrows very common. Feeding stations numerous. Very high degree of visibility. Disease, e.g., mange, may be present.
Rabbit (entire island)	Tracks and pellets frequent. 0-33% of forage browsed.	Trails well established. Numerous pellet groups. 33-66% of forage browsed.	Similar to Category 3. 66-100% forage browsed.
Deer (entire island)	Light browsing of preferred food. Trails lightly used, not well established by frequent travel.	Trails well established. Browsing moderate (33-66% of stems).	Trails heavily used. Browse lines apparent, stems 66-100% browsed. Numerous pellet groups. Secondary food plants being utilized.
Raccoon (entire island)	Tracks frequent, scats more numerous. Feeding stations present. One individual trapped.	Scats, tracks, and trails more frequent. Feeding stations frequent. Den trees present. 2-4 trapped, usually a family unit.	Trails well established into woods and around feeding stations, the latter numerous. Scats in piles at feeding stations, 2 or more den trees present. 2 or more family units, 5-7 individuals trapped.
Fox (entire island)	Frequently hunts while passing through. Use not intensive.	Regularly and systematically hunts while passing through. Old dens present, digging evident. Scats and tracks frequent.	Active den present. Heavy hunting use as evident from trails, digging, and kills.

where it is probable that the animal utilizes storm sewers, culverts, and similar artifacts for shelter, and human refuse for food. The transitional islands in the area have less suitable opossum habitat and greater likelihood of contact with dogs, other predators, and automobiles.

Woodchuck populations were well distributed, but low. This coincides with the current generally low populations throughout most of southeastern Wisconsin. Skunks appeared to be relatively infrequent at all study areas except Cedar-Sauk Forest, the largest rural site. Low abundance in the forest patches is not unexpected, as skunks are primarily animals of the edge and open.

White-tailed deer show relatively low levels in spite of documented reports that southeastern Wisconsin populations of up to six animals per square kilometer over winter are common. However, deer range includes both forest patches and other appropriate habitat such as abandoned fields and wetlands. Deer were infrequent to absent in the transition sites, probably because these islands are isolated, provide relatively little browse, and are susceptible to high levels of human and dog disturbance. A few deer penetrate to the urban islands, traveling along streams, and utility and railroad rights-of-way. In larger parks, deer find cover and sufficient food to maintain a limited population.

The white-footed mouse, present in all islands, usually occurred at high levels of abundance. Population sizes correlate well with those obtained in other studies, including a 10-year study at the UWM Field Station and other work within the Milwaukee metropolitan area (Matthiae, unpublished data). Two sites (15 and 20) in the transitional area showed very low populations, indicating either heavy local mortality or recent recolonization.

Variable use of islands was demonstrated by a number of species, especially the raccoon, gray squirrel, and white-tailed deer. Both raccoons and squirrels utilize den trees and mast (acorns) and are highly mobile (Oxley et al. 1974). It is not always possible to tell whether these mammals are visitors that do not breed in an area or whether the area does indeed support a breeding population. Ear-tagging of raccoon and opossum failed to indicate direct movement between adjacent and simultaneously trapped forest islands, although presumably movement occurs. However, several islands must be included within the home range of any one individual or family unit. One ear-tagged raccoon was killed while crossing a busy street halfway between two of the urban forest island locations. Another was killed while foraging in a residential garage two city blocks from the forest island most closely adjacent to its island of capture. In both cases, the islands were separated by 1.6 km. Knowledge of home range and habits of raccoon, squirrels, and opossum suggests that the absence of evidence for travel between the several sets of adjacent islands was a result of experimental design. Two five-day trapping periods and immediate postcapture trap avoidance by many species combined to make interisland movement difficult to assess. Raccoons are relatively abundant in urban sites (Cauley and Schinner 1975), where they not only utilize natural resources but sewers, other structures, and garbage.

The cottontail rabbit is not a forest animal, and it was not abundant in most of the forest patches studied. Rabbit use was confined largely to fall and winter when it was highest where openings in the canopy had stimulated a dense shrub understory. The pattern of seasonal use by rabbits agrees with that found by March (1976), who showed somewhat higher abundance levels, probably because his census included edge as well as the forest interior.

The chipmunk was common to abundant at most locations and did not respond to differences in the matrix. Chipmunk populations normally fluctuate as a result of mange and other diseases as well as habitat factors. However, once the chipmunk has been exterminated, repopulation of woodlots is difficult since the animal is both vulnerable to predators and sensitive to sunlight. Absence of natural corridors such as fencerows and rights-of-way appears to limit chipmunk and mouse movement (Wegner and Merriam 1979).

Abundance of foxes was determined from scats, tracks, and den sites. Red foxes were usually present in islands with several species of rodents. Rabbits were present in all but one of the islands frequented by foxes.

Island Size

As anticipated, there was some relationship between mammalian species richness and island size, although this relationship was not always clear-cut. This coincides with findings of others (MacArthur and Wilson 1967, Simberloff 1976a, Heatwole and Levins 1973, Forman et al. 1976), although most of their conclusions were based on studies of birds rather than mammals. In this study, the number of species increased with island size. No stand under 2.5 ha had more than eight species, nine species were present at 2.9 ha, 11 at 14.5, and 13 at 40 ha. Species richness is greatest at Cedar-Sauk Forest where the existing regional mammalian community of 22 different species is completely represented. This assemblage of mammals might once have been present in any of the forest islands under study.

Differences in relative abundance were also related to island size. In larger islands (seven islands over 2.47 ha) 69% of the species were in the common or abundant categories. Other factors are also involved. The six urban islands had high abundance values for the few species present, whereas only 39% of the species in the ten transitional islands were common or abundant. Urban islands act as refuges for species such as raccoon, opossum, and gray squirrel which concentrate in these isolated forest patches in numbers greater than carrying capacity. The islands are used for daytime cover, a type of urban behavior cited also by Cauley and Schinner (1973).

Gray squirrels and raccoons were more abundant in larger than in smaller islands. Red squirrel was present in the large islands associated with conifer stands. The fox squirrel requires the more diverse food source and abundant mast in large islands. Red fox dens were located in the larger islands, and fox tracks indicated that the animals cross islands during the winter but appeared to hunt at or near the edge. The fox may simply cross the larger woods to reach more desirable hunting sites. Deer presence (Fig. 5-3) was correlated with larger islands, but deer were frequently found in medium-sized islands in smaller numbers and are not uncommon in the fringe suburbs.

Rabbits were observed in the small and medium-sized islands (Fig. 5-3). This presumably reflects the high proportion of forest edge to island interior (cf. Ranney et al., Chap. 6) in smaller islands, which benefits edge species such as the cottontail.

Abundance of the white-footed mouse (Fig. 5-3) and the chipmunk showed no relationship to island size. The opossum also showed no strong correlation with island

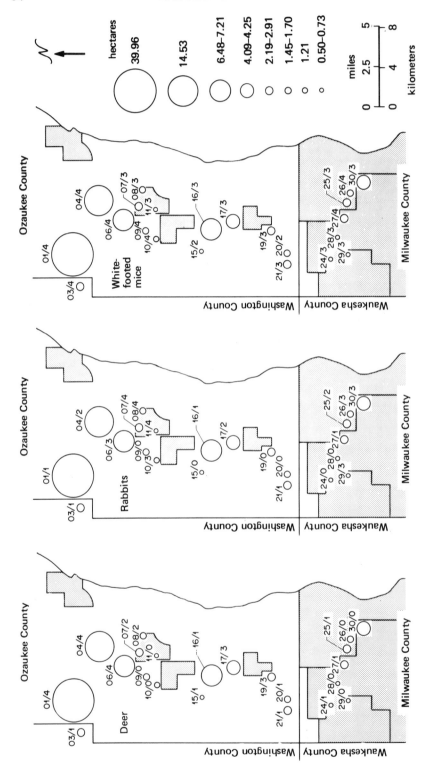

Fig. 5-3. Forest island size, matrix type, and abundance values for three representative mammals; large (deer), medium (rabbits), and small (white-footed mice). Numbers at the top of each circle are the site number/abundance value for the individual species.

size, occurring most frequently and in higher abundance in large rural and small urban islands. Skunk presence and abundance increased with island size, perhaps reflecting treatment as a pest species in the urbanized sites. Woodchucks appeared to prefer larger islands, but the sample was too small to show any clear relationship.

Relationship between Animals and Island Matrix

As noted, the presence and abundance of mammals differed considerably across the transect from rural through transitional to urban matrix. For many animals, the urban matrix may provide benefits, such as food and the possibility of moving safely through storm sewers and drainageways. Habitat diversity and food availability appear lowest in the transitional zone between city and country. The rural sites with lower human populations have islands of adequate size and adjacent habitats which provide the food resources necessary to support a diverse population.

Predators change along the transect. The automobile and the dog are major problems in the city, while fox, hawk, and owl are more common predators in the rural portion of the transect. Human disturbance also varies along the gradient and may affect animal distribution considerably. Those mammals that adapt to human noise, grazing, fire, etc., are capable of surviving in the urban and urban-rural transition areas while other species may be better adapted to the country. A dendrogram (Fig. 5-4) based on species occurrence and abundance suggests some relationships among mammalian groups. Species in Group A are most frequent in occurrence and highest in abundance, with a broad distribution over the transect. Species in Group B occur infrequently and are restricted in their general distribution. Each of these groups is capable of forming a food chain, and the skunk may associate with either group as an omnivore. The complete assemblage is not present in these islands, however, restricting full development of the two food chains.

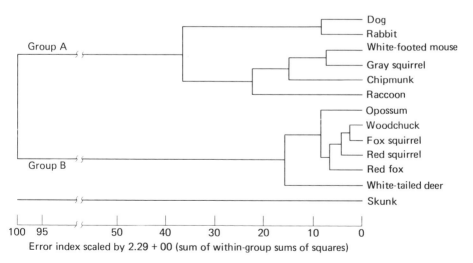

Fig. 5-4. Dendrogram based on cluster analysis showing relationships among mammals based on presence and abundance values in forest islands in southeastern Wisconsin.

Conclusions

Mammalian adaptation to changing land use practices continues even after 140 years of regional development. At the time of settlement there were habitat specialists such as woodland bison, moose, wolverine, black bear, elk, and lynx. All were extirpated by about 1850. The mammals in this study included only three true deciduous forest specialists, the chipmunk, raccoon, and white-footed mouse. Of these, the raccoon and mouse have extended their range beyond the forest. The other ten species are adapted to a broad range of habitats, although the forest may be preferred. These species tend to be associated with successional stands, whereas the chipmunk is strictly a mammal of the forest interior. Only five to seven forest specialists, including the chipmunk, raccoon, and white-footed mouse, and others not included in the study (flying squirrel and several bats), remain in the region (Fig. 5-5).

Both island size and the matrix in which the island is embedded influence the presence and abundance of certain species. The mammalian community changes in composition and abundance from rural to urban areas. The far rural sites are the most diverse, the urban islands serve as refuges for small rodents and larger nocturnal scavengers and omnivores, while islands located in the urban-rural transition have lower mammalian species richness and lower abundance. It is postulated that the lower diversity in the transition region results from greater isolation of islands and the absence of diverse adjacent habitat. Presence and abundance were positively correlated with island size among the islands in each of the three segments of the transect. Effects of home range size, habitat connectivity, human disturbance, and adjacent varied habitats require further investigation.

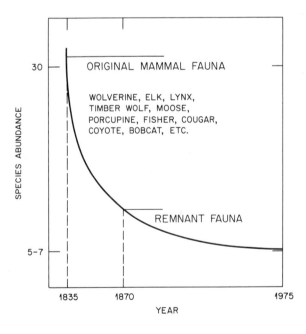

Fig. 5-5. An approximation of a negative exponential curve representing the rapid loss (regional extinction) of specialized forest mammals following settlement and agricultural development in southeastern Wisconsin.

6. The Importance of Edge in the Structure and Dynamics of Forest Islands

J. W. RANNEY
GRADUATE PROGRAM IN ECOLOGY
UNIVERSITY OF TENNESSEE-KNOXVILLE

M. C. BRUNER
DEPARTMENT OF BOTANY
UNIVERSITY OF WISCONSIN-MILWAUKEE

JAMES B. LEVENSON
DEPARTMENT OF BOTANY
UNIVERSITY OF WISCONSIN-MILWAUKEE

In much of the eastern deciduous forest biome, previously extensive forests have been reduced to widely dispersed woodlots, fencerows, and urban forests (Curtis 1959) as a result of clearing for agriculture, urbanization, and service rights-of-way. Such changes in the density, distribution, and composition of native tree species have altered population dynamics and forest development considerably. For example, some species encounter distance barriers in recolonizing isolated forest islands [e.g., sugar maple in southern Wisconsin (Auclair and Cottam 1971)] while others seem to be favored by such an arrangement [e.g., black cherry in the same area (Auclair and Cottam 1971)]. As fewer, more isolated forest islands remain, their condition and dynamics become more relevant to sustaining a region's natural forest resource and very importantly, a region's genetic base for tree species.

Two recent studies (Wales 1972 and Levenson 1976, Chapter 3) have shown that the edges of small (4-6 ha) forest islands may play a particularly important role in the development of individual forest islands. Such a role has far reaching implications for the population dynamics of individual tree species and for forest composition for a region. The study reported in this chapter relates forest characteristics to interactions between edges and interiors of forest islands and carries conclusions on to development within a landscape of forest islands. Effects of island size, orientation, isolation, site conditions, and rates of forest disturbance are discussed in order to relate the dynamics of forests on a regional scale. A number of analytic techniques, including computer simulations, are used to infer long-range trends. The data for this work were collected from forest islands in southeastern Wisconsin.

Concepts of Forest Edge

Concepts of forest edge have been based primarily on attributes of ecotones and sharp transitions between habitat types, rather than as distinct communities or habitat types (Leopold 1933, Allen 1962, MacArthur and Pianka 1966, Ranney 1977). To quantify effects of edge, say, for wildlife diversity or productivity, edges have been measured as linear structures, ratios of lengths to areas, and only rarely as two dimensional entities (Kelker 1964, Patton 1975, Ghiselin 1977).

Forest edge structures and dynamics in forest island landscapes are dependent on the way an edge is maintained by man (Fig. 6-1). Gysel (1951) noticed that some edges were maintained at the outer dripline of edge canopy trees (canopy dripline edges)

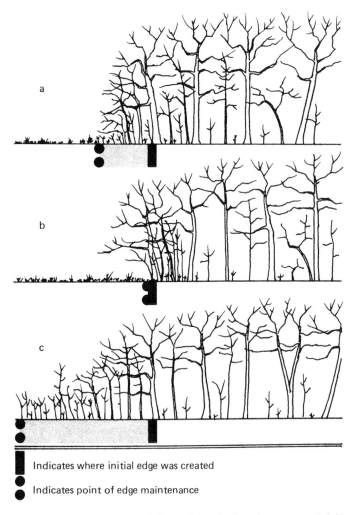

Indicates where initial edge was created

Indicates point of edge maintenance

Fig. 6-1. Three common structures of forest island edges (a, canopy dripline; b, cantilevered; c, advancing), dependent on where an edge was created and where it has been maintained.

while other edges were maintained either at the base of edge canopy trees (cantilevered edges) where barbed wire could be strung on trees, or beyond the tree dripline (advancing edges). This variation in edge maintenance has considerable influence on environmental conditions within an edge which affect the way vegetation responds and develops.

The complexity of the environment of forest island edges is suggested by the literature on edge effects, recently reviewed by Ranney (1977). Effects of wind and solar radiation are paramount. Wind buffets edge trees (Moen 1974), enhances seed dispersal, alters temperature fields (Crockett 1971), and changes soil moisture by increasing evapotranspiration. Solar radiation is controlled by the aspect of the edge (Wales 1967, 1972, Swift and Knoerr 1973) and by latitude (DeWalle and McGuire 1973). Solar radiation intensity at edges may at times exceed that of the horizontal canopy due to lateral exposure to direct and diffuse radiation (Hutchinson and Matt 1976a, 1976b, 1977), increasing evapotranspiration (Salisbury and Ross 1969, McConathy et al. 1976). In newly created edges, increased light intensity affects shade tolerant and intolerant species differently (Smith 1962), a feature that is reversed as edges mature (Wales 1972). In summary, forest edges generate microclimatic gradients which result in a physical environment that differs from both open field and interior forest. This difference is amplified by various plant responses to environmental gradients.

Forest edges have also been seen as locations of heavy use by wildlife. Edge has high cover density (Schreiber et al. 1976, Johnson et al. 1979) and represents the convergence of contrasting habitats (Odum 1959, States 1976). As a result, both seed predation and the opportunity for seed dispersal are enhanced (McDiarmid et al. 1977, Thompson and Willson 1978).

New concepts in niche quantification and studies of the effects of animals on tree propagule dispersal and mortality (Howe 1977, Reichman and Oberstein, 1977, Mitchell 1977, Bullock and Primack 1977, Vander Wall and Balda 1977, Thompson and Willson 1978) are coalescing to provide a holistic view of forest edge and forest development. However, the many factors influencing the dynamics and importance of edge ecotones in man-dominated landscapes have made the identification of edge processes difficult. Some of these issues have been addressed by field and simulation studies in this chapter.

Description of the Study Area

The study area of Milwaukee, Wisconsin, and its suburban and rural fringe includes parts of Milwaukee, Ozaukee, and Washington counties. The area contains numerous isolated forest islands, some of which have old edges (70 years and older). Interior vegetation of the islands has been studied (Levenson, Chapter 3; Hoehne, Chapter 4) and a computer model simulating forest growth has been developed and tested using tree species native to the island interiors (Ek and Monserud 1974). The combination of an island landscape pattern, the availability of supplemental data, and the existence of an appropriate local forest growth model made the area desirable for studying forest edges. The vegetation and environmental conditions of this region are described by Levenson (Chapter 3).

Methods

In order to minimize the number of factors encountered in this study, islands were selected according to strict criteria from the 43 islands of Levenson (Chapter 3). Only islands containing a high component of sugar maple and/or beech were selected to eliminate large variations in vegetation types between islands and to provide the best available conditions under which edge characteristics could be identified. Islands had to be over 3 ha in size so that relatively undisturbed interior conditions would be present. Stands were further examined for soil and relief differences, and visually checked for recent disturbances such as grazing, thinning, and fire. Finally, islands were selected which had linear edges facing cardinal directions (±10°) and which were long enough (140 m) to allow at least three sample transects to be taken for statistical replication. The transition from forest to open field had to be abrupt with well-developed edges (i.e., maintained at their existing location more than 10 years) and free from severe man-induced or natural disturbances. Table 6-1 lists the islands sampled and the number of transects taken according to aspect. Detailed descriptions of forest island conditions and history, and their locations are shown in Levenson (Chapter 3). Woody vegetation was sampled using a modified line-strip method (Lindsey 1955) (Fig. 6-2).

Three plot arrangements were oriented along transect centerlines to sample four vegetation strata; trees, saplings (2.5 cm ≤ dbh ≤ 10 cm), shrubs (less than 2.5 cm dbh and taller than one meter), and woody groundlayer species (woody vegetation less than 1 m tall).

Ages of all edges sampled were determined from increment borings of selected trees. The age of edges was used to determine whether vegetation structure and composition changed through time, and to elucidate the developmental processes of forest edge. Trees were chosen for coring based on evidence that they would indicate at least a minimum age of the edge, e.g., trees with barbed wire buried in the interior of the bole and with a form characteristic of edge growth. In addition, the structure or maintenance form (see Fig. 6-1) of each edge sampled was recorded.

Data were divided into two groups. The first was used to determine edge composition, structure, and function and included six islands containing 45 of the 76 tran-

Table 6-1. Numbers of Edge Transects Sampled, by Aspect, in Forest Islands in Southeastern Wisconsin

| Island no. | Island name | Aspect | | | | Total |
		North	East	South	West	
3	Fechter's Woods	4	–	3	4	11
4	Kurtz Woods	–	–	3	4	7
6	Buckskin Bowmen	–	–	4	4	8
16	Mee Kwon Park	3	–	–	–	3
17	Highland Woods	4	3	–	–	7
18	Wausaukee Woods	4	4	4	4	16
25	Brown Deer Park	–	–	4	4	8
41	Cudahy Woods (South)	–	4	4	–	8
45	Rawson Park	4	4	–	–	8
	Totals	19	15	22	20	76

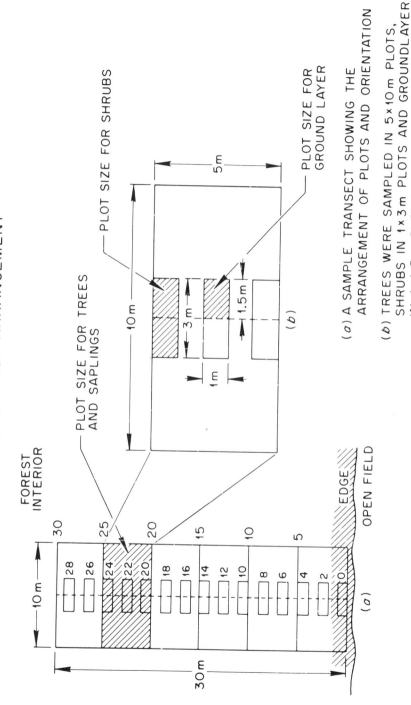

SAMPLE PLOT ARRANGEMENT

Fig. 6-2. Sample plot arrangement where (a) is a sample transect showing plots and orientation, and (b) shows the arrangement of plots for the sampling of shrubs and groundlayer vegetation.

sects. The second group was used to validate models constructed from the first data group. Plane table surveys, tree height measurements, and various structural tree measurements (crown asymmetry and dominance class) were made to check or parameterize models used in simulating forest edge dynamics.

Several important questions about the role of edges in forest island ecosystems remained unanswered after the direct analysis of field data. Can equilibrium in species composition be expected to develop along edges? What is the effect of edges on species composition in island interiors in contrast to similar interiors embedded completely in a large forest matrix? Is there a reverse effect of island interiors on species composition of forest edges? How is interisland seed rain affected when edges are more exposed to dispersal vectors than are forest interiors? Will interisland seed rain modified by the contribution of forest edges cause changes in development of other forest islands? Finally, can edge and interior forest development be related to island size or other factors which allow generalizations to be made concerning regional forest dynamics, i.e., can these factors be used to predict how forests may change under various regimes of management and modification?

A computer model, FOREST, which simulates forest growth and accepts propagule inputs (Ek and Monserud 1974), was used to answer these questions. The original model did not fulfill all requirements for simulation of edge dynamics and was modified for variable seed inputs by the staff at Oak Ridge National Laboratory (see Johnson et al., Chapter 10) to appropriately represent edge growth. Ek and Monserud (1974) developed and calibrated the model for specific application in southern Wisconsin. The model has worked well in simulating forest growth in northern and southern Wisconsin using simulation periods of 50 years.

FOREST contains important components for tree reproduction, growth, competition, and mortality. The generalized structure of the model (Fig. 6-3) demonstrates the interaction of these components with the main program. In the model, trees less than 2.5 cm dbh are tabulated according to height class and density and occupy subplots representing general growing conditions in portions of the simulated plot. In our case, the simulated plot was approximately 0.08 ha (1/5 acre).

All canopy trees are identified by coordinates and evaluated individually for height and diameter growth on the basis of potential growth for a dominant tree of the same height and species. Potential growth is then adjusted through the use of a competition index which is derived from measures of crown overlap and height, crowding, shade tolerance, crown size (a three-dimensional measurement), and sometimes effects of release (Ek and Monserud 1974, Monserud 1976). Canopy tree mortality is determined from predicted diameter (age), predicted diameter growth, and competition index. This determination is not species-specific.

Understory dynamics are simulated differently. Understory trees are considered in height classes rather than age classes. Growth is determined from canopy height and density and from understory density. Survival rates, determined by height-classes, were drawn directly from remeasured permanent plot data (Monserud 1976). Growth into taller height-classes was determined in a similar fashion.

Reproduction includes seed and sprout production, introduction of exogenous seeds, germination, and survival to a certain size class. Here, as well as in other parts of the model, processes are introduced to account for the natural variability known to occur

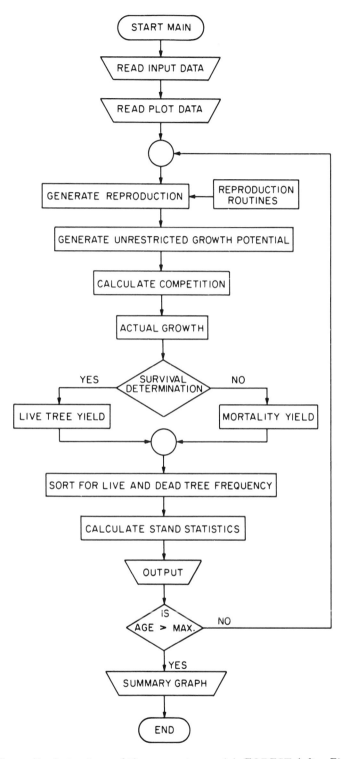

Fig. 6-3. Generalized structure of the computer model, FOREST (after Ek and Monserud 1974).

as forests develop. For example, good and poor seed crops are simulated to occur periodically and to depend on species characteristics.

In adjusting the model, the simulated annual height growth of edge trees was reduced at a rate inversely proportional to tree distances from an edge (up to 15 m). Crown diameter growth and bole diameter growth remained unadjusted to simulate the effects of lateral crown expansion and bole growth of edge trees. Thus, for edge trees, crown diameter and bole diameter growth progresses almost as if height growth was not adjusted, although height growth is reduced. Slowed height growth places the edge trees at a slight disadvantage with more interior companions since competition between adjacent trees is based, in part, on relative height. However, crown size is also a factor in determining growth. This appears to foster sufficient growth in edge trees to maintain vigor and compete with taller interior trees.

Results

Comparison of Islands

Before any computer simulations were run, basic statistics were calculated from field information. Continuum index values (Curtis 1959) were calculated using only the major species for each island. Continuum index values can range between 300 and 3000 for bur oak and sugar maple stands, respectively. Comparison of the interior index values with 22 values calculated by Whitford and Salamun (1954) in the same region shows that the values for the nine islands in the current study rank among their top third. If the range of continuum index values is seen as a successional gradient, the islands in this study are near the "terminal stages" of development. The occurrence and importance of some species betrayed the fact that the islands probably have not reached terminal state or, if they have, they contain a natural component of species considered somewhat shade intolerant. Comparison of importance values of edge and interior locations shows definite shifts for some species. Basswood and ash increase toward the edge while beech and sugar maple showed a definite decrease. In fact, no species with an adaptation number greater than 8.5 showed a greater importance value in island edges than in interiors. It will become apparent later that extreme edge plots (0-5 m in from an edge rather than 0 to 15 m) have much lower continuum index values than Table 6-2 indicates because of the increased importance of shade-intolerant species not tabulated.

General Edge Characteristics

Continuum index values suggest that island edges differ significantly from island interiors. A nonparametric sign test on continuum index values suggests ($p \leqslant 0.05$) that edges are more pioneer and less mesic in composition than forest interiors regardless of interior composition.

A structural analysis of forest edges demonstrates that basal area for trees and saplings decreases from the edge toward the island interior up to about 15 m (Fig. 6-4). The same is true for stem density. Beyond this distance, values are rather constant. The shrub and groundlayer strata show less distinct trends in density except for higher values in the first several meters. Typically, tree and sapling basal area and density nearest edges are more variable than interior zones. Gysel (1951), Bray (1956), Trimble and Tryon (1966), and Wales (1972) found similar structure in edge communities. They found that density and frequency of saplings in the border areas were much greater than the interior, but the density and frequency of larger trees were not greatly different, perhaps because of the limited stands they studied.

No abrupt discontinuity occurs in the edge-to-interior gradient that would serve to separate edge from interior. But, beyond the 10 to 15 m distance, the consistency of basal area and density measurements may indicate a shift from vegetation influenced by the edge conditions to vegetation adapted to interior conditions.

Multiple linear regression analysis was used to determine more precisely the depth of the edge, the variables that change significantly along the edge gradient, and how precisely these variables predict sample position in the edge gradient. The dependent variables used were total density of each structural stratum, total basal area of the tree stratum, and species richness of each stratum. Distance in from an edge was the independent variable and each aspect was analyzed separately.

Regression analyses were performed initially using the two distance intervals (0-5, 5-10 m) in from the exterior edge. The analysis was repeated with an additional distance interval being added with each new run. The grouping of distance intervals that gave the highest coefficient of determination (R^2) and the greatest number of variables with slopes differing significantly from zero (F test, $p = 0.05$) was used to approximate the depth of the edge zone. The group of distance intervals interior to those constituting the edge zone were then analyzed to determine if any significant variation occurred within them. The general regression equation was

$$Y = a - b_1 X_1 - b_2 X_2 \tag{1}$$

with Y as the expected distance interval. Table 6-3 shows that forest edge effects extend at least 15 m into a forest stand.

Trends in species richness and stem density have a pattern of general decrease through the outer 10-15 m of the edge, and a leveling off interior to that point. Decreases in variability of the mean richness and density values also suggest that the functional edge is 10 to 15 m in depth. The best results in the analyses were obtained on the north, south, and east edges with the 10-15 m depth. In contrast, on the west edge a 30 m or greater depth is indicated. Complex factors appear to be responsible for the increased depth in the west edges sampled. Land use effects on the forest edge may also contribute.

Mean tree and sapling richness values are highest along edges but are accompanied by a high degree of variability. Both richness and variability decrease as the forest interior is reached. The richness values are also indicative of the total number of species found in edges when individual plots are aggregated. This general pattern holds for the sapling and shrub strata, but is weaker for the tree stratum and nearly nonexistent for the groundlayer. In a limited survey, Gysel (1951) found 40 interior species and 59 edge species; however, 38 of the interior species also occurred in edges. Thus, a large portion of the interior trees were found in edges but the reverse is not true.

The distribution of several important tree species is directly related to the proximity of edges (Table 6-4). A general division of species into xeric, shade-intolerant species

Table 6-2. Species Importance Values in Nine Forest Islands in Southeastern Wisconsin[a]

CAN[b]	Species	Forest island no.									Average importance value
		25	6	45	3	17	41	18	16	4	
10.0	Acer saccharum	44	78	45	75	105	157	93	115	160	97
9.5	Fagus grandifolia	23	55	5	56	94	118	55	38	117	62
9.0	Ostrya virginiana	9	47	76	60	53	20	29	71	52	46
9.0	Carpinus caroliniana	12	23	27	9	35	6	14	—	40	18
8.5	Carya spp.[c]	19	61	36	43	34	9	54	15	43	35
8.0	Tilia americana	17	47	26	36	4	16	40	—	46	26
8.0	Ulmus rubra	16	9	16	8	5	5	9	28	—	10
7.5	Fraxinus spp.[d]	25	29	13	18	50	20	44	—	30	25
7.0	Acer rubrum	25	78	67	21	68	40	42	65	60	52
7.0	Juglans cinerea	—	9	—	—	—	—	—	—	—	1
7.0	Ulmus americana	4	—	—	4	—	—	11	11	—	4
6.0	Quercus borealis	26	11	27	66	9	56	8	55	15	29
5.0	Betula papyrifera	37	10	22	36	16	23	14	14	—	21
4.0	Quercus alba	35	16	28	3	24	5	—	—	7	13
3.5	Prunus serotina	13	36	37	8	14	34	5	10	8	18

Climax adaptation number[b]	Species										
2.5	*Crataegus* spp.	13/33	29/17	4/17	–/7	–/14	5/4	9/17	–/14	–/–	7/14
1.5	*Populus tremuloides*	–/–	–/3	–/–	–/–	–/–	–/–	–/–	–/–	–/–	–/14
1.0	*Quercus macrocarpa*	21/–	19/8	–/–	–/7	3/–	3/–	–/–	–/–	–/–	3/2
	Continuum index value	2240/1991	2262/2178	2410/2128	2511/2357	2574/2512	2577/2338	2612/2382	2621/2274	2802/2644	2512/2312

[a] The upper value of each pair is the response to interior conditions; the lower value is for the forest edge community.
[b] Climax adaptation number (Curtis 1959).
[c] Predominantly *Carya cordiformis* with some *C. ovata*.
[d] Predominantly *Fraxinus americana* with some *F. pennsylvanica*.

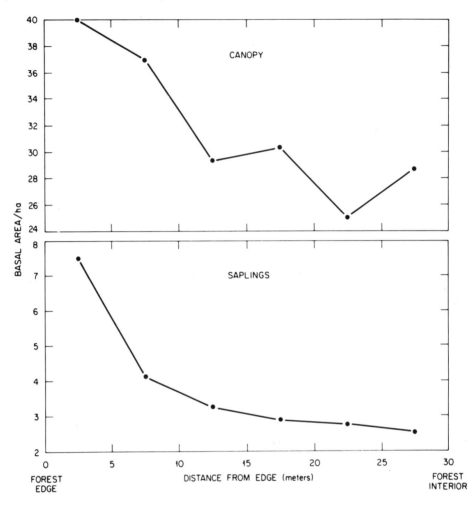

Fig. 6-4. Canopy and sapling basal area near forest edges expressed in square meters per hectare. Stem densities follow a similar pattern.

Table 6-3. Determination of Depth of the Edge in Forest Island Communities Using Multiple Regression Techniques

Aspect	Best calculated edge depth (m)	R^2 (%)	Important variables (R^2 %)	Interior variation
North	0-15	73	Sapling density (55) Shrub density (18)	Tree species richness
South	0-15	64	Sapling richness (36) Basal area (28)	None
East	0-15	79	Sapling density (59) Tree density (20)	None
West	0-30	77	Tree density (26) Shrub density (18) Tree richness (19) Sapling richness (7) Basal area (4)	(Not applicable)

Table 6-4. Relative Importance Values of Woody Species Near Forest Edges

Species	Distance from edge (m)					
	0-5	5-10	10-15	15-20	20-25	25-30
Group A (edge-oriented)						
Subgroup 1 (moderately oriented)						
Tilia americana L.	86.2	61.7	48.9	52.2	20.1	37.0
Fraxinus spp.	50.4	45.3	19.8	17.6	22.8	15.9
Quercus borealis Michx. f.	14.6	19.8	45.9	23.0	58.5	31.5
Ulmus rubra Muhl.	12.4	4.4	2.4	3.3	2.8	12.9
Amelanchier laevis L.	2.3	–	1.0	1.1	–	1.5
Ulmus americana L.	7.0	2.9	1.2	2.7	4.0	–
Prunus virginiana L.	–	2.2	2.1	1.1	1.2	–
Quercus alba L.	–	3.3	7.7	5.1	–	–
Cornus rugosa L.	–	1.4	2.1	1.9	–	–
Subgroup 2 (strongly oriented)						
Crataegus spp.	15.3	13.3	3.3	1.6	3.5	5.3
Rhamnus cathartica L.	5.7	3.7	8.1	2.6	1.2	–
Carya spp.	8.1	7.2	2.2	1.1	–	–
Populus tremuloides Michx.	7.1	2.5	3.0	–	–	–
Juglans cinerea L.	–	0.9	1.1	–	–	–
Quercus macrocarpa Michx.	2.6	1.1	–	–	–	–
Lonicera tatarica L.	1.9	–	–	–	–	–
Viburnum lentago L.	1.5	–	–	–	–	–
Salix spp.	1.2	–	–	–	–	–
Cornus racemosa Lam.	0.9	–	–	–	–	–
Acer negundo L.	0.8	–	–	–	–	–
Betula lutea Michx. f.	0.8	–	–	–	–	–
Prunus americana L.	0.8	–	–	–	–	–
Group B (interior oriented)						
Acer saccharum Marsh.	36.5	69.6	71.4	91.1	89.3	100.0
Fagus grandifolia Ehrh.	3.4	11.2	33.0	42.0	26.1	57.0
Acer rubrum L.	–	–	–	–	–	2.2
Group C (ubiquitous)						
Ostrya virginiana (Mill.) K. Koch	19.3	27.0	31.0	26.4	33.4	22.6
Prunus serotina L.	17.4	21.6	10.9	18.5	18.5	9.6
Hamamelis virginiana L.	4.1	1.1	2.1	–	4.3	3.8
Carpinus caroliniana Walt.	–	–	2.8	8.8	4.3	–

and mesic, shade tolerant species according to Curtis (1959) and Harper et al. (1970) is useful in elaborating the differences between forest edges and interiors. For example, dividing species into two groups based on adaptation values showed that the basal area of shade tolerant species having high adaptation values (7.0-10) increases away from edges. Poorly adapted species (shade-intolerant, xeric spaces having adaptation values below 6.5) were abundant near edges but were found in small quantities in forest interiors. Their occurrence in the interior is thought to be a result of gap-phase replacement that is facilitated by the proximity of seeds of edge species. A group of species having intermediate adaptation values make up considerable basal area in all locations but was more important along edges. Assuming that basal area corresponds directly to seed production, it can be deduced that there is at least twice as much seed production near edges for species with low and moderate adaptation numbers (0-7.0) compared to forest interiors. Similarly, seed production per unit area for mesic species is as much as 20-25% less along edges compared to interiors.

Basal area of intolerant species is 300% greater in edges than in forest interiors. The combination of intermediate and well adapted species showed an even distribution in basal area.

Patterns in Tree Species Distributions

Several earlier studies (Wales 1972, Weber 1975) have noted that edge composition varies along with aspect. However, compositional variability within the same aspect is also high compared to forest interiors. Edge maintenance is often so variable that definite compositional trends with aspect are difficult to identify. Extensive field data and historical records of edge treatments, the latter often not available, are needed to clarify the relationship between edge aspect and composition.

An evaluation of the species distribution within forest islands permitted division of species into characteristic groups. The groups are moderately edge-oriented, strongly edge-oriented, interior-oriented, and ubiquitous species (Table 6-3 and Fig. 6-5). In the edge-oriented group, importance values are highest in the outer portions of islands (Fig. 6-5a and b) but are reduced in the interior locations. Many of these species have propagules that are wind and bird dispersed. The moderately edge-oriented group (Fig. 6-5a) includes species which commonly dominate edges but may still make up a considerable part of the interior community. Ash, basswood, and elms are in this group. Almost all the species in this group are intermediate in shade tolerance and appear to be favored along north and east edges.

Members of the strongly edge-oriented group much less tolerant of shade and much more restricted to island edges (Fig. 6-5b) include hawthorns, most oaks, yellow birch, hickories, buckthorn, willows, poplars, and butternut. The group is most favored along south and west aspects. All exotic species fall in this group. The interior-oriented species group (Fig. 6-5c) is composed of beech and sugar maple, the two dominant equilibrium species in the study area. Beech is nearly nonexistent in edges but quite important in interiors. Sugar maple is less distinctly oriented toward interiors. The decrease in importance value toward the edge is more pronounced on south and west aspects. This group is most affected by island size where islands less than 2-3 ha may prevent the full development of interior forest conditions causing the group to decline in importance.

The ubiquitous group (Fig. 6-5d) is characterized by hop hornbeam and black cherry. Distributions within islands are relatively homogeneous. Species in this group are affected very little by island size, shape, and orientation. A Wilcoxon signed-rank test demonstrated that most species had moderate-to-significantly different importance values between edges and interiors. Only sugar maple and American beech contained significantly lower importance values along edges.

The preceding species groupings suggest that small forest islands having different shapes and orientations will vary significantly in forest composition as a result of forest edge development. For example, if islands were less than about 30 m wide in the study area (no matter what their length), the chances of beech surviving in the region as a self-sustaining population would be extremely low even if the islands were left

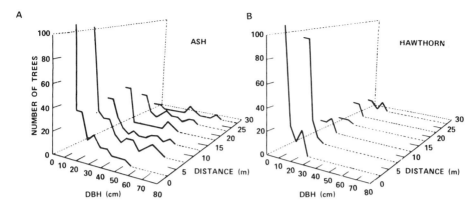

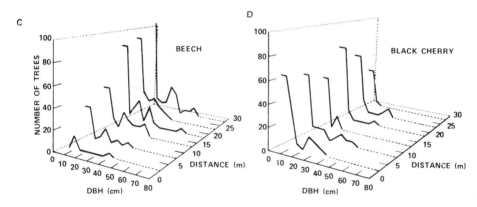

Fig. 6-5. Representative density-diameter distributions for species groups in southeastern Wisconsin. The ashes and hawthorn are edge species, beech is an interior species, while black cherry is representative of a group of ubiquitous species.

undisturbed. At the same time, species of intermediate shade tolerance would increase relative to more tolerant species (i.e., islands would be "all edge"). Island width can be a limiting factor in the distribution and importance of the interior-oriented species group, regardless of other features of island pattern, assuming propagule dispersal is not limiting. For this reason, hedgerows and very narrow forested corridors are believed to be unimportant as "safe" sites in the population dynamics of interior-oriented species. Judging from past island studies (Levenson 1976) and personal observation, corridors would have to be slightly over 100 m wide to permanently sustain beech populations and about 30 m to sustain sugar maple. The study area is at the extreme edge of the range for beech which may account for this species' unusual behavior near forest edges.

South and west island edges provide a permanent xeric habitat which allows species such as hawthorn, bur oak, hickories, and prickly ash (*Zanthoxylum*) to occur in otherwise mesic islands. The exposure that islands have along these aspects directly affects the importance values of these species. With other conditions equal, islands oriented northeast-southwest along their longest axis would minimize the importance of this species group and maximize the extent of moderately edge-oriented species (basswood, ash, etc.). Other orientations would not change the amount of area in "edge" but would affect the distribution and relative proportion of area in an island dominated by the two groups of edge-oriented species.

Forest Edge Structural Dynamics

To better understand how edges develop through time, a conceptual model was developed from field data and published literature (Gysel 1951, Trimble and Tryon 1966). We assumed that edges of different ages sampled at the same point in time can be chronologically arranged to represent edge change through time. Much of the variability between sites (soils, topography, climate, general vegetation type) has been eliminated through site selection criteria to permit use of such an assumption. Table 6-5 shows the temporal dynamics of stem density and basal area of forest edges for three edge maintenance types. In all cases, stem density for trees over 2.5 cm dbh peaked at about 20 years at a density of about 2600-2700 stems/ha. Stem density values in the interior consistently ranged from 46 to 67% less than edge densities (with the exception of young advancing edges which had not yet peaked). Basal area measurements demonstrate the effects of maintenance types on edge dynamics. Cantilevered edges continued to increase in basal area throughout successive age intervals with no plateau apparent. Individual trees along this type of edge theoretically experience the least amount of competition for light. The second structural category (edges maintained at the outer tree dripline) shows a maximum edge basal area value in the 20-40 year age class. Beyond this age, larger trees extending toward the edge are able to shade out individuals closer to the edge. Advancing edges are quite distinctive from the other structural types because they effectively begin from open field conditions, thus exhibiting very low basal area in the 10-20 year age class. Eventually this type of edge will take on the form of one of the other two types as it develops along a new boundary.

The structure of newly created forest edges changes during the first decade or two of development. The first stage (Fig. 6-6b) is dominated by the development of a dense understory of herbaceous perennials and woody seedlings and saplings. The

Table 6-5. Effects of Edge Structure and Age on Basal Area and Stem Density

Edge structure	Age class (yr)	Basal area (m²/ha)		Stem density (no./ha)	
		Edge	Interior	Edge	Interior
Cantilevered	20-40	41.8	26.5	2625	1343
	40-70	43.8	32.2	1865	1221
	> 70	67.4	34.5	1675	1125
Canopy dripline	10-20	37.9	26.5	2450	1244
	20-40	58.4	32.4	2700	1374
	40-70	31.6	27.1	2590	1451
Advancing	10-20	9.1	30.8	2033	900
	20-40	24.5	36.2	2625	1200
	40-70	43.2	31.8	1954	1147

canopy trees next to the edge respond to the increased side lighting by crown expansion toward the edge, epicormic branching on boles exposed to increased solar radiation, or basal sprouting. Near the edge, greater direct solar radiation (heat) reduces the availability of soil moisture. At the same time, the newly regenerated understory is making added moisture demands by increasing interception and evapotranspiration along an edge. The change in moisture regime may at times place an added stress on canopy trees.

Studies of understory vegetation along edges less than 20 years old show a large component of xeric species which are often considered pioneer in secondary succession (Curtis 1959). It is not known if the agent(s) responsible for this phenomenon involves dispersal capacities of propagules from pioneer species, environmental conditions affecting germination and growth of propagules and favoring more xeric-adapted species, or some combination of the two. Other factors enter into the composition of the edge understory. Pre-edge seedlings may survive edge establishment. For basswood, prolific base sprouting from larger trees already located at the edge contributes heavily to edge regeneration. Seed invasion from the overstory, which, if predominately mesic in species composition, contributes to the establishment of mesic species in edge regeneration. This and the dormant seed stock already existing on the forest floor may outweigh the high mortality these mesic species will experience in becoming established along edges. The dispersal and establishment of "edge species" is most apparent at this time when edge species do not occur within the stand (Ranney and Johnson 1977).

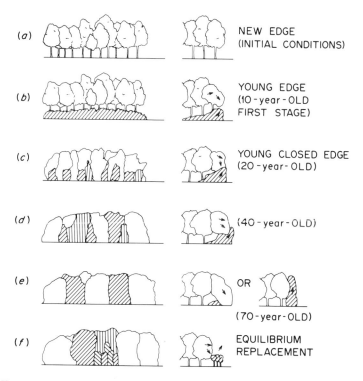

Fig. 6-6. Forest development along a newly created edge. Crosshatching illustrates development of the structure of edge-oriented species.

The edge overstory undergoes structural changes at a rate slower than that of the understory. The main change is in lateral growth of the crowns toward and beyond the edge. Some species have the capacity to adjust to lateral crown growth more rapidly than others (Trimble and Tryon 1966).

The second stage of forest edge development is characterized by a vegetative closing of the space between the canopy and the growing understory (Fig. 6-6c). Lateral canopy expansion beyond the forest edge is apparent due to asymmetrical crown development of individual edge trees. Competition for light between edge trees favors individuals capable of the fastest lateral limb growth (Trimble and Tryon 1966). Increased edge shading from the canopy and rank understory vegetation results in slight shifts of understory species composition and importance values. Canopy cover density is still apparently not sufficient to cause a decline in the density of the understory.

The vegetative closure of the edge causes microenvironmental changes in addition to shading. The island's interior becomes more insulated from external wind and allows more mesic conditions to develop. Because of altered wind conditions within the forest island and the reestablishment of the integrity of internal forest conditions, the pattern of invasion and ecesis of propagules within an island is altered and a conspicuous edge community develops.

The third stage of edge development (Fig. 6-6d) is marked by intensive competition between the understory and canopy for soil moisture and edge side lighting. Lateral growth rates of canopy trees and vertical and lateral growth rates of invading understory edge species determine which species survive. The numbers of dominant and codominant individuals are decreased and the general appearance of the edge is one of fewer and larger tree crowns with a few smaller individuals scattered along the edge margin.

The fourth stage is characterized by an edge dominated by a few large tree crowns (Fig. 6-6e) except where recent mortality has created gaps (Fig. 6-6e and f). At the age of 70 years, the dominant tree species of an edge will differ from those in the interior as a result of the different environmental conditions.

Studies which identify stable tree communities in forests seem to indicate that, based on species composition and size-class distribution, no edges examined in the selected literature were in a steady state. The persistence of xeric species in many size classes in the older edges seems to indicate that edges do reach an equilibrium under Zedler and Goff's (1973) definition although they differ in composition from forest interiors. In fact, edges appear to be a good example of Clements' (1916) disclimax where edge maintenance acts in much the same way as perpetual disturbance of forest interiors. No edges examined were of sufficient age, composition, and structure to verify edge equilibrium and the character of a "climax edge."

Five factors appear important in modifying the dynamics of edge development. These are (1) regional vegetation types, (2) successional stage of a forest island when an edge is created, (3) edge aspect, (4) activity of herbivores (Harper et al. 1970), and (5) the way an edge is maintained. The relative weights of the five factors listed above are not known and probably change through time and space. In order to gain some understanding of the interaction of these conditions or phenomena, we resorted to computer simulations.

Possible Edge Effects on Regional Forest Composition

In simulating forest development with effects from edge, it is necessary to estimate interisland propagule exchange as a background phenomenon to intraisland forest dynamics. Plants have often been thought to "saturate the surrounding environment with progeny" (Salisbury 1942) but this apparently does not hold true for many regions. When land is extensively cleared of original forests, proportionately more area in edges is created, i.e., the ratio of edge area to interior area increases (Curtis 1956, Levenson 1976). In such landscapes, forest edges, with their high exposure to propagule dispersal vectors, may contribute an important component to the declining propagule supply available for interstand transport. This contribution may eventually affect propagule invasion rates of various tree species and ultimately, the species composition of isolated forest stands.

Forest edges are assumed to produce more propagules per unit area than interior forests on the basis of substantiated higher edge primary productivity. If so, the propagules of edge trees would have a higher probability of dispersal (and perhaps subsequent caching) to other forest islands than interior trees. Sork and Boucher (1977) found this not to be true when studying individual trees; for hickories, the probability of removal remained constant per nut per week. Ghiselin (1977) suggests that there are also more animal species in edges where primary productivity is high, and infers that there may be more efficient predation (or dispersal) of tree propagules.

There may be a marked difference in animal predation on plants, seed consumption, propagule caching, and seed excretion by birds and mammals for many tree species between forest edges and interiors (Carpenter 1935, Smith 1975, Thompson and Willson 1978). Propagule invasion rates and survival to reproductive maturity is of considerable importance for forest development (Hett and Loucks 1971, Holt 1972). Models of seed dispersal by animals (DeAngelis et al. 1977, Bullock and Primack 1977) present a useful means for understanding some of the interrelationship between trees and animals.

Important variables in dissected landscapes are the spatial pattern of forest islands, the amount of forest edge (which is related to forest stand size and shape), the dispersal characteristics and agents of the tree species involved, and the normal sequences in succession (MacArthur and Wilson 1967, Cromartie 1975). McIntosh (1957) suggests that vegetation in a region may utilize several alternative but parallel avenues of succession. Changes in propagule invasion rates could favor one avenue of succession over another causing a shift in the composition of forests on a regional basis (Curtis 1956, Auclair and Cottam 1971).

The logistics of propagule dispersal and spatial patterns of forests may be more critical than the direct analysis of landscape patterns would lead us to believe. Forest islands may be potentially composed entirely of edge, an important consideration in propagule dispersal efficiency and in the arrangement of "safe sites" (Harper et al. 1961) for propagules of certain tree species within a given landscape. For example, the propagules of mesic, shade tolerant species would not have an equal opportunity for establishment within an entire forest stand but would be restricted to a smaller core area that excludes edge.

There is a size for forest islands below which mesic conditions cannot usually be found (Forman 1976, Levenson 1976). Forman (1976) believed that small islands of below about 2.3 ha and islands without dense, well-developed edge vegetation are permeable to wind and allow humus and litter layers to dry out in their interiors. Thus, were it not for the density of edge vegetation, the minimum critical size for maintaining forest interior conditions would probably be larger.

Edge vegetation provides propagules of a different species composition than would otherwise be available to interior forest tree replacement processes if edges were not present. The presence of seedlings of hawthorn (*Crataegus* spp.) and black cherry (*Prunus serotina* Ehrh.) in the interior, where mature seed-source trees are absent or infrequent, may be the result of extensive seed transport from edge trees to the interior. We believe that because of forest edges, the forest islands we examined have a greater component of shade-intolerant species in the interior than if the same sites were in extensive forest. For small forest islands the removal of one canopy tree may effectively alter the microclimate of the entire mesic interior (Levin and Paine 1974). Replacement by shade-intolerant species could lead investigators to identify the entire stand as edge. This review of past research on forest island edges raises a number of issues that require study in fragmented forest landscapes. These include:

1. How important is the animal vector in seed dispersal among forest islands and can this vector be manipulated and managed to assist in vegetation management?

2. Can propagule import to a forest island be measured and related to dynamics in spatial and temporal island composition? If so, how does it relate to concepts of island biogeography where the islands are habitat islands in a terrestrial environment?

3. Is there an arrangement of forest islands which can maintain the presettlement list of plant species given consideration of animal and wind propagule vectors and existing general land uses?

4. Can parallel courses of plant succession be documented for different forest islands as a result of propagule availability during critical stages of forest development?

5. If present trends are projected, what effects will edges have on forest island development in the future? Are there any basic principles which can be drawn from these effects?

Effects of Edge on Forest Island Development

The computer model FOREST was used to deal with the last question above by simulating forest edge and island interior conditions and then projecting possible forest conditions. Results from the projections were subsequently analyzed for basic principles applicable to forest islands and edges in the region where mixed hardwoods dominate the forest vegetation.

Model Validation. The ability of the modified FOREST model to simulate edge conditions and the contrasting conditions of interior and edge was tested by comparing simulations with field data. Data from a plane table survey (approximately 0.08 ha or 1/5 acre edge plot) of a cantilevered edge from Buckskin Bowmen Island were used to represent initial conditions for the model. The age of the sampled edge was 30 years. FOREST simulates this type of edge without any additional modifications. Exogenous seed input was set at an intermediate level. Forest development was simulated for 90 years and compared with actual conditions of edges and interiors having comparable ages and maintenance types. Basal area and density for edges and interiors were calculated as ratios. Ratios were computed as the average of five sample runs (five different random starts) and compared with ratios determined from field data. Actual and predicted basal area ratios were very close, indicating an accurate representation of simulated differences between edges and forest interiors (Fig. 6-7). Comparison of stem densities between edges and forest interiors also indicates that the model accurately represents the processes of competition, reproduction, and mortality.

The model's ability to simulate species composition dynamics was also tested. The initial conditions and simulation period were the same as above. The nine most important tree species were used in the model, and, again, field data were used as the control. Field data were drawn from several different forest islands, so composition on control plots was variable. Different site conditions (i.e., as represented by the controls) are second only to the distance from edges as variables which influence edge composition. Taking into account the differences between the initial conditions of the simulations (a relatively pure stand of sugar maple) and the composition of the controls, simulated trends in composition were remarkably similar to field data. A comparison was made of the relative basal area of the nine species in the model and the control data at 90 years using a Spearman rank correlation coefficient, r_s (Snedecor and Cochran 1967). The comparison (as used by Shugart and West 1977) required an ordering (numerical ranking from maximum to minimum) of percent values in both the control and simulation data sets. The r_s values resulting from the calculations were 0.59 for interior plots and 0.39 for edge plots, indicating similarity between control plots and simulations. In view of the variability and availability of the field data, the effects that initial conditions have on species basal area dynamics of interior and edge, and the

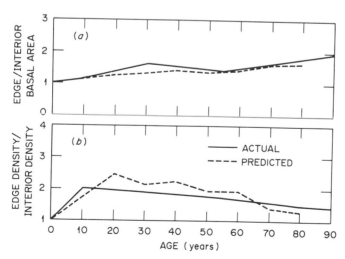

Fig. 6-7. Comparison of actual and predicted ratios of edge/interior basal area (a) and stem density (b).

variables of the environment, we believe that the test of the model was the most rigorous available. Since the model functioned acceptably, it was used to project island dynamics beyond the 90 year control period. In addition, simulations were run for edge and interior locations to identify long-term differences in development and equilibrium. Runs were also used to infer effects of edges on interiors.

Simulation Results. Two sets of simulations were run to determine what differences would develop between a forest interior and an edge. The simulation of edge dynamics was also used to further identify possible edge (0-10 m) development characteristics such as time to reach equilibrium and composition at equilibrium. Initial conditions for the edge simulations are from a 28.4 × 28.4 m (1/5 acre) plane table survey along the south edge of Cudahy Woods. The edge is cantilevered, and the island interior (and presumably the newly created edge at time zero) is composed primarily of sugar maple. Increment borings indicated that the edge surveyed was 45-60 years old. Since 11 species were tallied in the survey and only nine were simulated in FOREST, slippery elm (a species not in the FOREST model) was treated as northern red oak, and hornbeam was treated as sugar maple in the simulations. The substitutions were based on similarities in adaptation values (Curtis 1959). Initial conditions for the interior simulations were taken from the interior half of the survey plot and merely duplicated to derive a plot 0.08 ha in size. Five simulations were included in each of two sets. Each simulation began with a different random start and represented 20 years of growth. Exogenous propagule input was set at about 25,000/ha/species so seed source would not be limiting.

Comparison of interior simulations for both edge-bordered and deep interiors (Fig. 6-8) reveals effects of edges on interior composition. In both simulations, the relative

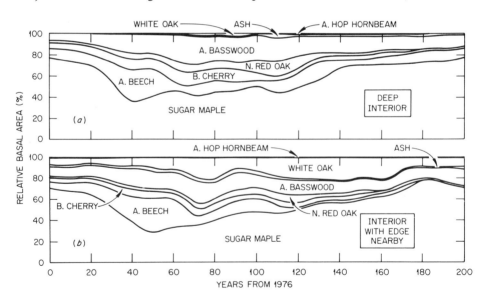

Fig. 6-8. Species basal area distributions resulting from model simulations of deep interiors (*a*), representing forest islands larger than two hectares, and edge-bordered interiors (*b*), representing interior development for islands of 1-2 ha.

basal areas of sugar maple, black cherry, and basswood are similar. However, beech is eliminated from edge-bordered interiors whereas white oak increases in basal area. The edge-bordered interiors reflect interior conditions for islands of 1-2 ha (i.e., islands of this size are not capable of sustaining beech populations for many generations). The other interior simulations show effects of further increases in island size.

The influence of the invasion of edge species propagules appears to be responsible for the increased time required for edge-bordered interiors to reach a general equilibrium (i.e., $\approx$ 140 years for deep interiors and $\approx$ 180 years for edge-bordered interiors). General lack of reproductive success accounts for the elimination of beech from edge-bordered interiors. Although beech propagules and root sprouts were simulated, conditions which represent small forest islands (1-2 ha) were beyond the species' tolerance limits. Eventually all older individuals died. Results were in agreement with field data which indicated that American beech could not establish itself in edge habitats through natural reproduction.

Average simulations of edges next to interiors (Fig. 6-9) demonstrate several changes that contrast with total interior plots (Fig. 6-8b), even with the continued invasion of "interior" propagules. By 200 years, edge species richness is greater than or equal to interior plots. Species richness appears to stabilize roughly 80 years after initial simulation conditions or when the edge is 125-140 years old (the edge used for initial conditions was 45-60 years old). Relative basal area fluctuates more widely in interior simulations but begins to stabilize at 180 years plus the age of the island interior at the beginning of the simulation (a total of approximately 300 years). The contrasts in relative composition between edges and interiors at the equilibrium points in Figs. 6-8 and 6-9 are distinct. Edges contain roughly equal proportions of sugar maple, basswood, and ash with most of the remaining basal area contributed by black cherry, northern red oak, and hop hornbeam. Beech is totally absent for the entire simulation period. The species was not present in the initial conditions for the simulation and failed to germinate and grow at any time. Edges maintain a small component of hickory whereas interiors do not.

Edge basal area stabilized at about 50% more than interiors (which strongly agrees with field data). This added basal area increases the importance of edges in producing propagules that invade and affect forest interiors. It also affects the component of "edge" species and quantity of "edge" propagules exposed to dispersal vectors. At the

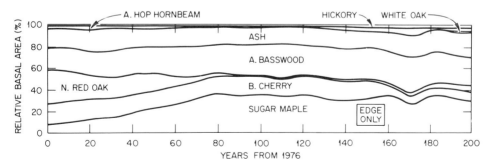

Fig. 6-9. Species basal area distributions resulting from model simulations demonstrating that a general equilibrium is reached about 80 years after initial conditions.

same time, interior species are insulated from dispersal vectors. Combining differences in basal area, composition, and propagule dispersal rates, edges contribute more to interisland propagule exchange than do interiors. This is especially true for older, smaller islands.

After evaluating the propagule production of edges and interiors and their probable contribution to interisland exchange, exogenous propagule input in the FOREST model was modified according to Table 6-6 and rerun for an island approximately 1 ha in size to determine effects on forest composition. The effect was minimal. Nearly all of the increase in sugar maple occurred in the island interior. Again, beech was eliminated. Were these propagules to be dispersed to forest islands just beginning secondary succession (with adjustment of acorn dispersal), it is likely that propagules would be limiting only for beech. However, increased island isolation would be expected to favor the successful dispersal of bird- and light wind-dispersed propagules which include black cherry, northern red oak, white oak, ash, and basswood. Poorer dispersal of other species which are predominantly shade tolerant, except for hickories, would likely result in a slightly slower rate of forest development toward the equilibrium stages demonstrated in the simulations. Although the "strategy" of beech does not require fast invasion of sites to successfully maintain a viable population in a forested environment, island isolation and especially island size (already small) may have scattered and reduced remaining populations past threshold levels necessary for adequately invading suitable sites.

The compositional development of forest islands is dependent upon forest island size, especially in the smaller islands. The source of this dependency is the ratio of edge area to interior area. Simulations of deep forest interiors, relative to edge-bordered interiors (within 20-30 m) show a lower component of shade-intolerant species and a lower species richness in equilibrium. The added propagule input from nearby edges is apparently enough to modify the course of forest development on smaller islands whose interiors necessarily are close to edges and whose edge/interior area ratios are greater than about 1.75-2.00 (i.e., island size is 2-3 ha). This is surprisingly similar to the 2.3 ha that Levenson (Chapter 3) independently determined as the island size of greatest species richness, containing both tolerant and intolerant species. Beyond this size, composition in island centers was dominated by tolerant, mesic species.

Figure 6-10 is an average of the results of 40 simulations showing species composition reached after 200 years. Initial conditions for the simulations were again taken from Cudahy Woods. Simulations for an edge plot and an interior plot were combined proportionately to represent different island sizes. For example, composition for islands less than about 1 ha was drawn completely from edge simulations whereas larger islands included increasing proportions of composition from interior simulations to account for size increases. There is an overlap of interior and edge species occurrence at the 2.5-3.5 ha island size as well as an absence of American beech in islands smaller than 2.5 ha. Although only nine species (plus a category for "others") are specifically considered, the results indicate an important principle for landscape management. Given a landscape of large forest islands, the change in forest composition can be determined if these islands were reduced to uniform size and allowed to equilibrate. But landscapes are usually not quite so organized; they have islands of various sizes. The results are still applicable to a region of many-sized islands if island size classes are weighted according to relative frequency and area.

Table 6-6. Factors Affecting the Amount and Composition of Propagules Contributed by One Forest Island to Another If Both Have 70-Year Old Edges[a]

Species	Propagule density[b]	Basal area		Mode of dispersal	Effect of distance	Exogenous propagules/yr
		Interior	Edge			
Acer saccharum	.3-13 × 10^6/ha	High	High	Wind	Moderate-severe	17,500
Prunus serotina	10,560/**kg**	Medium	Medium	Animal	Moderate-slight	25,000
Ostrya virginiana	?	Low	Low	Animal	Moderate	2,500
Quercus borealis	800-1600/tree	Low	Medium	Animal	Severe	250
Quercus alba	0-1900/tree	Low-medium	Low	Animal	Severe	250
Carya spp.	?	Low	Low	Animal	Severe	125
Fraxinus spp.	?	Low-medium	High	Wind-animal	Moderate	5,000
Tilia americana	6600-17,600/kg	High	High	Animal	Moderate	2,500
Fagus grandifolia	3520/kg	None-low	None	Animal	Severe	150

[a] Low, medium, and high basal area values corresponds to less than 1, 1-10, and over 10 m^2/ha. Most animal dispersal is by birds, although the oaks, hickory, and beech are distributed by mammals as well.

[b] Data from Fowells (1965). The silvics of forest trees. USDA Forest Service, Washington, DC.

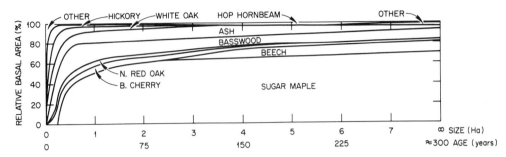

Fig. 6-10. Projected equilibrium composition for small forest islands (all considered to be square) based on island size and an approximate time scale. Although a linear relationship holds between size and time, this is not true for islands less than two hectares in size. Changes in equilibrium composition represent the continuum of vegetation recognized for southeastern Wisconsin. Equilibrium composition would differ for other regions.

The effect of forest island dissection is also apparent (Fig. 6-10). As islands are dissected within given size ranges, forest composition is shifted toward the left where the percent basal area of shade-intolerant species (hickories, oaks, etc.) is increased at the expense of more tolerant species (maple and beech). Change can be projected in somewhat specific terms using the continuum concept of forest composition. The exponential changes in composition occurring as island size changes represent shifts in the edge/interior ratio with accompanying composition interactions. Thus, equilibrium species composition differs for islands of different sizes depending on edge/interior ratios.

The forest composition of island landscapes is dependent not only on island size and equilibrium but also on the frequency and extent of island perturbations. In the study landscape, many islands were at intermediate and early stages of secondary succession. Accepted concepts of forest development for the area (Curtis 1959, McIntosh 1957, Whitford and Salamun 1954, Ward 1956) are remarkably similar to Fig. 6-10 if the island size axis is replaced by a time axis. Perturbations occurring at any frequency will generally tend to shift regional composition to earlier stages of development from the equilibrium for a particular island size. If frequency of perturbations is transformed into an expected time that islands may remain free from disturbance, the actual shift in forest island landscape composition can be projected. A simple linear model would suffice to accomplish such a task.

Conclusions

Clear differences exist in structure between forest island edges and interiors. Edges, which are about 15 m wide, support up to 50% more basal area than interiors. Simulations indicate that the increased edge basal area is a permanent condition, whereas tree density is transient. Greater edge productivity suggests greater propagule production. Density-diameter distributions along a gradient from edge to interior strongly support the hypothesis of Goff and West (1975) that diameter distributions vary from negative exponential form toward a rotated-sigmoid form as canopy-understory interactions increase. In young edges, this interaction is minimal because of side-lighting resulting in a negative exponential curve (R^2 = 0.96), whereas in older edges (over 70 years) and interior locations, the curve was of rotated sigmoid form. Consequently, the form of the diameter distribution can serve as an indicator of edge age along forest islands having well-developed interiors.

Clear differences in species composition also exist between edges and interiors. Edges are richer and contain a higher component of species that are intermediate to intolerant of shade. Although limited sampling indicates that more tree species are found in edges, other studies (e.g., Frei and Fairbrothers 1963) indicate that interiors contain more rare species that are not identified in local studies. For all but the youngest islands in a landscape, continuum index values (Curtis 1959) are significantly lower in edges than in adjacent interior locations. This points to the fact that edge dynamics are responding to side lighting and wind penetration (drying effects) along forest edges.

High tree densities, particularly in the shrub layer, are characteristic of disturbed forest islands and smaller forest islands. The high densities of the forest edge again suggest that forest edge and disturbed forest interiors are similar. This similarity may be critical in planning for the preservation of mesic forest stands. Mesic forest edge is different from undisturbed mesic forest interior. Small islands or disturbed islands will not support mesic forest. Instead they will support a successional forest similar to edge.

Spatial distributions of tree species in forest islands fall into three distinct categories. The first group is favored by interior conditions and declines in importance value near edges. These are the most mesic of the species in the region (sugar maple and beech). The second category is the reverse of the first and may be further subdivided according to the compass aspects on which species are most abundant. Species on south- and west-facing edges (hickories, hawthorns, bur oak, and others) are the most xeric and rarely extend into mesic forest interiors. Species on north- and east-facing edges (e.g., basswood, ash) are more shade tolerant and may extend into island interiors. The third category is ubiquitous and includes hop hornbeam and black cherry. The proportions of these categories (except for the last) is dependent on the shape, orientation, and minimum width of forest islands. Small and narrow islands reduce the proportion of interior-oriented species. Islands having considerable edge exposure to the southwest will contain a relatively high proportion of species in the most xeric, edge-oriented group. Irregularly shaped islands will have a high component of all edge-oriented species. Further dissection of forest islands in southern Wisconsin will result in the extirpation of beech in the region as the species does not maintain itself in edges for more than 200-400 years. However, the species is at the edge of its range in the study area.

A mathematical model adapted for simulating forest edge dynamics predicted a number of attributes of forest islands. Simulated edges reached equilibria at consistently lower continuum index values than interiors. Simulations also suggested that island interior composition is dependent on the presence and composition of edge forests, especially for islands smaller than about 5 ha. Based on composition, the rates at which both edges and interiors developed were comparable but edges consistently equilibrated at earlier successional states than interiors. Where forest islands were less than 5 ha, the influence of edges on interior development caused interior composition to reach equilibrium at a slower rate and at lower continuum index values. Studies of edge/interior ratios led to the conclusion that between the extremes of 0.5 and 5 ha, as island size increases, so does the level of forest development at equilibrium through the local continuum of vegetation. Also, there is a range of island sizes (roughly 2-4 ha) within which the rate of interior development is slowed below rates of deep interiors due to the mass invasion of edge propagules into the interior. At the same time, edges will generally take less time to reach equilibrium since forest development does not progress as far along the vegetation continuum. When large mesic islands are reduced in size below 4-5 ha, they are placed at a point on the vegetation continuum which may be above what the island can sustain at equilibrium. Forest composition will change eventually to a less mesic state at an undetermined rate.

In the well-developed islands of this study, newly created edges were dependent on propagule input from other islands. Once established, edges themselves act as a source of propagules for edge-oriented species and invade adjacent forest interiors far in excess of interisland dispersal levels for the same species (i.e., interisland propagule dispersal becomes less important while intraisland propagule exchange becomes more important). Simulation indicated that the interaction of edges with interiors is a function of island size and increases the proportion of species of intermediate shade tolerance in interiors. This occurred even when additional "interior" propagules were added to the seed input from other islands. This does not rule out the possibility that continued interisland propagule exchange will allow occasional interior or edge species to be added to existing lists for individual islands since invasion is more or less continuous. For the mesic equilibrium species, the question still remains as to whether the island landscape in the study area is sufficiently saturated with "interior" propagules to insure their survival.

Forest edges provide suitable sites for the maintenance of intermediate and shade-intolerant species. In forest island landscapes, edges (combined with other forms of island disturbance) lead to an ample supply of "edge" propagules to sustain these species. However, when islands become small, edge species may replace some mesic species and deny them potential sites for establishment. This process may cause the extirpation of species such as American beech and limit the invasion potential for other mesic species. A logical application of the results of small island dynamics would be to study adjusted propagule inputs into islands just starting secondary succession.

Results of this study are directly applicable to land management and environmental impact assessment. Edges are artifacts of man-modified landscapes which are permanent, yet dynamic, and are highly associated with the universal impacts of urbanization in forested regions. It has been conclusively shown that there is a selective effect on tree composition and forest island dynamics when edges are created. Some species are favored by the dissection of the forested landscape while others decline in importance. With the knowledge of such effects, landscapes can be managed and planned to obtain desired results in forest composition on a regional basis founded on the dynamics of individual woodlots. It should be realized that the faults of many regional plans with respect to large scale conservation efforts have resulted from their inability to directly connect broad scale planning goals with objectives that are meaningful to the small land owner and yet heed critical ecological phenomena. Perceiving regional changes in forest composition from the perspective of naturally functioning but artificially maintained habitat units (forest islands) can connect human intent with ecological wisdom in landscapes already dominated by man.

Results of this study should also help foresters in assessing the present and future hardwood timber resources of forest island landscapes. Effects of further forest island dissection on species composition can be assessed, as can the effects of land consolidation.

Predictability of vegetation development and equilibrium also contributes to wildlife management in particular habitats. Results of edge analyses could serve the wildlife manager in assessing existing and potential habitat available for species of concern. For example, future locations of habitats needed by endangered species requiring mesic interior forest conditions could be projected. From this, appropriate measures could be implemented to prevent local species extinction.

An important consideration in the management of forest islands and their edges is that the creation of new edges or the disruption of an intact edge will lead to a regression from mature, mesic conditions to dryer, pioneer conditions in the interior. If the objective of a landscape management plan is to preserve patches of mesic forest in the mature state, the protection of the edge from disturbance is at least as important as the protection of the interior.

If disruption of forest edges or interiors is unavoidable, strategies may be employed to minimize the disruption of the forest island. Plantings along a newly created edge will speed the development of the buffer vegetation. The sooner an edge shifts from an encroaching to a maintained state, the less the interior will be damaged. The best species to plant along a new edge are fast growing species that produce dense foliage. In southeastern Wisconsin, native species characteristic of pioneer stage forests and forest edges, such as basswood, hawthorn, and black cherry are preferred. Another alternative is to plant introduced, but seminaturalized species often found in edges, such as buckthorn (*Rhamnus catharticus*) or the introduced bush honeysuckle (*Lonicera* spp.). The third option is to plant species not characteristic of southern mesic forests. Conifers at the edges of a forest island create an unnatural edge, but one that appears to be a highly effective buffer. Islands were observed which had a belt of conifers planted in three rows at the edges that served as an effective wind and light barrier for the forest interior. These planting strategies using native, seminatural, or introduced species can minimize the damage caused by the creation of a new forest edge.

Acknowledgments. Research supported in part by the Eastern Deciduous Forest Biome, US-IBP, funded by the National Science Foundation under Interagency Agreement AG-199, BMS69-01147 A09 with the U.S. Department of Energy under contract W-7405-eng-26 with Union Carbide Corporation, and in part by NSF grant DEB 78-11338 to Southern Illinois University. Contribution No. 347, Eastern Deciduous Forest Biome, US-IBP. Publication No. 1636, Environmental Sciences Division, Oak Ridge National Laboratory.

7. Biogeography of Forest Plants in the Prairie–Forest Ecotone in Western Minnesota

MICHAEL J. SCANLAN
DEPARTMENT OF BIOLOGY
VIRGINIA COMMONWEALTH UNIVERSITY

The equilibrium model of island biogeography (MacArthur and Wilson 1963) interprets the number of species present on a given island as a function of two counteracting processes, immigration and extinction. The rates of immigration and extinction are directly related to separate environmental parameters. Immigration is considered to be inversely proportional to the distance from source regions: distant islands are impoverished, a "distance effect." Extinction is viewed as a stochastic process with probability of extinction falling off as island area increases: the "area effect."

MacArthur and Wilson suggested that these relationships should apply to the entire biota of islands, and to continental as well as oceanic islands. However, the hypotheses are most commonly applied to the avifauna of oceanic islands. For example, in an excellent review of the island biogeographic literature, Simberloff (1974) refers to at least 31 papers of this kind.

That the hypotheses are appropriate for entire biota is exemplified by their application to a wide range of organisms including Hemiptera (Leston 1957), bats (Koopman 1958), invertebrates (Simberloff and Wilson 1970), zooplankton (Schoener 1974), plants (Johnson and Simberloff 1974), and primate intestinal protozoans (Freeland 1979).

In this chapter, I examine the floristic richness of planted and natural woodlands in the tall grass prairie region of west-central Minnesota (Fig. 7-1). The woodlots were established during the past century on the prairie uplands, usually near farmsteads. The trees that dominate the woodlots were initially transplanted as saplings from the natural forests of the prairie–forest margin by the early European settlers. Now that 80 years have passed since the woodlots were planted, most of the prairie

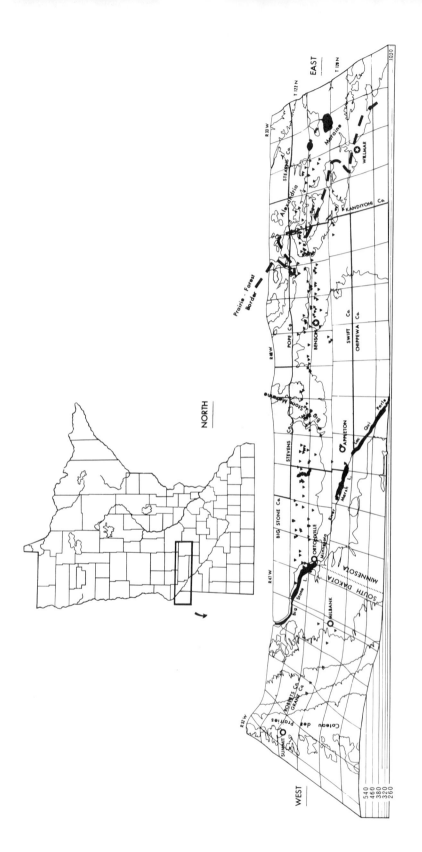

Fig. 7-1. Block diagram of the transect area in western Minnesota showing sample sites and a county map of the state outlining the transect location. Contours are in meters above mean sea level.

species that once existed in the understory of these stands should have been replaced by forest plants immigrating from the nearby natural forests. Assuming that the immigration and extinction rates of forest plant species on these "continental" islands have reached equilibrium, then species richness should be a function of island size and isolation. The results suggest that this relationship exists.

Description of the Study Area

The study area is located in four contiguous counties, Kandiyohi, Swift, and Big Stone counties in Minnesota, and Grant County in South Dakota (Fig. 7-1). Two topographic highs are present in the area, the Coteau des Prairies on the west with an elevation of 610 m, and the Alexandria Moraine complex on the east which reaches a maximum of 411 m. Between these two landforms lies a level plain centered in Swift County that has a general elevation of 320 m with a very gentle slope to the southwest. Most of the planted forests that were sampled are located on this plain.

The Minnesota River, forming the boundary between Minnesota and South Dakota, fills two segments of its valley to form Big Stone Lake and Marsh Lake Reservoir. Two other rivers, tributary to the Minnesota, are the Chippewa River, which drains the eastern three-fourths of Swift County and the northwestern corner of Kandiyohi County, and the Pomme de Terre, which enters the Minnesota River at Marsh Lake Reservoir near Appleton, Minnesota, and drains western Swift and eastern Big Stone counties.

The large lakes of the area include Green Lake, Mud Lake, and the Norway-Middle-Games Lakes complex in the northern part of Kandiyohi County, as well as Artichoke Lake near the Big Stone-Swift County line. Numerous smaller lakes are present in Big Stone and Kandiyohi counties. Although lakes are absent from most of Swift County, the central area was mainly swampy before settlement by European man and the water table is still above the surface in several places.

History of the Vegetation

The woodlots examined in the study area did not exist prior to settlement by European man. The Sioux Indians who then occupied western Minnesota commonly set fire to the existing tall grass prairies (Murphy 1931). As an example of the extent of these fires, the prairie east of the Pomme de Terre River was set on fire in October, 1878. Spreading at a rate of 10 km/hr, it burned an area of 80 km^2 (Monitor and News 1878). Frequent fires of this kind would have destroyed all upland stands of timber on the prairie other than those sheltered by lakes, rivers, or topographic features. Fires would also have prevented woodland shrub and herb species from growing on the uplands.

Marschner described the presettlement vegetation along the prairie–forest border in the study area as a thin strip of savanna (Heinselman 1974). However, the "Surveyor's Subdivision Notebooks," written by the first government surveyors, indicate that the ecotone was a broad area with scattered bur oak and occasional large stands of bur oak forest. Based on these records, savanna was certainly more extensive than indicated by Marschner. The natural stands of trees on the prairie–forest border that were found by the original surveyors in the period from 1857 to 1870 are presented in Fig. 7-2.

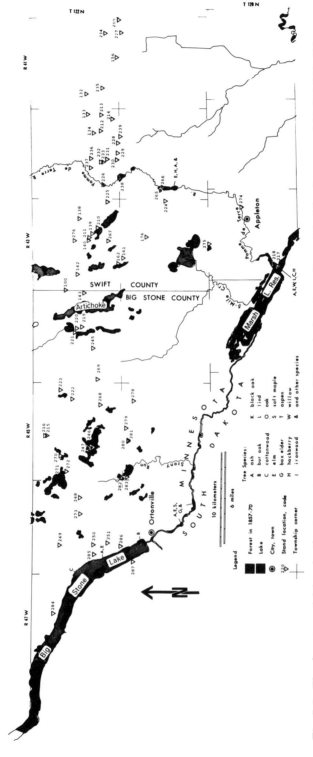

Fig. 7-2. The forests and lakes in the transect area in western Minnesota based on the original land survey of 1857-1870; this page: western half, facing page: eastern half.

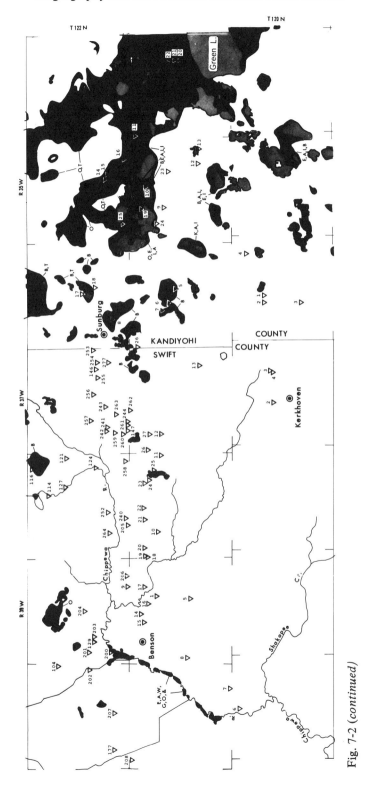

Fig. 7-2 *(continued)*

With the plowing and destruction of the western Minnesota prairies in the late 1800s, fire ceased to be an important force in determining the vegetation. The absence of fire no doubt led to the closing of the canopy of oak openings and savannas on the prairies and prairie margins (Cottam 1949). This change would also have provided more hospitable sites for the establishment of woodland plant immigrants, especially those that were sensitive to fire.

It is apparent (Fig. 7-2) that proximity to lakes was essential for the persistence of at least those forests outside of the main body of deciduous forest. These forest outliers along rivers and lake shores, as well as the main forests of the ecotone, were the potential sources of understory immigrants for woodlots. They were also the sources of the trees that were planted. The settlers west of Benson, Minnesota, hauled trees, often in carts, from the Pomme de Terre River terraces (personal communication from two landowners in Swift County). East of Benson the most likely source of trees would have been the forested area on the Alexandria Moraine. West of the Pomme de Terre River the source of trees would have been the Minnesota River Valley.

The physiognomy of the study area is in general the same today although the area occupied by forest has been much reduced (cf. Figs. 7-2 and 7-3). Furthermore, the large nonforest area in Fig. 7-2 that was prairie has been converted to pasture and cropland.

The Planted Forests

Based on information from the landholders, data concerning original land sales (Minnesota State Records Department), aerial photographs dated 1938 and 1961, and the notes of the original surveyors, the tree stands in the area can be separated into those that were present prior to European settlement and those that were planted by the early settlers (Tables 7-1, 7-2, and 7-3).

The woodlots were planted for wind and snow breaks, firewood sources, game sources, livestock shelter lots, timber, and aesthetics. The age of these stands can be estimated by means of settlement history and legal documents. Settlement of the Swift County area began in 1866 (Appleton Press 1973, Monitor and News 1970). Therefore, no woodlot should predate 1866. The locations first selected by settlers were forested sites where tree plantations were probably not necessary. Upon completion of the railroad to Benson in 1870, settlement of the Swift County prairie intensified. Therefore, tree planting on the prairie probably began around 1870. This sets a maximum age for planted stands at slightly more than 100 years. As settlement of the prairie generally proceeded in a westerly direction in the transect area, the majority of the planted stands should be 108-110 years old or younger. Many more stands should date from the large immigration that occurred in 1876 and 1877, and from the period of the Timber Culture Act, 1873-1891, when the Federal Government made the planting of trees on the prairies financially attractive (see references to the 42nd and the 43rd Congress). Of the 94 planted stands listed in Tables 7-1, 7-2, and 7-3 at least 13 originated as patented Timber Claims (original land sales records in the U.S. Land Office Tract Books), five in the Alexandria Moraine archipelago (Table 7-1), three in the Pomme de Terre region (Table 7-2), and five in the Minnesota River stand area (Table 7-3).

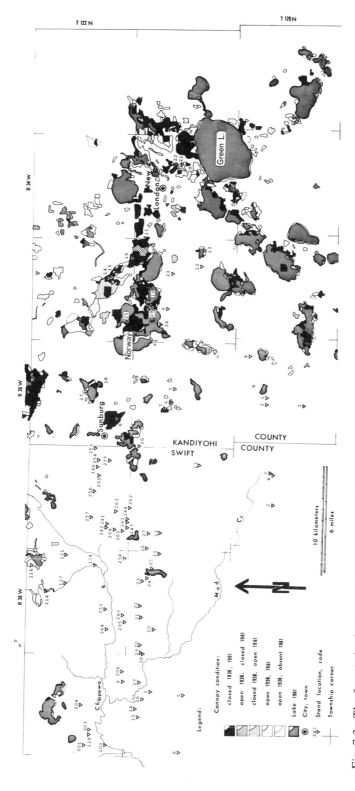

Fig. 7-3. The forests in the transect area in western Minnesota in 1938 and 1961 (1963) as recorded on aerial survey photographs.

Table 7-1. Alexandria Moraine Stand Areas, the Distance from Major Source Areas and from Nearest Neighbor, the Number of Species in Each Stand, and Estimates of Stand Age[a]

Stand code	Mean area (ha)	Distance (km)			No. species		Age (yrs) (in 1972)
		D_w	D_n	D_{nb}	Total	Natural	
Kandiyohi County							
20*	5.80	0.0	0	0.50	82	69	
21*	7.70	0.0	0	0.50	63	57	
22*	11.42	0.0	0	0.50	53	47	
11*	6.92	6.8	0	2.15	56	52	
16*	75.20	8.8	0	2.15	48	42	
15*	8.28	10.3	0	0.36	39	36	
14*	8.28	10.5	0	0.36	56	50	
19*	12.27	13.6	0	1.99	33	30	
25*	775.22	15.1	0	1.99	59	56	
10*	5.27	11.3	0	2.44	59	56	
17*	3.69	21.3	0	1.48	46	43	
18*	7.48	20.6	0	1.48	16	14	
8*	16.49	23.6	0	3.16	66	57	
7*	7.83	22.8	0	0.86	68	59	
26*	1.22	26.1	0	3.18	42	34	
23*	16.31	9.8	0	2.50	61	53	
24*	6.21	14.8	0	1.66	28	23	
13*	4.41	8.0	0	1.04	42	40	
12*	1.06	9.0	0	1.04	38	26	
6*	3.27	22.1	0	0.86	17	15	
5*	1.60	21.1	0	1.02	35	30	
9	1.25	13.3	1.3	1.66	29	16	68
4	1.82	17.6	3.3	3.94	28	17	34
3	2.05	21.8	5.8	3.51	34	21	85[b]
1	1.44	21.1	6.5	1.00	16	6	>49
2	1.39	22.1	7.3	1.00	22	10	>51
Swift County							
116*	30.66	36.9	0	3.20	50	43	
114*	6.71	39.4	0	1.39	52	44	
121	2.20	36.1	1.5	2.28	40	22	55
253	1.10	26.6	1.8	1.12	29	15	46
277	1.21	27.6	2.0	1.00	31	11	
254	1.58	27.6	2.8	0.50	29	17	
146	1.80	28.1	3.0	0.50	13	5	
255	1.27	28.6	3.3	0.50	22	10	
256	1.07	30.4	4.8	1.77	29	11	42
262	0.82	31.6	4.5	1.32	27	10	
244	1.79	32.9	4.5	0.50	33	15	65
243	1.15	31.6	5.3	1.28	20	8	82
263	0.58	32.4	5.3	1.12	24	7	
257	1.06	32.9	6.0	1.66	35	11	>33
13	2.64	27.9	6.3	6.07	21	11	
261	1.08	33.4	6.3	0.20	31	12	83
259	0.85	33.9	6.6	0.73	36	21	64
147	1.33	33.6	7.1	0.20	36	21	94
250	0.36	34.1	7.1	0.71	27	13	86
258	0.74	36.4	8.1	2.25	15	6	

Table 7-1 (continued)

Stand code	Mean area (ha)	Distance (km)			No. species		Age (yrs) (in 1972)
		D_w	D_n	D_{nb}	Total	Natural	
27	0.30	34.1	7.6	1.30	14	7	86[b]
12	2.88	34.1	7.3	1.30	20	11	
26	0.53	35.6	8.6	1.24	13	5	
11	1.21	35.9	9.8	1.24	25	13	
25	0.24	37.4	10.6	0.85	13	4	
23	0.82	38.4	11.3	0.54	15	5	
24	0.45	38.2	11.8	0.54	18	7	
252	1.07	41.2	6.6	1.77	26	9	
240	0.64	41.7	9.3	0.58	21	7	
264	0.79	42.9	8.8	1.66	20	8	41
205	1.90	42.4	9.8	0.58	25	14	83
22	1.36	40.7	10.6	1.00	28	14	88[b]
21	1.60	41.7	10.8	1.00	22	11	
10	3.80	42.9	12.3	1.77	31	21	96[b]
20	3.72	44.4	11.8	0.71	27	12	96[b]
19	0.34	45.2	12.6	0.76	20	7	
18	0.69	44.9	13.6	0.71	12	4	
3	1.51	28.6	13.4	0.30	12	7	
4	0.44	28.6	13.6	0.30	18	9	
2	1.58	31.4	14.4	2.80	27	14	

[a] Asterisk denotes natural stands; all others are planted. D_w is distance west of the most eastern stand; D_n is distance to the nearest natural stand; D_{nb} is distance to the nearest stand, natural or planted.

[b] Timber claim.

Methods

Location of the Sample Transect

In the spring of 1972, an east-west transect in western Minnesota was established along which the sample stands were selected. The transect is approximately 170 km long by 10 km wide. On the east end the transect begins at a point 24 km north of Willmar, Minnesota, and extends 120 km west to the Minnesota River at the Minnesota-South Dakota border. It extends an additional 50 km into South Dakota to an area just south of Summit, South Dakota. During the survey, 171 stands were sampled. The location of each stand is mapped in Fig. 7-1. However, for the present paper, only the data from the 134 stands listed in Tables 7-1, 7-2, and 7-3 will be used. The remaining 37 stands were excluded. One of the 134 sampling sites is in South Dakota. The remaining stands include 30 in Big Stone County, 77 in Swift, and 26 in Kandiyohi.

The location of the transect offered advantages in testing several hypotheses. In order to assume that dispersal of forest species from the forest margin had been occurring continually since the woodlots were planted, a large forest area on the prairie—forest border was essential. The aerial photographs of the region revealed the existence of a number of large forests on the forest margin, and a large number of woodlots that

Table 7-2. Pomme de Terre Stand Areas, the Distance from Major Source Areas and from Nearest Neighbor, the Number of Species in Each Stand, and Estimates of Stand Age[a]

Stand code	Mean area (ha)	Distance (km)			No. species		Age (yrs) (in 1972)
		D_w	D_n	D_{nb}	Total	Natural	
Swift County							
156	5.09	68.8	12.4	2.57	30	21	98[b]
235	1.50	72.0	8.0	2.15	24	10	49
132	1.13	72.8	7.0	2.15	17	9	70
214	1.80	75.3	5.5	1.04	25	9	67
213	0.52	75.0	5.2	1.04	16	5	55
133	3.19	75.1	4.9	1.70	18	9	83
239	0.91	77.3	3.7	0.85	27	10	85
212	0.85	76.8	3.7	0.76	20	4	
228	0.80	77.1	3.1	0.85	26	8	80
134	2.13	77.1	2.9	0.76	23	12	95[b]
231	0.66	79.8	2.3	0.42	34	17	81
229	0.78	79.1	2.2	0.51	20	9	
233	0.38	80.1	1.8	0.36	22	12	45
232	0.84	79.8	1.7	0.36	27	11	104
230	0.52	79.6	1.7	0.51	17	7	
236	1.14	79.6	0.9	1.00	20	10	65
237	0.92	80.6	0.4	1.00	29	10	62
238*	4.21	81.8	0	1.58	38	30	
226*	2.19	82.3	0	1.58	31	28	
266*	5.19	82.8	0	0.30	22	19	
265*	7.95	83.1	0	0.30	54	42	
274*	5.70	84.8	0	5.87	51	33	
224	1.81	84.3	−1.3	1.24	37	23	92[b]
225	0.88	84.1	−1.3	1.83	24	11	45

[a] Asterisk denotes natural stands; all others are planted. Negative distance indicates stands *west* of the source areas. D_w is distance west of the most eastern stand; D_n is distance to the nearest natural stand; D_{nb} is distance to the nearest stand, natural or planted.

[b] Timber claim.

could be colonized by migrants. Many woodlots were required because of the expected high rejection rate and the expected low richness of each stand. Finally, the generally level topography of the area west of the Alexandria Moraine (Fig. 7-1) essentially eliminated topography from the list of major environmental variables.

Stand Selection and Measurements

The stands in Minnesota were selected from U.S. Soil Conservation Service aerial photographs taken in 1961 and 1963, while those in South Dakota were selected during a field reconnaissance in 1973. The stands had to consist of trees with a canopy more than two trees wide; that is, trees in fence rows were not acceptable. The canopy had to be continuous and the stand had to appear undisturbed on all aerial photographs from the oldest to the most recent.

Table 7-3. Minnesota River Stand Areas, the Distance from Major Source Areas and from Nearest Neighbor, the Number of Species in Each Stand, and Estimates of Stand Age[a]

Stand code	Mean area (ha)	D_w	D_n	D_m	D_{nb}	Total	Natural	Age (yrs) (in 1972)
Swift County								
138	1.78	86.1	20.6	20.6	1.92	18	10	
210	0.96	87.1	18.1	18.1	0.71	27	12	>102
139	1.21	87.6	18.5	18.5	0.36	24	12	81
211	0.37	87.9	18.5	18.5	0.36	23	15	74
267	0.82	88.6	16.6	16.6	1.99	15	6	63
140	2.06	88.4	18.5	18.5	0.58	25	14	90
276	1.36	88.4	19.9	19.9	1.20	25	11	
161	2.14	90.4	14.2	14.2	0.71	32	20	96[b]
162	2.69	90.0	14.3	14.3	0.71	29	13	86
142	1.89	91.9	17.9	17.9	1.92	25	14	90[b]
100	2.91	93.4	16.6	18.5	1.84	29	15	94[b]
176	1.35	88.1	13.2	13.2	3.93	23	8	90
275	0.99	88.9	7.3	7.3	5.76	26	9	
Big Stone County								
143	1.65	94.1	15.8	16.2	1.84	25	15	75
221	1.37	98.1	11.9	16.2	0.28	20	5	
220	4.10	97.9	12.1	16.4	0.28	14	10	> 89
219	1.08	97.6	12.3	15.7	0.85	13	5	60
223	1.93	104.2	6.4	15.7	1.64	29	10	85
222	1.87	105.2	5.0	14.4	1.64	24	12	45
216	4.16	108.7	4.7	14.1	0.10	28	15	80
215	3.25	108.7	4.7	13.9	1.64	39	22	79
245	4.17	99.9	10.0	13.8	2.51	22	13	73
269	1.29	103.2	6.7	12.1	2.70	27	8	
270	0.83	110.9	2.3	11.5	1.68	24	15	
268	0.74	105.9	4.0	11.2	2.70	33	11	
272	1.75	112.5	1.0	9.7	0.99	30	17	77
279	1.09	108.2	1.0	7.9	1.02	34	13	69
278	0.60	105.4	3.8	7.7	2.89	31	10	
248	2.99	115.2	2.8	6.8	1.50	38	21	96[b]
273	2.22	116.7	4.0	5.1	1.50	34	12	77
249	2.62	120.0	3.1	3.1	3.50	27	17	95[b]
250	2.58	120.0	1.2	1.4	1.22	30	17	81
247*	5.07	109.9	0	11.4	0.50	40	30	
246*	4.81	109.9	0	11.2	0.50	42	32	
271*	3.81	113.2	9	9.1	0.99	32	16	
280*	2.77	109.2	0	7.4	0.20	30	12	
281*	0.24	109.4	0	7.4	0.20	32	15	
282*	1.04	113.2	0	6.9	0.50	37	20	
283*	1.21	113.2	0	6.5	0.50	47	40	
251*	2.10	120.0	0	0.9	1.32	37	24	
284*	10.89	127.3	0	0.5	7.34	58	39	
285*	1.58	121.2	0	0.1	1.22	45	37	
286*	2.07	120.2	0	0.3	1.32	58	50	

Table 7-3 (continued)

Stand code	Mean area (ha)	Distance (km)				No. species		Age (yrs) (in 1972)
		D_w	D_n	D_m	D_{nb}	Total	Natural	
South Dakota								
287*	8.30	122.2	0	−0.3	2.24	57	47	

[a] Asterisk denotes natural stands; all others are planted. Negative distance indicates stands *west* of the source area. D_w is distance west of the most eastern stand; D_n is distance to the nearest natural stand; D_{nb} is distance to the nearest stand, natural or planted; D_m is distance to the Minnesota River.

[b] Timber claim.

During the field survey any stand found to contain livestock or with a tree canopy of less than 20% cover was rejected. Interviews with the owners were conducted to aid in the evaluation of disturbance. After a stand was accepted in the field, a list of all overstory and understory species of nonaquatic angiosperms in the entire stand was compiled.

In many of the woodlots a bole core was taken to the center of the green ash (*Fraxinus pennsylvanica*) with the greatest girth. *Fraxinus pennsylvanica* was selected as the standard because it was present in most stands, the growth rings are very well defined, and the tree is not subject to heart-rot. The number of annual growth rings was recorded to establish the approximate year in which the trees were planted (Tables 7-1, 7-2, and 7-3). Also, based on coring data, the species longevity exceeds 108 years in the sample area, longer than the predicted age of the oldest woodlots.

The area of each stand was measured using both the oldest (1938) and the most recent (1961, 1963) available aerial photographs. A transparent plastic grid was used to measure area to the nearest 0.02 ha. The "mean area of the stand" was calculated by averaging the oldest and most recent canopy area sizes.

Delineation of Habitat Island Groups

A source area is defined as an aggregate of natural forest stands that contains a relatively rich understory from which propagules are released for the potential colonization of planted stands. The canopy in each stand is closed and the stands are all more than 100 years old. In the study area there are three distinct source areas. The first and most important source area includes the forests of the Alexandria Moraine (Table 7-1), which marks, in general, the westward limit of natural upland forests. The Minnesota River forests constitute the second most important source area (Table 7-3). Here the natural forests are located on the floodplain of the river or in bordering ravines. The final important source area consists of the forests along the Pomme de Terre River (Table 7-2).

In order to examine the relationships between the isolation of a planted stand from a source area and the size of the stand to its total or understory species richness, the 134 stands in the main study area were separated into three groups centered on the source areas. These groups are: (1) the Alexandria Moraine group, 43 planted stands and 23 natural stands; (2) the Pomme de Terre River group, 19 planted and 5 natural stands; and (3) the Minnesota River group, 32 planted and 12 natural stands. This subdivision was based on the assumption that each source area had a domain of primary influence. The woodlots closest to the source area should receive most of their species from the species pool of that source area.

Certain historical and geographic factors also had to be accommodated in delineating the archipelagos of woodlots. For example, the planted stands in the zone between the two eastern archipelagos were excluded from both groups because the stands are close to the original location of Six Mile Grove, a large, natural forest that once bordered the Chippewa River (Fig. 7-2). This forest was destroyed in the early 1900s during dredging operations. Several stands near its original location are floristically richer than can be explained if the forest had not existed.

The selection of the western terminus of the Pomme de Terre River transect is based on the fact that the river and the natural stands on its margin are located at the foot of the abrupt east-facing bank of the river valley. If wind is an important dispersal agent in the area, then anemochores are more likely to be dispersed eastward by the prevailing winds. The eastern limit of the Minnesota River transect was determined after realizing that the natural forests on the Minnesota River are much larger than those on the Pomme de Terre River and, given the prevailing west winds, even stands close to the Pomme de Terre River are more likely to have received their colonists from the Minnesota River source stands. Finally, most South Dakota stands were deleted from consideration because they were quite young compared to the woodlots in Minnesota. At the conclusion of the discrimination processes 37 of the original 171 sites were excluded from further consideration.

Selection of Dependent Variables

Two dependent variables were selected to test the application of MacArthur and Wilson's model to the planted stands. The first obvious variable was total species richness (S_t) (Table 7-4). Because this quantity may include plant species that have originated from several sources, it was necessary to find the component that had immigrated from the deciduous forest source areas, the natural species (S_n). The definition of natural species rests on the assumption that the frequency of occurrence of a species indicates its preferred community type. Natural species were defined as those species that occurred in 13 or more ($> 48\%$) of the natural forests at the Alexandria Moraine, or that were found more often in natural than in planted forests (Scanlan 1975a).

The resulting species numbers for each stand are listed in Tables 7-1, 7-2, and 7-3. In support of the identification process, the classification of species agrees well with their habitats quoted in Fernald (1970).

The results of this classification for natural species are given in Fig. 7-4A. During the study, 287 species were observed in the sample stands, but only 273 were positively identified and used in Fig. 7-4. Of these 89 were restricted to the natural forest source areas. An additional 90 species occurred in both forest types but with an equal or higher frequency in the natural forests.

It is surprising to find just over half of the 179 species that typically occur in the natural forests of the prairie—forest ecotone have colonized the new habitat islands in the last century. The 94 species of planted forests are slightly more mobile than the natural forest species because 57 (61%) were found in both forest types.

Table 7-4. List of Symbols Used in Regression Analyses of Forest Islands in Western Minnesota

Age	=	Age of stand (based on age of the largest trees)
A_{38}	=	Stand area in 1938 aerial photographs
A_{61}	=	Stand area in 1961 aerial photographs
A_{mn}	=	Mean stand area, $(A_{38} + A_{61})/2$
D_m	=	Distance to the Minnesota River (for Minnesota River archipelago)
D_n	=	Distance to the nearest natural forest
D_{nb}	=	Distance to the nearest neighboring stand
D_w	=	Distance west of the easternmost natural stand in the study area
S_t	=	Total species richness
L_{St}	=	$\text{Log}_{10}\ S_t$
S_n	=	Total natural species richness
L_{Sn}	=	$\text{Log}_{10}\ S_n$

It is useful to view the species quantities in Fig. 7-4A as if they had resulted from a random distribution of species individuals among the sampled stands. If we also assume that each species had a frequency of 2% and that the two stand types were equally numerous, then we would expect 25% of the species to be restricted to woodlots, 25% restricted to natural forests, and 50% present in both forest types. The observed percentages differ from expected particularly with respect to species restriction. Thus, one (or more) of the three assumptions appears incorrect.

Frequency is not consistent for the sampled species, but 26 of the forb species do have a stand frequency of 2%. If the remaining assumptions apply to these species then they should show the predicted percentage restrictions and co-occurrences. However, among the 26 forb species an unexpected 73% are restricted to one or the other of the forest types. Of the 27% that occur in both forest types, all are more common in the natural forests. In addition, for the total species observed, the average frequency is equal to or greater than 5% and the probability of restriction rapidly decreases as species frequency increases above this value. Although the precise probabilities for stand type restriction or co-occurrence require complex calculations, the observed percent of species restriction is far above the values that would be predicted assuming random dispersion.

The forest types were not equally sampled. Of the original 171 sampled stands only 44 (25.7%) are natural stands. For that reason more of the species should have occurred in the planted stands if species occurrence is random and related to sample size. The extremely underrepresented flora of the planted forests and the overrepresented flora of the natural forests indicate that the distribution pattern is not related to the proportion of forest types sampled.

Therefore, after allowances are made for differences in species frequencies and in stand type frequency, there is still an unexpectedly large number of species restricted to each of the two stand types. Conversely, fewer species than expected co-occur in the two stand types. These considerations and the relatively rich flora of the natural forests lead to the hypothesis that the natural forests at the prairie—forest ecotone are indeed operational source areas for plant species in the woodlot islands.

Fig. 7-4. (A) Venn diagram of the number and percent of species restricted to each forest type and that occur in both types. (B, C, and D) The percent of the flora categorized as forbs (f), graminoids (g), shrubs (s), or trees (t), and n, the total number of species used. The diagrams show (B) the subdivision of all identified species, (C) those classified as natural forest species, and (D) species that have colonized the planted forests from the source areas.

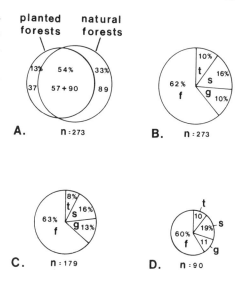

In the results section, the nonrandom species distributions are examined in terms of the geographical features of forest size and isolation to determine if these features can explain the distribution patterns. However, it is first necessary to determine if there is reason to separate the plant species by category. In Fig. 7-4B, the pie diagram shows the 273 identified species separated into the four basic life-form categories. The graph for the 179 species of natural forests (C) and that for the 90 natural forest species that are colonists (D) are both very similar to Fig. 7-4B. Therefore, the species were analyzed as a single group rather than as subsets of life-form categories.

Selection of Independent Variables

Woodlot richness could depend on a number of independent variables. Those that were initially expected to be important were stand area on the 1938 aerial photographs (A_{38}); stand area on the 1961 photographs (A_{61}); mean stand area $A_{mn} = (A_{38} + A_{61})/2$; distance to the nearest natural forest (D_n); distance west of the easternmost natural stand in the study area (D_w); distance to the nearest neighboring stand natural or planted (D_{nb}); and, for the Minnesota River archipelago, distance to the Minnesota River (D_m). The measurements are presented in Tables 7-1, 7-2, and 7-3.

An initial scatter plot of total species (S_t) and total natural species (S_n) against the area and distance variables indicated that their variances were not constant over the range of independent variables, hence linear regression could not be used. Therefore, a logarithmic transformation was employed (S_t and S_n become L_{St} and L_{Sn}, respectively, after transformation) (Table 7-4). The correlation coefficients between species richness (L_{St} and L_{Sn}) and each independent variable were used as an indication of the importance of the independent variables.

As a result of MacArthur and Wilson's use of the Arrhenius power function to model the relationship between species richness and island area, many researchers have assumed that this is the appropriate model. However, based on an examination of 100 data sets, Connor and McCoy (1979) found no single best-fitting model. Consequently, they advised that the model chosen should depend on the observed form of the species-area curves. Based on correlation coefficients, my "logspecies/area model" (Connor and McCoy 1979 terminology) fit the Minnesota vascular plant data slightly better than the power function and was therefore used in the analyses.

For each of the three data sets, a distance measurement had the highest correlation with the log of total richness (L_{St}), although none of the correlations for the Pomme de Terre River island group were significant (Tables 7-5, 7-6, and 7-7). Distance to source area was the variable with the highest significant correlation in both the Alexandria Moraine and Minnesota River island groups. Distance west (D_w) also had a significant correlation with total richness in the Minnesota River group, and two area variables were highly significant in the Alexandria Moraine group. There was a low correlation between L_{St} and the distance of the nearest neighboring stand (D_{nb}), in all cases.

Based on the correlation results, I used regression analysis to determine the significance, and to describe the relationship between independent and dependent variables. L_{St} was used for regressions on each of the distance measurements except D_{nb}, and stand area was also used in the simple regressions. I used A_{mn} instead of A_{38} or A_{61} because neither of these latter two measurements had a consistently higher correlation in all three cases, and because the three area variables were highly correlated.

The area variables had the highest correlations with L_{Sn}. Because D_n for the Alexandria Moraine archipelago and D_w for the Minnesota River group also appeared to be related to L_{Sn}, they were included in the regression analyses for all three transects. Nearest neighbor distance had no significant correlation with L_{Sn} for any group and was excluded from further consideration.

A final variable with a potentially important effect on L_{Sn} and L_{St} is stand age. The importance of stand age was tested only for the Minnesota River transect where the age of a large number of stands was known (Table 7-3). The low correlation coefficients between age and the dependent variables suggest that the range of stand ages encountered had very little to do with richness. Consequently, age was not used as an independent variable in the regressions.

Table 7-5. Correlation Matrix for All Potential Regression Variables for 43 Planted Stands in the Alexandria Moraine Archipelago[a]

	D_n	D_w	D_{nb}	A_{38}	A_{61}	A_{mn}	L_{St}
D_w	.60***						
D_{nb}	-.19	-.30					
A_{38}	.05	.09	.30*				
A_{61}	-.15	-.15	.32**	.76***			
A_{mn}	-.05	-.03	.33**	.94***	.93***		
L_{St}	-.44***	-.15	.17	.24	.40***	.34***	
L_{Sn}	-.34***	-.25	.25	.43***	.51***	.50***	.90***

[a] Symbols are given in Table 7-4. Statistical significance is indicated for the 99% (***), 95% (**), and 90% (*) levels.

Table 7-6. Correlation Matrix for All Potential Regression Variables for 32 Planted Stands in the Minnesota River Archipelago[a]

	D_m	D_n	D_w	D_{nb}	A_{38}	A_{61}	A_{mn}	Age	L_{St}
D_n	.84								
D_w	−.78***	−.89***							
D_{nb}	−.38**	−.18	−.04						
A_{38}	−.02	−.05	.15	−.23					
A_{61}	−.13	−.23	.38**	−.24	.80***				
A_{mn}	−.08	−.15	.28	−.24	.95***	.95***			
Age	.01	.17	.14	−.01	.30	.10	.21		
L_{St}	−.50***	−.48***	.46***	.02	.14	.05	.10	.12	
L_{Sn}	−.25	−.24	−.36**	−.21	.46***	.34*	.42**	.16	.69***

[a] n = 23 for age coefficients. Symbols are given in Table 7-4. Statistical significance is indicated for the 99% (***), 95% (**), and 90% (*) levels.

In addition to analyzing the planted stands, the effects of area and isolation on the richness of the natural, or unplanted, forests in the Alexandria Moraine and Minnesota River transects were examined. The natural forests of the Pomme de Terre transect were not used because of the small data set (n = 5). The same analytic pattern used for planted stands was followed, and the same initial variables were used except age (which was unknown) and D_n, which was inappropriate.

For the forests of the Alexandria Moraine source area, the only significant correlation was between L_{Sn} and D_w (Table 7-8). Because there was also a high correlation between L_{St} and D_w, the relationship between both measures of richness and D_w was tested using linear regression. For the Minnesota River forests, all distance variables were significantly related to total richness, and D_m and D_w were significantly related to L_{Sn} (Table 7-9). Thus, the dependence of L_{St} and L_{Sn} on both D_m and D_w was tested in regression analysis. Results for the three transects follow.

Table 7-7. Correlation Matrix for All Potential Regression Variables for 19 Planted Stands in the Pomme de Terre Transect[a]

	D_n	D_w	D_{nb}	A_{38}	A_{61}	A_{mn}	L_{St}
D_w	−.93***						
D_{nb}	.74***	−.57**					
A_{38}	.70***	−.54**	.68***				
A_{61}	.68***	−.59***	.62***	.95***			
A_{mn}	.70***	−.57**	.66***	.99***	.99***		
L_{St}	−.07	.27	−.02	.24	.18	.22	
L_{Sn}	−.09	.14	.23	.52**	.40*	.47**	.75***

[a] Symbols are given in Table 7-4. Statistical significance is indicated for the 99% (***), 95% (**), and 90% (*) levels.

Table 7-8. Correlation Matrix for All Potential Regression Variables for 23 Natural Stands in the Alexandria Moraine Archipelago[a]

	D_w	D_{nb}	A_{38}	A_{61}	A_{mn}	L_{St}
D_{nb}	-.26					
A_{38}	-.06	.13				
A_{61}	-.13	.30	.70***			
A_{mn}	-.07	.14	1.00***	.73***		
L_{St}	-.32	.10	.17	.23	.17	
L_{Sn}	-.41*	.12	.20	.28	.21	.97***

[a] Symbols are given in Table 7-4. Statistical significance is indicated for the 99% (***), 95% (**), and 90% (*) levels.

Results

The Alexandria Moraine Archipelago

Based on a sample of 43 planted stands, total richness significantly decreases with increasing isolation from the natural forests on the Alexandria Moraine. The regression results for these and other data in which L_{St} is the dependent variable are found in Table 7-10. For this set of planted forests, over 19% of the variation in L_{St} is explained by stand isolation. To examine the effect of area on richness, regression analyses of L_{St} on stand area were performed. Approximately 11% of the variation in the dependent variable is explained by stand area. Because the total richness of the planted stands was significantly related to both D_n and A_{mn}, both variables were used in a multiple regression equation to predict L_{St} (Table 7-12):

$$L_{St} = 1.41 - 0.02D_n + 0.06A_{mn}; p(F) \approx 0.001 \tag{1}$$

The equation accounts for about 30% of the variation in L_{St}. Therefore, for the planted stands in the Alexandria Moraine transect, the number of species in the stand

Table 7-9. Correlation Matrix for All Potential Regression Variables for 12 Natural Stands in the Minnesota River Archipelago[a]

	D_m	D_w	D_{nb}	A_{38}	A_{61}	A_{mn}	L_{St}
D_w	-.90***						
D_{nb}	-.53*	.79***					
A_{38}	-.26	.52*	.70**				
A_{61}	-.29	.60**	.85***	.87***			
A_{mn}	-.28	.58**	.81***	.96***	.97***		
L_{St}	-.62**	.75***	.60**	.47	.45	.48	
L_{Sn}	-.51*	.63**	.40	.34	.27	.31	.95***

[a] Symbols are given in Table 7-4. Statistical significance is indicated for the 99% (***), 95% (**), and 90% (*) levels.

Table 7-10. Simple Regression Analysis for the Planted Stands of the Three Transect Areas in Western Minnesota, Where $y = L_{St}$ (see Table 7-4)

Transect	x	Intercept	$b \pm SE$	F	R^2 (%)
Alexandria	D_n	1.489	-0.017 ± 0.0055	9.69	19.12
Moraine	D_w	1.457	-0.003 ± 0.0030	0.97	2.30
($n = 43$)	A_{mn}	1.281	0.060 ± 0.0260	5.24	11.33
Minnesota	D_n	1.491	-0.008 ± 0.0027	9.10	23.27
River	D_m	1.550	-0.011 ± 0.0037	8.48	22.04
($n = 32$)	D_w	0.909	0.005 ± 0.0018	8.50	21.37
	A_{mn}	1.386	0.011 ± 0.0198	0.31	1.01
Pomme de	D_n	1.377	-0.002 ± 0.0084	0.08	0.45
Terre River	D_w	0.812	0.007 ± 0.0062	1.35	7.37
($n = 19$)	A_{mn}	1.342	0.020 ± 0.0215	0.82	4.62

can be successfully described by an equation that includes both isolation and size of the woodlot as the dependent variables.

Because L_{St} may include species that have migrated both from the natural forests and from fields and roadways adjacent to the woodlot, the relationship between L_{St} and the two variables A_{mn} and D_n is difficult to explain without knowing whether L_{Sn} is also dependent on the same variables.

For the Alexandria Moraine archipelago, L_{Sn} was significantly dependent on distance to the nearest large source area. The regression equation is

$$L_{Sn} = 1.150 - 0.019D_n \qquad (2)$$

The complete regression results for planted forests where L_{Sn} is the dependent variable are listed in Table 7-11. In this case, about 12% of the variation in L_{Sn} is ac-

Table 7-11. Simple Regression Analysis for the Planted Stands of the Three Transect Areas in Western Minnesota, where $y = L_{Sn}$ (see Table 7-4)

Transect	x	Intercept	$b \pm SE$	F	R^2 (%)
Alexandria	D_n	1.150	-0.019 ± 0.0081	5.38	11.60
Moraine	D_w	1.229	-0.007 ± 0.0041	2.62	6.00
($n = 43$)	A_{mn}	0.847	0.123 ± 0.0334	13.63	24.94
Minnesota	D_n	1.138	-0.006 ± 0.0043	1.83	5.75
River	D_m	1.185	-0.008 ± 0.0057	1.98	6.19
($n = 32$)	D_w	0.534	0.005 ± 0.0026	4.42	12.84
	A_{mn}	0.957	0.064 ± 0.0255	6.34	17.44
Pomme de	D_n	0.980	0.006 ± 0.0148	0.15	0.88
Terre River	D_w	0.511	0.006 ± 0.0112	0.32	1.83
($n = 19$)	A_{mn}	0.900	0.075 ± 0.0345	4.67	21.61

counted for by D_n. Concomitantly, the mean area of the woodlots accounts for 25% of the variation. The regression line is described by

$$L_{Sn} = 0.847 + 0.123A_{mn} \tag{3}$$

When both A_{mn} and D_n are incorporated into a multivariate model the relationship is

$$L_{Sn} = 0.984 + 0.119A_{mn} - 0.017D_n \tag{4}$$

The variables account for about 35% of the variation in L_{Sn} (Table 7-12). The relationship between L_{Sn} and the two independent variables A_{mn} and D_n is indicated graphically in Fig. 7-5.

The regression coefficient and the F value were not simultaneously significant in either the regression of L_{St} or of L_{Sn} on D_w. Because D_w approximately locates each stand on the east-west macroclimatic gradient (Scanlan 1975a), the decrease in richness does not appear to be caused by macroclimatic changes. Thus, the colonization of woodlots by species from the Alexandria Moraine source areas appears to be controlled by the distance that the immigrants have had to travel and by the size of the recipient forest islands.

Table 7-12. Stepwise Multiple Regression Analysis with Selected Independent Variables for the Planted Stands of the Three Transect Areas in Western Minnesota[a]

Transect	y	b_i	Value ± SE	F	R^2 (%)
Alexandria Moraine ($n = 43$)	L_{St}	Intercept	1.41		
		D_n	−0.02 ± .0053	9.69**	19.12
		A_{mn}	0.06 ± .0236	5.62*	9.96
	L_{Sn}	Intercept	0.98		
		A_{mn}	0.12 ± .0315	13.63**	24.94
		D_n	−0.02 ± .0070	6.16*	10.01
Minnesota River ($n = 32$)	L_{St}	Intercept	1.48		
		D_n	−0.01 ± .0028	9.10**	23.27
		A_{mn}	0.003 ± .0180	.03	0.07
	L_{Sn}	Intercept	0.58		
		A_{mn}	0.05 ± .0260	6.34*	17.44
		D_w	0.004 ± .0026	2.38	6.24
Pomme de Terre River ($n = 19$)	L_{St}	Intercept	0.10		
		D_w	0.02 ± .0068	1.35	7.37
		A_{mn}	−0.05 ± .0236	4.54	20.49
	L_{Sn}	Intercept	0.93		
		A_{mn}	0.12 ± .0462	4.69*	21.62
		D_n	−0.03 ± .0176	2.45	10.40

[a] Symbols are given in Table 7-4. Statistical significance is indicated for the 99% (**) and 95% (*) probability levels.

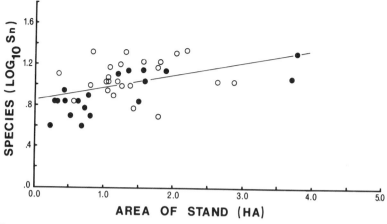

Fig. 7-5. The relationship between mean stand area and the number of natural species per woodlot in the Alexandria Moraine archipelago. Woodlots less than 7.6 km from the source area are indicated by open circles; those at a distance of 7.6 km or more are indicated by closed circles.

Based on the high correlation between the measure of westward location of source stands, and the two measures of stand richness, L_{St} and L_{Sn}, I performed two regressions using D_w as the independent variable. Only the regression of L_{Sn} on D_w was significant (Table 7-13). Therefore, with increasing distance westward, the number of species that are common in natural forests decreases, while the total number of species in the stands does not change significantly. The relationship between L_{Sn} and D_w is not significant in the case of the planted stands. The different relationships may be due to the somewhat different species pool in each case. Included in L_{Sn} for the source stands are many species that are restricted to the Alexandria Moraine, because of dispersal problems, the westward deterioration of the climate, or a westward decrease in some unmeasured factor necessary for their existence. The restriction of these species may be the reason that D_w is an important variable in explaining L_{Sn} in the source stands.

Table 7-13. Simple Regression Analysis for the Natural Stands of the Alexandria Moraine and Minnesota River Source Areas in Western Minnesota[a]

Transect	y	x	Intercept	$b \pm SE$	F	R^2 (%)
Alexandria Moraine ($n = 23$)	L_{St}	D_w	1.719	$-0.004 \pm .0026$	2.89	10.35
		D_m	1.682	$-0.006 \pm .0028$	5.03*	16.75
Minnesota River ($n = 12$)	L_{St}	D_w	0.174	$0.013 \pm .0035$	12.66**	55.86
		D_m	1.695	$-0.014 \pm .0057$	6.22*	38.34
	L_{Sn}	D_w	-1.030	$0.021 \pm .0083$	6.66*	39.99
		D_m	1.561	$-0.024 \pm .0126$	3.59	26.44

[a] Symbols are given in Table 7-4. Statistical significance is indicated for the 99% (**) and 95% (*) levels.

The Minnesota River Archipelago

In the Minnesota River group, two distance measures (D_m and D_n) were used to differentiate between the influence of the source areas in the Minnesota River Valley (D_m = distance to the Minnesota River) on the richness of the planted stands, and the effect of all source areas in the group including upland source areas (D_n = distance to nearest source area). Although the upland source areas near the Minnesota River are probably as old as the stands in the valley proper, many of the shallow lakes of the area dried up during the drought in the 1930s. Because the upland outliers are usually located on lake shores, the microclimate in the stands may have deteriorated during the drought resulting in reduced floristic richness. Also, these upland source areas may have had open canopies prior to the arrival of European settlers (Cottam 1949). As a result of the open canopy, prairie species may have dominated the understory. This possibility, in conjunction with the effects of the drought, may have diminished the significance of the outliers as source areas and requires the distinction between D_m and D_n in regression analysis.

For the planted stands near the Minnesota River, the total richness was regressed against D_n, the distance from the nearest source area; D_m, the distance from the Minnesota River; D_w, the distance west; and A_{mn}, the mean stand area. L_{St} was significantly related to all of the distance measures (Table 7-10).

Based on a sample of 32 planted stands, total richness significantly decreases with increasing isolation; as either D_m or D_n increases. It is important to note that the sign of the regression coefficient in the equation for L_{St} (and L_{Sn}) on D_w is negative. As there is a gradient of decreasing precipitation westward in the sample area (Scanlan 1975a), the significant positive coefficient for D_w in the Minnesota River transect suggests either that macroclimate is not important in the distribution of plant species in the woodlots, or that the effects of distance outweigh those of macroclimate. Proximity to the source areas is the important determinant of species richness.

When D_n and A_{mn} are incorporated into a multivariate equation, A_{mn} accounts for an insignificant 0.07% of the variation in L_{St} while D_n accounts for 23%. Therefore, in this case mean area of a woodlot is not an important predictor of L_{St} either by itself or when incorporated into a multivariate equation.

While the distance variables are most important in predicting L_{St} in the planted stands of the Minnesota River archipelago, both A_{mn} and the distance variable, D_w, are important in predicting L_{Sn} (Table 7-11) in the following manner:

$$L_{Sn} = 0.95 + 0.06A_{mn} \tag{5}$$

and

$$L_{Sn} = 0.53 + 0.005D_w \tag{6}$$

A_{mn} explains 17% of the variation in L_{Sn} while D_w explains 13% of the variation. If the two variables are included in a multivariate equation, only the coefficient for A_{mn} and the amount of variation accounted for by A_{mn} are significant (Table 7-12). This is apparent when L_{Sn} is graphed against A_{mn} and D_n is approximated by open or closed symbols (Fig. 7-6).

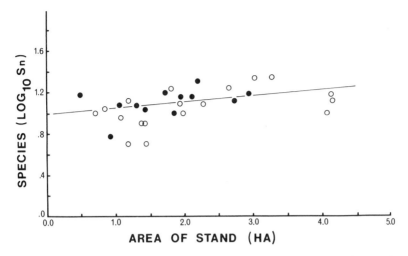

Fig. 7-6. The relationship between mean stand area and the number of natural species per woodlot in the Minnesota River archipelago. Woodlots less than 13.7 km from the source area are indicated by open circles; those at a greater distance are indicated by closed circles.

In the Minnesota River archipelago, 12 source stands were sampled. Among the independent variables, both D_w and D_m are significantly related to the two dependent variables, L_{St} and L_{Sn}. The regression results (Table 7-13) demonstrate the positive effect of increasing western location and the negative effect of increasing isolation from the Minnesota River on the abundance of natural forest species in the source stands. The two distance variables account for 40% and 26% of the variation in L_{Sn}, respectively. The effects of D_w and D_m on total richness are similar but there is less deviation from the regression relationship in each case ($\%R^2$ is higher).

The relationship between richness and the two distance variables is expected because the upland source stands are more easterly and at a greater distance from the Minnesota River. If these stands are considered forest islands, as the planted stands have been, then the species richness in the upland stands should be less. The insignificant relationship between richness and mean area, and the significant dependence of L_{St} on nearest-neighbor distance, however, are unexpected. Possibly the wide range of the area variable is important in the relationship between L_{St} or L_{Sn} and A_{mn}. If there is a maximum number of species available and if that maximum number is present in all stands above some threshold size, then the relationship between A_{mn} and L_{St} (or L_{Sn}) may not appear. On the other hand, the significant dependence of L_{St} on nearest neighbor distance seems to be artificial. In the first place, D_{nb} has not been found to be significantly related to either of the dependent variables prior to this. Second, the river valley supports the primary source stands which contain approximately twice as many plant species as the upland stands. However, the primary source stands are separated from themselves and from other stands in the transect by almost six times the average distance between the upland stands and their nearest neighbors. Therefore, the relationship simply indicates that there is a greater distance separating the richer source stands in the river valley than the upland source stands, and that it does not provide information applicable (in general) to the dependence of stand richness on the distance between adjacent islands.

The Pomme de Terre River Archipelago

For the planted forests in this archipelago, the dependence of the abundance of natural-forest species (L_{Sn}), on mean area is the only significant relationship. The equation for the regression line (Fig. 7-7) is

$$L_{Sn} = 0.90 + 0.08A_{mn}; \ p(\beta = 0) < 0.05 \tag{7}$$

which accounts for nearly 22% of the variation in the dependent variable. Although D_n is not significantly related to L_{Sn}, many of the stands with the lowest richness in Fig. 7-7 are also the stands that are, on the average, more isolated from natural forest areas. The lack of significance may be due to the much larger and more abundant natural forests of the Minnesota River and the Alexandria Moraine source areas; these areas may contribute to the richness of the woodlots in the Pomme de Terre group. Also, Six Mile Grove was within 8 km of the transect stand that is most distant from the Pomme de Terre River. This may be important because the influence of Six Mile Grove as a source area can still be seen today in the greater richness of stands adjacent to it.

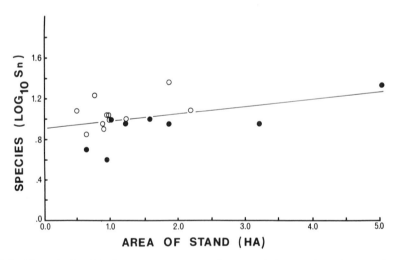

Fig. 7-7. The relationship between mean stand area and the number of natural species per woodlot in the Pomme de Terre River archipelago. Woodlots less than 3.7 km from the source area are indicated by open circles; those at a distance of 3.7 km or more are indicated by closed circles.

Discussion and Conclusions

The planted forests in the tall grass prairie and the stands of natural deciduous forest in the prairie—forest ecotone in western Minnesota have provided a unique situation in which to test the hypotheses of island biogeography of MacArthur and Wilson (1967) for plant species in habitat islands. Heretofore, these hypotheses have been tested almost exclusively in an oceanic setting, and with birds. In this study, the time for immigration is known and the scale is shorter, the distance from the source areas to the farthest island is much shorter, and the island areas are small. Therefore, the applicability of the hypotheses has been tested under several new conditions. The significant results indicate that the hypotheses are more generally applicable than has been previously demonstrated.

The natural forests of the prairie—forest ecotone at the Alexandria Moraine, along the Minnesota River, and on the banks of the Pomme de Terre River are the source areas. The nonnatural forests on the prairie uplands adjacent to each of these source areas are considered to be clusters of forest islands in a "sea" of farmland. The trees in these forest islands, or woodlots, were planted during the last century and since that time, plant species of the deciduous forest in the source areas have migrated to the woodlots. The distribution of these immigrants is related to two island parameters, the distance to the nearest source area, and the mean area of the forest island.

For the two largest sample areas, the Alexandria Moraine and the Minnesota River valley, total woodlot richness has the most significant relationship with isolation of a woodlot from the source areas. Also in the Alexandria Moraine, total species richness significantly increases with an increase in stand area. In many island studies, certain "weedy" or nonnatural species are excluded from the calculation of island richness. I objectively calculated the number of species in the woodlots that were more common in the source stands and, therefore, have probably immigrated from the source areas. In all three areas the richness of these natural-forest species in the woodlot islands is signficantly dependent on woodlot area. Therefore, the second hypothesis of MacArthur and Wilson is germane: richness decreases with a decrease in island area.

Niering (1963) has suggested, based on his results for the flora of the Kapingamarangi Atoll, that the species-area relationship may not exist for islands smaller than a certain size—1.4 ha in his case. The range in area of the woodlots studied spans Niering's 1.4 ha size limit, and I find a consistently significant area effect. Also, the size range of the islands is very broad compared to that of Niering (1963). My stands are not only below the lower limit of the usual size range, but they differ in size by only a factor of ten. In spite of these unusual circumstances, there is a significant area effect in the woodlots.

Usher (1979) demonstrated that after approximately ten years on Yorkshire nature preserves, the mathematical relationship between plant species richness and preserve area changed when previously unnoticed species are included in the calculations. Because my woodlot islands were censused, I do not expect this kind of sampling variation in the calculated relationships. However, the relationship for natural forests may change because they could not be completely censused.

Long-term studies of the individual plant species in censused woodlots of several different sizes would be extremely useful in supporting the results of this study. As

Diamond and Marshall (1977) point out, the species composition of an island is variable through time. Although we do not know what the extinction rate is for plant species on habitat islands, long-term studies of woodlots of different sizes would provide this information. Based on the insignificant correlation between age and richness, the planted forests appear to have reached species equilibrium within 100 years. Thus species turnover may be a process measurable in years or decades for plant species in the study woodlots.

Some work has already been done on the change in plant populations within bounded areas. Weaver's extensive work on the Nebraskan, Iowan, and Kansan prairies during the drought of the 1930s (Weaver and Albertson 1936), Tamm's long-term studies (1972), and Dyrness' data (1973) demonstrate that within small, fixed areas, a species can dramatically increase or decrease in numbers, become extinct, or can colonize an area within only a few years. Because these strong population fluctuations appear to be a necessary force leading to equilibrium species numbers on islands (e.g., Freeland 1979), island species are expected to be occasionally very abundant. In western Minnesota, unusually large populations of certain species are present in certain woodlots. For example, *Galium aparine, Osmorhiza longistylis,* or *O. claytoni* almost completely dominate the understory in selected stands in the Alexandria Moraine area. The reason for this overabundance may be an absence of competitors. Although the influence of competition in structuring island communities is difficult to demonstrate (Connor and Simberloff 1979), an examination of population sizes and species composition on different islands may yield information in this respect. In any case, the high density populations observed are examples of a situation that may occur in bounded areas. Studies of secondary succession have repeatedly demonstrated that whole communities may replace one another in a matter of years in the initial stages of succession. In summary, neither plant populations nor plant communities are stable. Because a fixed area has a limited amount of inhabitable space, the instability of the plant populations (or the community) may result in the species' extinction in the fixed area. That is, a plant species may be a common member of a community that is much larger than the species' population size but, in fixed areas approaching some lower size limit, extinction may be an increasingly common phenomenon.

Various mathematical models have been used to describe the dependence of species richness on land area. Included among them is the "logspecies/area model" (terminology of Connor and McCoy 1979) which I used to relate plant species richness to forest area. The slope and intercept statistics for the various models differentially depend on the area scale (Connor and McCoy 1979) making it difficult to compare my results with other observations. The Arrhenius power function model, $s = cA^z$, where s is the number of species and A is the land area occupied, has been recently applied to vascular plant data. It was also employed by MacArthur and Wilson (1967) who presented empirical boundary conditions for z. Therefore, for comparative purposes I have calculated the power function statistics c and z for the natural species in my three archipelagos (Table 7-14).

For the three source areas, the exponent z is smaller and the coefficient c is larger than for the planted forest groups. The difference in z values is expected since isolation tends to increase island values above those of source areas. MacArthur and Wilson (1967) presented the empirical z ranges of 0.12 to 0.17 for source areas and 0.20 to

Table 7-14. The Power Function Coefficient c and Exponent z Calculated for the Natural Plant Species Richness of the Planted Forest Islands and Source Areas in Each of the Three Forest Island Archipelagos in Western Minnesota

| | Power function statistics | | | |
| | Source area forests | | Planted forests | |
Archipelago	c	z	c	z
Alexandria Moraine	31.4	0.107	9.9	0.402
Minnesota River	23.2	0.187	10.6	0.245
Pomme de Terre River	21.0	0.220	9.8	0.283
Mean	25.2	0.171	10.1	0.310

0.35 for islands. The z values for the natural forests and woodlot islands are in reasonable range of these biogeographic rules of thumb.

It is not unusual that the MacArthur and Wilson boundaries are exceeded. For example, for vascular plant species in Yorkshire nature preserves, Usher (1979) found z values ranging from 0.276 to 0.314. Usher was satisfied with the values because they were within the reported range for islands. It would seem, however, that nature preserves should be viewed as source areas so that Usher's values are higher than expected. Similarly, Dony (1977) reported z values for vascular plant species in Purwell meadow and in the whole British Isles as 0.21 and 0.1873, respectively. Thus, although three of the z values calculated for Minnesota are not within the empirical ranges of MacArthur and Wilson, such deviations have been observed for other plant distributions.

The larger c values measured for the forest source areas reflect the larger species pool there; both colonizing and noncolonizing species occur in the natural forests. As the total pool of natural species is approximately twice the number that occur in planted forests, the average value of 25.2 for c in source area forests, 2.5 times greater than that in planted forests, is expected.

The first hypothesis of MacArthur and Wilson (1963, 1967) states that islands more distant from a source area are less rich because the immigration rate is lower. This hypothesis is supported by results for the Alexandria Moraine and by the high negative correlation between L_{Sn} and the two distance variables D_m and D_n in the Minnesota River region. In light of these results, I recommend that in future studies of succession, the source of the immigrants must be considered. Researchers have mentioned in a few studies that certain species are absent from a successional surface because the potential source area is downwind or at too great a distance. In view of these comments, observations of the pattern of tree encroachments on prairies, the studies of seed longevity in soil, and the dependence of woodlot richness on the proximity of source areas, it is quite clear that source areas must be considered in successional studies in order for the results to be meaningful.

In an applied sense, it seems that the size and isolation of habitat islands should receive considerable attention. Examples of subject areas in which these considerations would be useful include natural area preservation, agriculture, forestry, and regional planning. In the establishment of natural areas for the preservation of plant and animal species, there is certainly a minimum size requirement for the natural area. The size threshold would be a function of the life histories of the species involved as well as the site's proximity to other areas in which the species occur. In agriculture and forestry, where pest organisms are a problem, there is probably an upper limit to the size of the managed units. Small areas of crops or trees that are isolated should be less susceptible to pest attack and contain fewer pest species than large areas nearly adjacent to one another. Finally, in the planning of land use in urban areas, island biogeographic aspects should be considered. A high quality urban environment includes seminatural park areas. However, if these parks are to retain their natural qualities, attention must be given to their size and proximity to one another.

Acknowledgments. I thank Dr. Edward J. Cushing for suggesting that I apply MacArthur and Wilson's hypotheses of island biogeography to the natural forests and woodlots at the prairie—forest border. Field research was supported by a Carolyn M. Crosby Fellowship, an Alexander P. and Lydia P. Anderson Summer Fellowship and the Dayton Natural History Fund. Analysis of the data was made possible through a number of grants from the University of Minnesota Computing Center.

8. Effects of Forest Fragmentation on Avifauna of the Eastern Deciduous Forest

R. F. WHITCOMB
PLANT PROTECTION INSTITUTE (USDA)

C. S. ROBBINS
FISH AND WILDLIFE SERVICE (USDI)

J. F. LYNCH
SMITHSONIAN INSTITUTION

B. L. WHITCOMB
10271 WINDSTREAM DRIVE
COLUMBIA, MARYLAND

M. K. KLIMKIEWICZ
FISH AND WILDLIFE SERVICE (USDI)

D. BYSTRAK
FISH AND WILDLIFE SERVICE (USDI)

It has long been recognized that islands support fewer species of animals and plants than equivalent areas of mainland habitat (Lack 1942, Van Balgooy 1969, Carlquist 1974). Preston (1962) and especially MacArthur and Wilson (1963, 1967) provided a theoretical interpretation for this basic empirical observation of island biology, and subsequent experimental field research (Wilson and Simberloff 1969, Simberloff and Wilson 1969) has for the most part supported what has become known as the MacArthur-Wilson Theory of Island Biogeography (reviewed by Simberloff 1974, Strong 1979). According to the theory, the low diversity of insular biota reflects a dynamic equilibrium between rates of extinction and rates of colonization of individual species populations. The equilibrium number of animal species for a particular island is thought to depend on the island's productivity; on its vegetational, topographic, and climatic diversity; and, perhaps most importantly, on its size and isolation from sources of potential colonists. Some investigators have emphasized the importance of island size (e.g., Hamilton and Rubinoff 1963, 1964, 1967; Hamilton et al. 1964); others regard isolation as more important (e.g., Power 1972); still others point to habitat diversity as a crucial determinant of animal diversity in insular settings (Johnson 1975, Mühlenberg et al. 1977a).

Although the "island effect" has been rather extensively documented for oceanic islands, the effects of insularity are less well understood for isolated patches of habitat in terrestrial situations. Much of the published research on mainland "islands" concerns sharply circumscribed patches that differ from their surroundings climatically as well as structurally. Such patch-types include caves (Culver 1970b, Vuilleumier 1973) and montane habitats (Brown 1971, Vuilleumier 1970, Thompson 1978, Fritz 1979).

Others extended the island concept to include host-parasite relationships (Janzen 1973, Opler 1974, Dritschilo et al. 1975, Cornell and Washburn 1979). The present study concerns the relationship between the insularity of patches of eastern deciduous forest and the composition of forest-associated bird communities. Although MacArthur and Wilson (1967) recognized the relevance of their theory to the fragmentation of the eastern deciduous forest, biologists have only recently begun to assess the biogeographic implications of deforestation.

The influence of forest tract size on the composition of the breeding bird community in the eastern deciduous forest was first demonstrated by Bond (1957) in a study of the factors governing the distribution of birds in southern Wisconsin. Later, Linehan et al. (1967) found exceptionally high densities of birds in urban woodlots in Delaware, and stressed the importance of such woodlots to bird conservation. In Europe, Oelke (1966) found bird density to be an inverse function of woodlot size. In a study of British birds, Moore and Hooper (1975) concluded that species richness increased with forest area at least over a 0.5-50 ha size range, and that species differ markedly in their ability to utilize small woodlots. Galli et al. (1976) and Forman et al. (1976) studied small to medium-sized (0.1-24 ha) forest islands in the New Jersey Piedmont, and concluded, as had Moore and Hooper (1975), that bird species differ in their minimal area requirements. None of the cited studies addressed forest fragmentation as a regional phenomenon, nor was the degree of isolation of forest islands considered as a variable.

In 1974, we analyzed the avifaunal composition of a set of small islands of deciduous forest in the Maryland Piedmont. From these initial surveys, it was clear that entirely different avifaunal assemblages occurred in small isolated forests and in extensive forests. The effect was statistically significant at high levels, and furthermore, examination of published Breeding Bird Censuses convinced us that these differences were characteristic of deciduous forests throughout eastern North America. In 1975, we examined two extensive forest systems and two fragments within the systems, minimally isolated from large mainland forests, by the point survey and spot-mapping methods (MacClintock et al. 1977, Whitcomb et al. 1977). Results from these studies suggest that isolation, as well as size, is an important determinant of avifaunal composition. Subsequent analyses have focused on the life history features of bird species that adjust poorly to forest fragmentation. Such studies were made possible by the accumulation of vast amounts of data, over many years, by local bird students. From these analyses, it became clear that two principal life history features, neotropical migration and habitat preference, were powerful determinants of the ability of bird species to tolerate forest fragmentation. In particular, neotropical migration strategy has important implications in reproductive effort, dispersal strategy, behavior, and nest type and nest height. Such results suggest that fragmentation may impact forest-inhabiting bird species through any one or a combination of life history features. For example, migration itself, involving, as it does, the abandonment and recolonization of territories on an annual basis, may have powerful implications for the tolerance of a species to fragmentation. The lesson of these studies for conservation suggests that the multiplicity of factors generated by habitat fragmentation precludes, absolutely, any combination of small forest fragments acting as avifaunal preserves of eastern deciduous forest. Such reserves must be large. This chapter is a detailed exposition of our case for this conclusion.

Methods

Description of Study Area

Geographic Location. Field work was conducted in central Maryland, primarily in Prince George's, Montgomery, and Howard Counties. The study area (Fig. 8-1) lies within the "middle zone" of Maryland (Shreve et al. 1910). The eastern portion is within the Coastal Plain province, whereas the western portion is in the Piedmont province. In the eighteenth, nineteenth, and twentieth centuries the area was principally agricultural, but in recent years suburban development has become increasingly prevalent (Fig. 8-2).

Virtually the entire region was covered with oak-hickory-chestnut forest (Shreve et al. 1910) in presettlement times, but systematic destruction of the primeval forest began soon after the arrival of Europeans in the mid-seventeenth century. At present only about 22% of the study area is wooded, and most of the existing forest is in early or middle-stage second growth less than 50 years old (Table 8-1). The fraction of forested land probably was substantially less than 20% in the late nineteenth and early twentieth centuries, and much abandoned agricultural land has reverted to forest throughout the area. We estimate that about 18% of the original forest was bottomland, largely in the form of streamside forest strips. Additional bottomland areas of seepage forest, sinkholes, etc., would not elevate this figure substantially.

According to the vegetation map prepared by Brush et al. (1976, 1980), two major upland deciduous forest assemblages occur in the study area: the Tulip-tree association and the Chestnut Oak-Post Oak-Blackjack Oak association. The former association is the more extensive and is composed of the following common tree species in addition to the characteristic tulip-tree (*Liriodendron tulipifera*): red maple (*Acer rubrum*), flowering dogwood (*Cornus florida*), black gum (*Nyssa sylvatica*), white oak (*Quercus alba*), sassafras (*Sassafras albidum*), black cherry (*Prunus serotina*), mockernut hickory

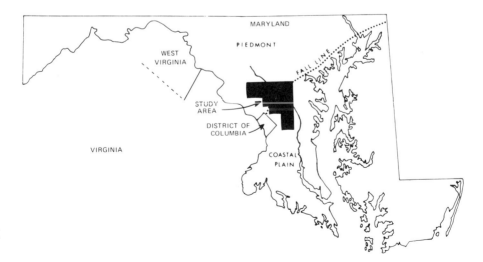

Fig. 8-1. Location of the Maryland study area in relation to the District of Columbia and the fall line between the Piedmont and Coastal Plain provinces.

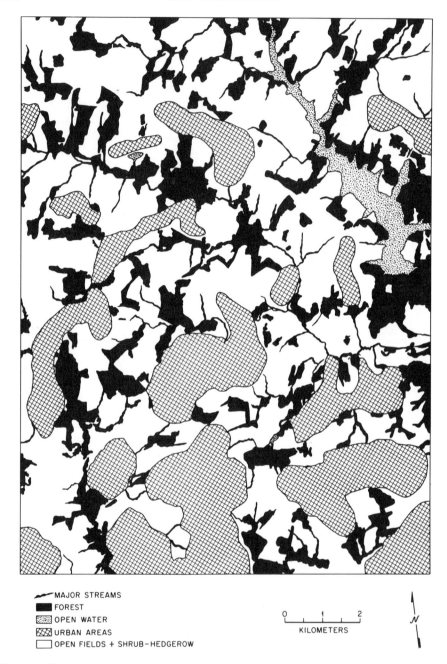

MAJOR STREAMS
FOREST
OPEN WATER
URBAN AREAS
OPEN FIELDS + SHRUB-HEDGEROW

0   1   2
KILOMETERS

N

Fig. 8-2. Forest fragmentation in an expanding urban region: Sandy Spring quadrangle of central Maryland. Forest islands are isolated by a combination of urban and agricultural developments.

Table 8-1. Land Use in the Maryland Study Area[a]

Land use	Miniroute sample[b]	Beltsville (south half)[c]	Sandy Spring	Kensington (northeast)	Gaithers- burg	Mean (weighted by area)
Water	.005	.005	.023	.000	.002	.010
Forest						
Bottomland	.108	.076	.077	.041	.055	.066
Upland	.198	.223	.135	.196	.133	.156
Residential						
Suburban	.065	.093	.023	.172	.017	.047
Subdivision	.073	.067	.062	.127	.089	.079
Rural	.178	.194	.060	.123	.062	.091
Field-hedgerow	.280	.233	.621	.341	.630	.528
Industrial	.093	.110	.000	.000	.012	.024

[a] Data are fractional areas for each land use category in the indicated sample source.
[b] Sample of eight Miniroutes (200 points) in Prince George's County in the Beltsville and Lanham quadrangles.

[c] The close correspondence of habitat distribution of this set of points with the distribution in the Beltsville quadrangle (typical of the region as a whole) suggests that the Miniroutes adequately sampled the habitat types in the subregion.

(*Carya tomentosa*), pignut hickory (*C. glabra*), black oak (*Q. velutina*), American beech (*Fagus grandifolia*), and northern red oak (*Q. rubra*). The less common chestnut oak-post oak-blackjack oak association typically occurs on substrates where availability of water is limited (e.g., fragipan soils, gravels, serpentine soils). Indicator tree species are chestnut oak (*Q. prinus*), post oak (*Q. stellata*), and blackjack oak (*Q. marilandica*). Common trees include red maple, black gum, white oak, sassafras, Virginia pine (*Pinus virginiana*), black oak, American beech, flowering dogwood, sweetgum (*Liquidambar styraciflua*), scarlet oak (*Q. coccinea*), Spanish oak (*Q. falcata*), mockernut hickory, black cherry, and pignut hickory.

Current Land Use. Current land use in the region was estimated by color-coding the habitat types on topographic maps, and cutting out and weighing the map fragments. The 7.5 min USGS quadrangles for Gaithersburg, Sandy Spring, Beltsville (southern half) and Kensington (NE quarter) were used for this analysis. Seven habitat types were provisionally identified and later confirmed by field surveys:

1. Bottomland forest: This habitat usually occurred along stream valleys. The upper limit of bottomland forest was no more than 15 m above the stream bank and was operationally recognized by the presence of significant numbers of red maple in the overstory.

2. Upland forest: This category encompassed forest stands at higher elevations between the dendritic stream valleys. In the Piedmont, deciduous species almost invariably comprise the major, if not sole, fraction of the trees present in mature upland forest. In the Coastal Plain section, varying percentages of Virginia pine, pitch pine (*Pinus rigida*) or, rarely, shortleaf pine (*P. echinata*) or loblolly pine (*P. taeda*) were present, particularly in younger stands.

3. Suburban with canopy: Older housing subdivisions may have a significant amount of canopy cover, and some new developments retain fragments of uncleared forest. Lawns are almost always present. Parks also fit in this category.

4. Suburban without canopy: This category encompases most newer housing developments. The isolated large trees, hedges, and bushes typical of suburban developments provide nesting and foraging habitat for a limited number of bird species.

5. Rural residential: This includes the low-density habitations along secondary roads in rural areas. Houses are located on large lots (generally 0.5-1.0 ha), which usually contain various combinations of lawns, trees, shrubs, and outbuildings. In some instances, more or less canopied woodlots may adjoin such residential areas.

6. Open field with hedgerow: In the study area, nearly all open fields are bordered or segmented by hedgerows.

7. Industrial: This category includes institutional building complexes, light industrial parks, shopping centers, apartment complexes, and similar developments. Lawns and planted shade trees (usually small) are typically present.

Choice of Forest Islands for Study. Twenty-five forest islands (Table 8-2) with the following characteristics were selected for study in 1974: (1) The tracts were upland stands of forest dominated by oaks, or oaks and hickories, with tulip-tree (generally an indicator of mesic conditions in this region) at most a minor component of the canopy. (2) The tracts were completely isolated from other forest by agricultural land, pasture, or suburban development. However, presence of isolated trees or hedgerows in adjacent habitat was permitted. (3) Permanent streams were absent. (4) The tracts were large enough to have well-developed forest interior. (5) The age of the tracts was at least 40 years. (6) The tracts showed no evidence of recent fire, heavy grazing, or other forms of intense disturbance. With respect to the last criterion, it must be pointed out that virtually all forest fragments in the study area have been subject to some disturbance, including persistent intrusion by people, their pets, or domestic stock. In the absence of obvious structural damage to the forest, a modest level of human activity was deemed acceptable. Figure 8-3 illustrates a segment of the 7.5 min USGS Gaithersburg Quadrangle, in which four of our study islands were located.

After our 1974 surveys, we grouped the 25 forest tracts that had been studied into three size classes: small (1-5 ha); medium (6-14 ha), and large (more than 70 ha). No stands between 14 and 70 ha were available for study. Because only five islands could be classified in the "large" category, we supplemented this group in 1975 by adding five comparable point surveys in extensive woodlands. Selection of suitable forests in the 70+ ha category was difficult. Although large, well-drained areas occurred within the primitive forest, most of them have long since been cleared. Remaining large forest tracts inevitably contain mesic and bottomland forest in addition to drier upland vegetation. We attempted to minimize the effect of habitat diversity in the proximity of points in this size class in two ways. First, we attempted to choose interior survey points that were maximally distant from streams and located at the center of patches of upland oak-hickory forest. This approach was employed in the extensive coastal plain forest in the Beltsville area. An alternative approach was the selection of upland forest artificially isolated from bottomland, e.g., situations in which artificial reservoirs covered the adjacent floodplain forest.

Measurement of Island Isolation. The initial experimental design, intended to assess the influence of forest area on bird populations, ignored the degree of isolation of our forest islands. This occurred largely because we knew of no generally accepted method for measuring this variable, given the complex configuration of islands, fragments, and riparian strips present in much of the study area. However, it was possible to assess isolation in retrospect. Sullivan and Shaffer (1975) suggested a "gravity" model for assessing isolation. We modified this to compute an isolation coefficient (I_I):

$$I_I = \frac{1}{\displaystyle\sum_{d=0.1}^{d=3} \frac{A_i}{d_i^2}} \tag{1}$$

in which A_i is the area of each forest island i between 0.1 and 3 km from the study island I, and d_i is the margin to margin distance between islands I and i. These isolation values are given in Table 8-2.

Bird Survey and Census Methods

Point Survey Method. The point survey technique we used is a modified version of the IPA method used by European workers to estimate relative abundances of bird species (Blondel et al. 1970, Ferry 1974). The occurrence of singing males audible at the point is recorded and mapped (Fig. 8-4). We located the approximate center of each forest tract with the aid of topographic maps and aerial photographs, and by on-site inspection. In certain large tracts, several survey points were selected; these were located at the approximate center of large oak or oak-hickory patches, without regard to the location of the geometric center of the tract.

Three 20-min visits were made to each point between 1-30 June. In 1974 each plot was visited once at an early (0530-0610), middle (0610-0650), and late (0650-0730) time period. This schedule enabled the observer to survey three sites per day. In 1975 we enlarged the early, middle, and late periods to 0530-0700, 0700-0830, and 0830-1000, respectively. This revised schedule permitted observers to survey up to nine sites in a morning.

Our use of 60 min total survey time per sample point was based on analysis of the results of six preliminary surveys, each of 2 hr duration. This analysis revealed that about 50% of the species found in a 2 hr survey had been detected within the first 20 min, and that about 88% of the species were detected within the first hour. The 60 min of total observation time were divided among three separate visits to increase the probability of encountering species whose vocal activity fluctuates seasonally or during the early morning hours.

Detectability of Species. Comparison of the results of point surveys with those of standard mapping censuses performed in the same forests revealed that most vocalizations recorded in point surveys involved birds located within 100 m of the survey point (Fig. 8-4). Under ideal conditions, however, individuals of certain species are audible at distances greater than 100 m. Based on our extensive experience in territorial

Table 8-2. Number of Bird Species Recorded in Point Surveys in the Interior of Oak-Hickory Forest Islands in Central Maryland

USGS[a]	Latitude	Longitude	Surrounding habitat[b]	Size (ha)	Isolation factor[c]	Habitat preference[d]				Total no. species
						Forest interior	Both	Edge	Field edge	
B	39.00.42	76.57.00	A	1.1	201.6	0	12	4	4	20
B	39.01.24	76.56.33	A	2.2	737.2	0	11	3	3	17
S	39.10.58	77.02.57	A	2.2	694.9	2	13	6	2	23
B	39.01.34	76.56.27	A	2.4	940.8	1	9	5	5	20
B	39.01.05	76.54.44	U	2.7	186.4	0	10	5	3	18
K	39.04.54	77.00.19	U	3.5	264.7	0	12	2	3	17
G	39.14.49	77.09.28	A	3.8	528.0	1	14	2	3	21
S	39.11.15	77.02.53	A	3.8	199.2	3	13	5	2	23
K	39.04.42	77.02.40	U	3.8	49.9	1	12	3	3	19
B	39.01.37	76.55.35	U	4.2	376.6	1	9	5	4	19
B	39.00.38	76.56.49	U	6.2	139.2	1	12	5	3	21
G	39.11.20	77.11.08	A	6.3	494.0	0	15	2	4	21
K	39.05.06	77.00.30	U	7.2	289.6	2	12	4	4	22
K	39.05.24	77.01.40	U	7.4	323.9	2	14	3	4	23
G	39.13.08	77.10.41	A	8.8	344.6	1	14	1	4	20
S	39.12.44	77.02.35	A	11.3	308.0	3	15	4	2	24
L	39.01.00	76.49.18	A	11.7	14.9	8	13	2	4	27
B	39.03.00	76.57.05	U	13.3	114.9	0	12	4	3	19
G	39.12.44	77.09.55	A	13.8	121.7	1	14	2	2	19
K	39.07.18	77.00.13	A	14.0	80.0	3	14	4	4	25
G	39.11.31	77.09.44	A	73.7	16.3	6	14	2	5	27
K	39.03.30	77.01.49	U	97.9	141.0	6	12	1	3	22
C	39.12.27	76.59.40	A	219	14.2	9	12	1	2	24
C	39.07.53	76.55.27	A	283	5.7	5	16	2	3	26
C	39.08.04	76.55.43	A	283	5.7	7	14	4	3	28
C	39.07.53	76.56.14	A	283	12.4	10	13	2	2	27
C	39.09.36	76.52.49	A	357	43.9	7	14	0	3	24
L	39.02.11	76.49.30	U	905	1.5	7	11	0	2	20

| L | 39.02.06 | 76.50.17 | U | 905 | 1.5 | 4 | 14 | 2 | 1 | 21 |
| L | 39.02.29 | 76.49.00 | U | 905 | 1.4 | 6 | 12 | 1 | 3 | 22 |

[a] U.S. Geological Survey quadrangles: B, Beltsville; C, Clarksville; G, Gaithersburg; K, Kensington; L, Laurel; S, Sandy Spring. Latitude and longitude given as degrees, minutes, seconds.

[b] A, agricultural; U, urban.

[c] Computed as described in text. Units are $(km^2/ha) \times 10^{-5}$.

[d] Number of bird species of given habitat preference recorded in 60-min point surveys (three 20-min intervals) taken at geometric center of small islands, or within upland oak-hickory patches of major woodlands.

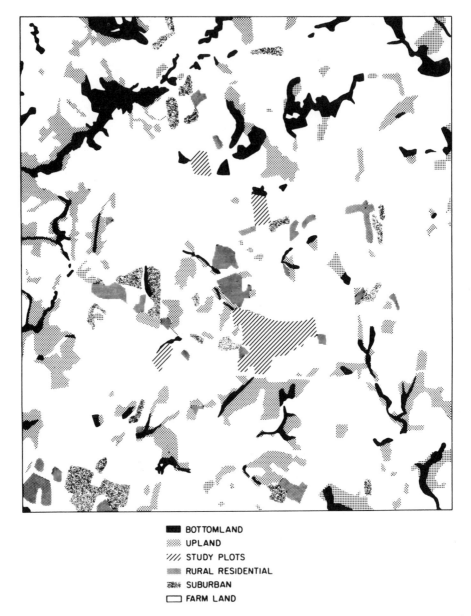

■■■ BOTTOMLAND
▒▒▒ UPLAND
/// STUDY PLOTS
▓▓▓ RURAL RESIDENTIAL
▦▦▦ SUBURBAN
☐ FARM LAND

Fig. 8-3. Forest fragmentation in a predominantly agricultural area: North portion of the USGS Gaithersburg Quadrangle of central Maryland.

mapping in bird census work, estimates of the maximum radius of vocal detectability in forest habitats are given in Table 8-3.

To estimate detectability we constructed "discovery curves" for each species by compiling data from successive 5-min intervals. This permitted computation of a detectability coefficient for each species (Appendix A). These coefficients made it possible to take interspecific differences in detectability into account in estimates of species abundances.

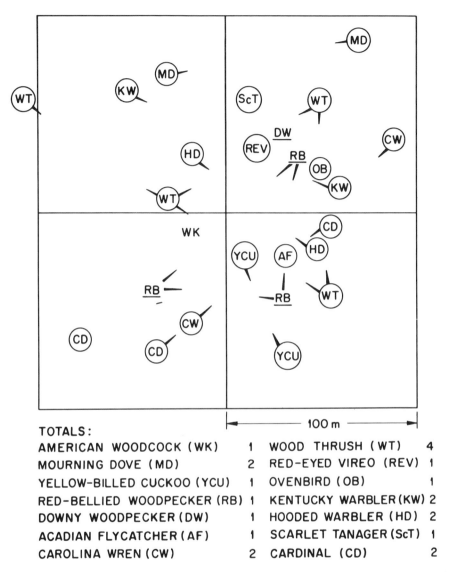

TOTALS:

AMERICAN WOODCOCK (WK)	1	WOOD THRUSH (WT)	4
MOURNING DOVE (MD)	2	RED-EYED VIREO (REV)	1
YELLOW-BILLED CUCKOO (YCU)	1	OVENBIRD (OB)	1
RED-BELLIED WOODPECKER (RB)	1	KENTUCKY WARBLER (KW)	2
DOWNY WOODPECKER (DW)	1	HOODED WARBLER (HD)	2
ACADIAN FLYCATCHER (AF)	1	SCARLET TANAGER (ScT)	1
CAROLINA WREN (CW)	2	CARDINAL (CD)	2

Fig. 8-4. A typical point survey visit (redrawn from a field sheet). Observer mapped singing (circled) territorial males from centerpoint. Most observations are within 100 m of the point; therefore the technique surveys somewhat less than 4 ha. The actual surveyed area varies for each bird species, but recorded territories may be compared between points. In mapping, simultaneous registrations are used as the basis for assigning multiple territories. Note: In international standard mapping methods (International Bird Census Committee 1970) such circles would be connected by dotted lines; on maps such as these with many simultaneous registrations, such a system is impractical. Birds calling (e.g., DW: downy woodpecker) and not singing may be recorded as territorial if such an observation is biologically reasonable. Birds seen but not heard may also be recorded (e.g., American woodcock). If males and females cannot be distinguished vocally (e.g., red-bellied woodpecker), three simultaneous registrations are necessary to document two territories. The number of territories is estimated on the field sheet immediately after the visit.

Table 8-3. Estimates of the Maximum Radius of Vocal Detectability for Selected Birds in Eastern Deciduous Forest Habitats

Maximum radius (m)	Species
240	Common flicker, red-bellied woodpecker, wood thrush, scarlet tanager
200	Indigo bunting
180	Mourning dove, hairy woodpecker, blue jay, yellow-throated vireo, rufous-sided towhee
165	Tufted titmouse, brown thrasher, Kentucky warbler, cardinal
150	Yellow-billed cuckoo, great crested flycatcher, house wren, white-eyed vireo
140	American robin, ovenbird
120	White-breasted nuthatch, veery, starling, common grackle
100	Acadian flycatcher, red-eyed vireo, pine warbler, hooded warbler, northern oriole, American goldfinch
90	Downy woodpecker, common yellowthroat
75	Eastern wood pewee, Carolina chickadee, gray catbird, black-and-white warbler, worm-eating warbler, brown-headed cowbird
60	Blue-gray gnatcatcher

Comparison of Point Survey Method with Spot-Mapping Censuses. No one survey or census method is optimal for all purposes (Robbins 1978a), but the spot-mapping method (Williams 1936) is the commonly accepted standard for comparison. Standard spot-mapping censuses were performed as recommended by Hall (1964). The locations of singing males were mapped on each of 8 to 14 visits to the sites, which varied in size from 6 to 14 ha. Observers traveled grid lines located 100 m apart. The results of these censuses have been published elsewhere (Whitcomb et al. 1975, MacClintock et al. 1977, Whitcomb et al. 1977, L. MacClintock et al. 1978, N. MacClintock et al. 1978).

We compared the results of the point survey method with those of the spot-mapping method by conducting point surveys in 15 plots that were also censused by spot-mapping. The results of two of these comparisons are given briefly in Appendix B. The main points can be summarized as follows: The 60 min devoted to each point in a point survey slightly exceeds the time that normally would be allotted to the same point during the standard course of eight visits in a spot-mapping census. Therefore, point surveys have a high probability of detecting nearly all forest interior species audible from the point. Coverage of tracts > 15 ha, in which species occurrences may be patchy, is improved by establishing two or three widely spaced survey points instead of the usual single point. Such use of multiple points gives an even closer correspondence between the results of the two methods (Appendix B). The overriding advantage of the point survey method is an enormous saving of time, a feature that permits the observer to gather quantitative data for a large number of tracts during a single breeding season. The point survey method does have some disadvantages relative to the more intensive spot-mapping technique. For example, point surveys do not always distinguish territorial singing males from transitory singing nonbreeders. In comparison, the 8-10 visits typically made during a spot-mapping census allow a better estimation of the breeding status of singing males. However, in our view the drawbacks of the point survey method do not negate the advantage of greater coverage, and, whichever method is employed, consistent trends in species occurrence should be evident.

Breeding Bird Atlas Projects. Data from the Breeding Bird Atlas projects (Klimkiewicz and Solem 1978) were utilized retrospectively to assess the degree of patchiness in contemporaneous distribution of bird species in the study region. The goal of these projects was to map breeding birds on a quadrat-by-quadrat basis over countywide areas. Standard 7.5 min USGS topographic maps were divided into six quadrats, each covering 2.5 min of latitude and 3.75 min of longitude or approximately 25 km^2. One or more volunteer observers were assigned to each quadrat or block, and searched the area for the presence of bird species. Various degrees of certainty of breeding presence were designated, but in the present paper we accept all three categories, "possible," "probable," and "confirmed" breeding, i.e., all species found in suitable nesting habitat during the breeding season. In the context of our analysis, this treatment is conservative, because we are interested in documenting the *absence* of species during the breeding season. The number of species absent will be underestimated if mere presence of the species, without actual demonstration of reproductive activity, is accepted as sufficient evidence for breeding. On the other hand, unequivocal proof of the absence of a species from a large area is virtually impossible, and presents a particular problem in the detection of ephemeral colonizations and extinctions or marginal persistence of rare species (Lynch and Johnson 1974). To minimize the impact of such artifacts, a special effort was made by independent experienced observers to search optimum habitat in each block for expected species that had not been encountered.

Miniroute Surveys. A second method that was used retrospectively for obtaining estimates of relative abundance and distribution of bird species was the Miniroute survey (Bystrak 1980). This variant of the standard Breeding Bird Survey method (Robbins and Van Velzen 1974) was used to supplement the Breeding Bird Atlas surveys in Howard and Prince George's counties (Klimkiewicz and Solem 1978). The Miniroutes were transects consisting of 25 stations each. The stations were spaced at intervals of about 800 m along secondary roads. Unlike Breeding Bird Survey routes, which require the observer to record birds encountered at randomly selected points, Miniroute points were adjusted in some instances to maximize coverage of habitats that were otherwise underrepresented by strictly randomized sampling. Although such adjustments were not made quantitatively, our analyses (Table 8-1) suggest that the Miniroutes adequately sampled existing habitat. Each Miniroute was surveyed twice (once in each direction) during the period 1-30 June. Coverage began 30 min before sunrise, and about two hours were required to visit the 25 stations. At each of the 25 stations, observers recorded the identity and abundance of all bird species heard or seen during a 3-min interval. Special note was made of singing males, and the maximum number of such individuals observed at each station during the two visits was used in the analysis.

Avian Life History Features

Migratory Status. We defined *permanent residence* as the maintenance of year-round home range on the breeding grounds, with at most local "drift" into nearby habitats. *Short-distance migration* involves movement of the breeding population a few hundred kilometers to a different wintering area. In some instances, northern populations of a given short-distance migrant species overwinter in habitat vacated by more southerly populations when the latter move still farther south. *Neotropical migration* is the movement of most or all individuals of a given species to the neotropical (or in some cases subtropical) region for the winter.

Body Weight. Mean body weights (grams) of banded birds from files of the Migratory Bird and Habitat Research Laboratory of the U.S. Fish and Wildlife Service at Laurel, Maryland, were used. Most weights (Table 8-4) were records from birds banded in the northeastern United States during autumn migration under the cooperative Operation Recovery program. Although such weights may average slightly more than breeding season weights, the differences were not considered sufficient to bias the results of the analyses. For most species, sample sizes exceeded 100.

Longevity. Maximum longevity records (Table 8-4) of unrestrained banded birds taken from Kennard (1975) were supplemented with unpublished records (C. S. Robbins), records from the Bird Banding Laboratory files, and manuscript notes compiled by Roger Clapp et al. (in preparation).

Territory Size. Schoener (1968), among others, has emphasized the importance of territory size in avian life history patterns. We estimated territory size from maximum breeding densities (territorial males per km^2) that were obtained from computer files of the Breeding Bird Censuses that have been published in *American Birds, Audubon Field Notes,* and *Audubon Magazine* (Table 8-4). These files are maintained at the Migratory Bird and Habitat Research Laboratory of the U.S. Fish and Wildlife Service at Laurel, Maryland. Plots smaller than 6 ha were excluded from calculations of maximum densities because of sampling bias.

Nest Form and Location. Nest form (open or cavity) and location are well known for all eastern forest species (Bent 1946, 1968). Mean nest height for each species was estimated from Maryland Ornithological Society (MOS) records (Table 8-4). Sample size exceeded 20 nests for all but a few species.

Reproductive Traits. Nesting data gathered by the MOS were used to compute mean clutch size for each breeding species (Table 8-4). Data for at least 20 clutches were available for all but a few species. The number of clutches produced by individual females per breeding season is more difficult to estimate, but existing MOS records, together with published life history studies (Bent 1968, Brackbill 1977, Graber and Graber 1963, Henny 1972) permitted a reasonable approximation. An estimate of mean annual reproductive effort was then computed as mean clutch size multiplied by the mean annual clutch number. Although we recognize that this measure greatly oversimplifies the measurement of reproductive effort (Stearns 1977), we believe it is adequate for the purposes intended in our analyses.

Habitat Utilization. Several workers have attempted, by one means or another, to quantify descriptions of avian habitat utilization. James (1971), for example, developed ordination techniques for habitat description. More recently, Robbins (1978b) used parameters derived from habitat descriptions in spot-mapping censuses. To classify habitat utilization, we defined four habitat associations for breeding birds: (1) *forest-interior specialists* nest only within the interior of the forest and tend to avoid edge habitats (e.g., hooded warbler, Fig. 8-5a); (2) *interior-edge generalists* may have territories located entirely within the forest, but also can utilize forest edge (e.g., cardinal, Fig. 8-5b), or may in some instances integrate more than one discrete forest fragment into a single territory; (3) *edge species* organize their territories primarily or exclusively at the border of forest (e.g., indigo bunting, Fig. 8-5c); and (4) *field-edge species* may nest at the margins of forest, or even in the forest interior, but require fields or other open habitats for foraging (e.g., common crow). Classification of birds into these categories was largely based on spot-mapping data.

Measures of Species-Specific Isolation

Each of the 93 bird species in our regional pool is subject to a certain degree of isolation. This isolation is a result of (1) patchy occurrence of suitable habitat, even in continuous forest, and (2) inability of some species to locate suitable but isolated habitat. Thus, the degree of isolation to which each species is subjected does not conform to the geometry of forest islands, and is highly species-specific. Five methods were developed to estimate this type of isolation. Estimates of regional abundance derived from Miniroute data offer an indirect means of measuring isolation. Data from Atlas blocks offer another means of estimating isolation indirectly. Three other estimates were developed, discussed below.

Miniroute Data. Miniroute data from Prince George's County, taken in 1976, were used to establish general patterns of habitat utilization. Nine Miniroutes were selected as representative of the study area as a whole (Table 8-1), and each of the 225 survey points along these routes was inspected and classified according to the major habitat it sampled. These major habitats reflect the contemporaneous reality of land use in our study area, rather than subdivisions of the primitive forest. To increase the number of bottomland forest sites, 26 additional survey points in this habitat were selected. Of a total of 251 survey points, 185 were found to sample a single habitat. Data from this subset of points were used to compute fractional utilization of the 7 major habitat types (Appendix C).

Bridge Utilization. Forest islands may be isolated by a mosaic of open fields, housing developments, hedgerows, and early successional stands of various structural and vegetational composition. Some of these habitats may provide "bridges" that can be used by some species for breeding. The Miniroute data were used to compute an index of isolation for each species (Table 8-5) which we term *bridge utilization* (B_i):

$$B_i = \frac{D_i n_i}{N} \qquad (2)$$

where n_i is the number of Miniroute points at which the species (i) was recorded, D_i is the "detectability coefficient" for that species (see Appendix A), and N is the total number of Miniroute points sampled (= 251 in the present example).

Table 8-4. Life History Traits of the Regional Avian Species Pool in the Maryland Study Area[a]

Species	Body weight (g)	Feeding strategy	Nest type	Nest height (m)	Territorial density (males/km²)	Maximum longevity (months)	First clutch size (no. eggs)	No. broods/ year	Reproductive effort
Forest interior species									
Permanent residents									
Pileated woodpecker	310.0	I	H	10.4	1	242	2.00	1.0	2.00
Hairy woodpecker	67.0	I	H	8.8	11	268	4.08	1.0	4.08
White-breasted nuthatch	21.1	I	H	6.1	20	102	6.22	1.0	6.22
Short-distance migrants									
Pine warbler	12.6	I	O	10.4	76	111	3.80	1.0	3.80
Neotropical migrants									
Acadian flycatcher	12.6	I	O	2.8	68	189	2.80	2.0	5.60
Veery	31.1	O	O	1.0	42	164	3.33	1.0	3.33
Black-and-white warbler	10.7	I	O	0.0	27	226	6.17	1.0	6.17
Worm-eating warbler	13.6	I	O	0.0	26	302	4.14	1.0	4.14
Cerulean warbler	13.5	I	O	10.7	83	147	3.33	1.0	3.33
Ovenbird	19.9	I	O	0.0	114	133	4.67	1.0	4.67
Louisiana waterthrush	20.8	I	O	0.0	16	193	5.20	1.0	5.20
Kentucky warbler	13.8	I	O	0.0	36	103	4.12	1.5	6.18
Hooded warbler	12.1	I	O	0.9	63	133	3.43	1.0	3.43
American redstart	9.0	I	O	6.1	71	153	3.56	1.0	3.56
Scarlet tanager	31.2	O	O	6.4	27	173	3.12	1.5	4.68
Forest interior and edge species									
Permanent residents									
Red-bellied woodpecker	74.6	O	H	8.8	29	151	4.50	1.0	4.50
Downy woodpecker	26.1	I	H	9.1	21	115	4.06	1.0	4.06
Blue jay	86.7	O	O	7.0	40	120	4.55	1.0	4.55
Carolina chickadee	9.7	I	H	1.5	39	125	5.95	1.0	5.95
Tufted titmouse	21.8	I	H	3.4	56	131	5.10	1.0	5.10
Carolina wren	18.8	I	O	1.5	59	79	4.67	2.0	9.34
Cardinal	41.6	O	O	2.0	96	141	3.30	3.0	9.90

Species								
Short-distance migrants								
Common flicker	134.0	I	5.8	H	27	153	6.30	1.0	6.30
Eastern phoebe	19.2	I	2.3	O	15	127	4.57	2.0	9.14
Gray catbird	38.5	O	2.3	O	198	92	3.80	2.5	9.50
White-eyed vireo	12.5	I	1.0	O	40	118	3.45	2.0	6.90
Common yellowthroat	10.6	I	0.2	O	111	109	3.75	2.0	7.50
Rufous-sided towhee	40.8	O	0.4	O	68	124	3.75	2.0	7.50
Neotropical migrants								
Yellow-billed cuckoo	56.3	I	4.0	O	17	106	2.55	1.0	2.55
Whip-poor-will	55.6	I	0.0	O	13	151	1.78	1.5	2.67
Ruby-throated hummingbird	3.3	O	4.0	O	15	183	2.05	2.0	4.10
Great crested flycatcher	34.0	I	2.3	H	17	285	4.80	1.0	4.80
Eastern wood pewee	13.6	I	6.1	O	24	177	2.75	1.0	2.75
Wood thrush	51.8	O	2.6	O	125	104	3.80	2.0	7.60
Blue-gray gnatcatcher	6.4	I	8.8	O	28	166	4.75	1.5	7.13
Yellow-throated vireo	14.5	I	8.8	O	25	182	3.12	1.5	4.68
Red-eyed vireo	18.2	I	2.8	O	138	131	3.25	2.0	6.50
Prothonotary warbler	12.3	I	1.6	H	40	127	4.08	1.5	6.12
Northern parula warbler	8.8	I	5.1	O	47	147	3.20	1.0	3.20
Forest edge and scrub + field-edge (F) species								
Permanent residents								
Bobwhite	169.9	O	0.0	O	5[b]	64	14.53	2.0	29.06
Common crow (F)	482.8	O	10.4	O	n[b]	154	4.60	1.5	6.90
Mockingbird (F)	49.1	O	1.7	O	28	178	3.83	2.5	9.58
Starling (F)	78.9	H	4.0	H	n	138	5.30	2.5	13.25
Short-distance migrants								
Mourning dove (F)	121.0	H	2.4	O	n	74	2.00	3.5	7.00
House wren	11.1	I	2.3	H	100	76	5.59	2.0	11.18
Brown thrasher	68.5	O	1.3	O	34	126	3.88	2.0	7.76
American robin (F)	81.5	O	6.1	O	122	102	3.80	2.5	9.50
Eastern bluebird (F)	30.5	I	2.0	H	n	90	4.32	2.0	8.64
Common grackle (F)	99.2	O	5.5	P	n	145	4.45	1.0	4.45
Brown-headed cowbird (F)	41.4	O	1.6	O	42	121	4.50	1.5	6.75
American goldfinch	12.9	H	3.0	H	21	99	4.50	1.5	6.75

Table 8-4 (continued)

Species	Body weight (g)	Feeding strategy	Nest type	Nest height (m)	Territorial density (males/km^2)	Maximum longevity (months)	First clutch size (no. eggs)	No. broods/ year	Reproductive effort
Chipping sparrow (F)	12.3	O	O	2.2	90	91	3.06	2.0	6.12
Field sparrow	12.8	O	O	0.4	80	75	3.67	3.0	11.01
Song sparrow	20.8	O	O	1.0	109	88	4.32	3.0	12.96
Neotropical migrants									
Eastern kingbird (F)	43.6	I	O	8.5	17	208	3.38	1.0	3.38
Blue-winged warbler	8.7	I	O	0.0	47	237	3.78	1.0	3.78
Yellow warbler	10.3	I	O	1.2	63	129	4.00	1.0	4.00
Prairie warbler	8.0	I	O	1.0	85	123	3.56	1.5	5.34
Yellow-breasted chat	26.5	I	O	1.1	36	121	3.73	1.0	3.73
Orchard oriole	22.7	I	O	4.7	29	128	4.14	1.0	4.14
Northern oriole	34.8	I	O	8.9	10	152	4.59	1.0	4.59
Blue grosbeak	30.0	O	O	1.3	4	217	3.33	2.0	6.66
Indigo bunting	15.2	O	O	0.9	52	122	3.23	2.0	6.46

[a] Feeding strategy is coded: I, insectivorous; H, herbivorous; O, omnivorous. Nest type is coded: H, hole-nesting; O, open-nesting; P, brood parasite. Reproductive effort is clutch size times number of broods per year.

[b] Not territorial in forest.

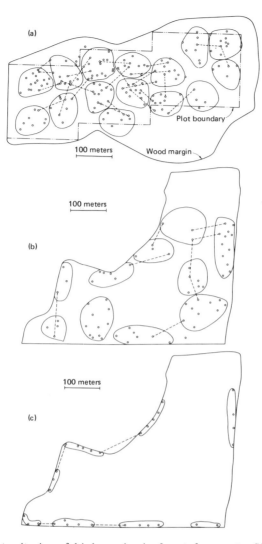

Fig. 8-5. Mapped territories of bird species in forest fragments. Simultaneous registrations are connected by dotted lines. (a) Territories of hooded warbler (a forest
interior resident) in a mature, selectively logged tulip tree-oak forest fragment in Maryland (R. F. Whitcomb et al. 1977). The 14.2 ha plot was located within the 21.5 ha
island; at least one additional territory lay completely outside the plot in the lower
right portion of the figure, but was not recorded. All the area of defended territories
was located entirely within the forest interior, even if, as in this case, dense packing
resulted. (b) Territories of cardinal, a species that can utilize either forest interior or
edge. Such species often are able to integrate several isolated patches of habitat into
a single territory. Six of the ten territories in a tulip tree-oak fragment (R. F. Whitcomb et al. 1977) were organized at the edge, but four territories were entirely within
the forest interior. (c) Territories of an edge species (indigo bunting). Although such
species may occupy territories in forest interior, major gaps, such as those generated
by large treefalls, are required. Most territories of such species are organized along
wood margins, hedgerows, or similar habitat.

Table 8-5. Tolerance to Fragmentation of Forest Bird Species and Measures of Their Isolation

Species	Island size[a]			Tolerance to fragmentation[b]	Habitat utilization			Regional abundance[f]			Regional distribution[g]
	1-5 ha	6-14 ha	>70 ha		Forest habitat utilization[c]	Bridge utilization[d]	General habitat utilization[e]	Howard	Pr. George's	Total	
Forest interior species											
Black-and-white warbler	0	0	9	.00	.55	.13	.12	4	20	24	.40
Worm-eating warbler	0	0	8	.00	.10	.02	.01	4	2	6	.26
Pileated woodpecker	0	0	2	.00	.14	.06	.03	0	3	3	.70
Ovenbird	0	2	21	.10	.84	.18	.14	26	147	173	.62
Hooded warbler	1	1	4	.25	.44	.10	.08	12	41	53	.29
Kentucky warbler	0	2	6	.33	.53	.12	.11	28	54	82	.74
Scarlet tanager	5	6	14	.43	.88	.40	.35	65	177	242	.87
Acadian flycatcher	4	4	9	.44	.58	.23	.18	97	203	300	.91
Hairy woodpecker	0	1	2	.50	.35	.10	.05	9	11	20	.83
White-breasted nuthatch	0	2	2	1.00	.20	.09	.02	2	7	9	.81
Forest interior and edge species											
Blue-gray gnatcatcher	0	0	6	.00	.28	.20	.11	30	36	66	.61
Yellow-throated vireo	0	0	2	.00	.14	.45	.32	9	21	30	.60
Yellow-billed cuckoo	1	1	7	.14	.50	.52	.37	85	62	147	.90
Red-eyed vireo	6	8	23	.35	.93	.61	.57	226	762	988	1.00
Wood thrush	14	14	23	.61	.91	.65	.46	556	724	1280	1.00
Tufted titmouse	10	13	15	.87	1.00	.63	.45	227	409	636	1.00
Mourning dove	6	8	9	.89	.34	1.29	1.30	862	818	1680	1.00
Blue jay	8	10	10	1.00	.88	.63	.48	204	301	505	1.00
Red-bellied woodpecker	7	9	9	1.00	.75	.56	.56	130	217	347	1.00
Cardinal	14	13	11	1.18	.80	1.07	1.01	715	1289	2004	1.00
Eastern wood pewee	6	12	10	1.20	.66	.39	.38	181	211	392	1.00
Carolina chickadee	11	11	9	1.22	.77	.66	.49	148	306	454	1.00
Downy woodpecker	8	10	8	1.25	.40	1.11	1.02	51	96	147	1.00
Rufous-sided towhee	9	12	9	1.33	.78	.64	.59	222	468	690	0.99
Great crested flycatcher	2	10	6	1.67	.49	.17	.13	55	39	94	0.95

Carolina wren	9	16	9	1.78	.81	2.19	2.07	269	654	923	1.00
Common flicker	9	11	6	1.83	.26	.62	.37	99	59	158	1.00
Gray catbird	14	7	3	2.33	.22	.85	.78	537	328	865	1.00
Field-edge species[h]											
Common crow	10	9	9	1.00	.63	1.41	1.53	1034	1078	2122	1.00
American robin	8	5	3	1.67	.35	1.46	1.98	1198	1397	2595	1.00
Common grackle	6	6	3	2.00	.29	2.19	2.07	3037	2526	5563	1.00
Starling	11	10	3	3.33	.05	1.40	1.32	2989	2650	5639	1.00

[a]. Numbers of presumed territories in 60-min point surveys on the 10 islands of each given size class.
[b] Number of presumed territories per point sampled on the 6- to 14-ha islands divided by the number per point sampled on the >70-ha islands.
[c] Calculated as described in the Methods section. Not corrected for detectability.
[d,e] Calculated as described in the Methods section. Corrected for detectability.
[f] Numbers of birds counted on Miniroutes in Howard and Prince George's counties.
[g] Fraction of 25 km² Atlas blocks in Howard and Montgomery counties in which possible breeding status was reported.
[h] No edge species were considered to be true island inhabitants. The four species in the field-edge category, however, were sufficiently abundant that their residence in islands could not be excluded.

General Habitat Utilization. Bridge utilization does not take into account unequal occurrence of the various habitat types. A better measure of general habitat utilization could be developed (Table 8-5) if habitat use was weighted according to habitat availability:

$$B'_i = D_i \sum_{k=1}^{7} h_k p_k \tag{3}$$

where h_k is the available fraction of habitat type k (from Table 8-1) and p_k is the fraction of points sampling habitat k where species i was in fact encountered in the Miniroute surveys.

Forest Habitat Utilization. Habitat designated as "forest" can also be classified in terms of subhabitats, and the tendency for a species to occur in different types of forested habitat can then be computed. We pooled results of 27 points in the Beltsville forest (Appendix D) and 18 point surveys in the Seton Belt woods (Appendix B) for this portion of the analysis. The sites and their associated birds are described in detail elsewhere (MacClintock et al. 1977, R. F. Whitcomb et al. 1977). Each survey point was classified according to the following three-way system: (1) upland vs. bottomland, (2) predominantly deciduous vs. predominantly coniferous, and (3) early (10-20 yr), middle (25-60 yr), or late (100+ yr) successional maturity. All of the 12 possible habitat combinations except bottomland-coniferous were represented by at least one of the 45 sample points. A coefficient of forest habitat utilization (F_i) for each bird species (i) was then computed as follows:

$$F_i = \frac{D_i \sum^{c} P_{ci}}{N} \tag{4}$$

where P_{ci} is the fraction of survey points within habitat c where territories of species i occurred, N is the total number of habitat types, and D_i is the detectability coefficient for species i (Appendix A).

Statistical Procedures

χ^2 Analysis: Taxonomic, Ecological, and Migratory Classification. A preliminary analysis was performed to search for evolutionary and ecological properties of bird species that were highly correlated with tolerance to fragmentation. We classified all taxa according to the following discrete attributes: *major habitat association* (Appendix E, Table 8-6: forest interior, forest interior and edge, forest edge-scrub, or field-edge); *migratory strategy* (Appendix E: neotropical, short-distance, permanent resident); *taxonomic affiliation* (19 families); *nest location* (ground, shrub, canopy, hole); *species-specific isolation (ability to utilize nonforest "bridges"*: Table 8-5); and tolerance to fragmentation (tolerant or intolerant, depending on value of the index) (Table 8-5 and Bond 1957). Utilization of the data of Bond and our own data for fragmentation tolerance in a single determination offered significant advantages that offset, to some extent, the disadvantage of neglecting the continuous nature of this variable.

Table 8-6. Regional Species Pool of Forest Birds[a]

Residence	Migratory strategy			
	Permanent resident	Short-distance migrant	Neotropical migrant	Total
Species analyzed				
Forest interior	3	1	15	19
Forest interior and edge	6	10	10	26
Edge and scrub	1	6	8	15
Field-edge	5	7	1	13
Total	15	24	34	73
Species not analyzed				
Water species	0	3	1	4
Raptors	5	5	1	11
Upland game birds	3	0	0	3
Crepuscular species	0	1	1	2
Total	8	9	3	20
Grand total	23	33	37	93

[a] An annotated list is presented in Appendix E.

We constructed separate contingency tables for each of these discontinuous attributes, utilizing data for the 43 species whose tolerance to fragmentation could be unambiguously classified. χ^2 tests were used to test the significance of apparent associations among attributes.

Correlation Coefficients. A matrix of correlation coefficients was computed to relate each of the life history variables to the Maryland indices of tolerance to fragmentation and to show which of the variables were closely related to each other (SAS Pearson product-moment correlation program, Barr et al. 1976). We recognize that the use of discontinuous variables with a small number of character states (e.g., migratory status) is not strictly in accord with the prerequisite for formal correlation analysis. However, because these character states in fact reflect essentially continuous underlying distributions, and because the states bear an interval relationship to one another, we included them in the analysis. The following variables were included in the analysis: migratory strategy (classification: permanent resident = 0, short distance migrant = 1, neotropical migrant = 2) or migratory distance (distance between study area and center of winter range of species); habitat utilization strategy (field-edge = 0, edge-scrub = 1, forest interior and edge = 2, forest interior = 3) or linear ordination values of habitat selection from James (1971); nest type (cavity = 0, open = 1); nest height; clutch size, number of broods and reproductive effort (clutch size × number of broods); longevity (maximum longevities from banding data); territory size (computed from data on packing density); body weight; bridge utilization (B_i); general habitat utilization (B_i'); Breeding Bird Atlas data (fraction of 94 blocks in which species i was present); regional abundance (total individuals of species i observed on 50 Miniroutes in Prince George's and Howard counties; forest habitat utilization (F_i); and indices of tolerance to fragmentation from our island set (Table 8-5). Because some of the variables were not linearly correlated with tolerance to fragmentation, we also examined matrices using square root and square transformations of certain variables.

Stepwise Regression. The results from the correlation matrices were used to select variables for stepwise regression. In the instances in which two or more estimates of the same life history feature were available, the estimate that showed the greatest correlation with tolerance to fragmentation was chosen. Because several life history features were highly correlated, sets of variables were chosen that showed low intercorrelation. Sets of independent variables were used in the stepwise regression analyses (SAS stepwise regression program with STEPWISE option; Barr et al. 1976), with tolerance to fragmentation as the dependent variable. These variables (Appendix F, Tables 8-4 and 8-5) included migratory strategy, major habitat association, reproductive effort, general habitat utilization (B'_i), longevity, nest type, nest height, body weight, territorial size, and forest habitat utilization (F_i).

Results

Species Composition of Forest Islands

Tolerance to Fragmentation. Bird species differed markedly in their tolerance to habitat fragmentation. We obtained an index of the tolerance of each species (Table 8-5) by computing the ratio of the mean number of perceived territories per survey point in 6-14 ha islands vs. 70+ ha islands. This ratio approached unity for species unaffected by island size, exceeded unity in species that were more abundant in fragmented forest systems, and decreased to zero for species that were not found breeding on forest islands as small as 14 ha (Table 8-5). Data from forest islands 1-5 ha in size were not used for this computation because the distance between the survey points and the forest margin in such islands was less than the radius of vocal detectability of many bird species (Table 8-3). The indices obtained from these computations formed a basis for later statistical analyses involving life history features.

Species Richness. Avifaunal composition of forest islands varied according to island size. Figure 8-6 shows the regression of the number of forest-interior, edge, and ubiquitous species on forest area. For this computation, edge and field-edge species were combined in a single category. A strong negative correlation ($r = -0.713, p < 0.001$) was found between the number of edge species and the size of the forest island. Conversely, there was a strong positive correlation ($r = 0.527, p < 0.005$) between the number of forest-interior species and island area (Fig. 8-6). Because these positive and negative trends tended to compensate, there was no significant change in overall species richness with increasing island size ($r = 0.128, p > 0.1$). These results permit the rejection of a null hypothesis (Strong et al. 1979, Grant and Abbott 1980) that forest island area is unrelated to the composition of the forest interior avifauna. It should be emphasized that the point surveys summarized by these regressions estimate the density of species at a given location. They do not represent the total species count for entire tracts, as do conventional species-area curves.

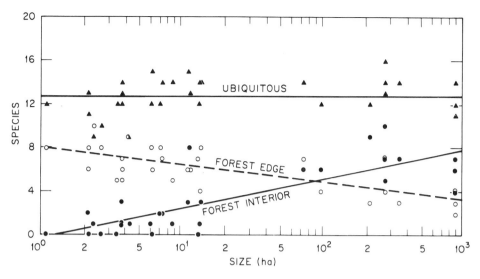

Fig. 8-6. Point survey data plotted vs. forest island size. The regression is *not* a species-area curve. Forest interior species richness is positively correlated with island size ($y = -.267 + 2.678 \log_e x$; $p < .001$). Species that can utilize edge as well as forest interior (ubiquitous) occur with equal frequency on large and small forest islands. Species richness of edge species is an inverse function of island size ($y = 8.005 - 1.559 \log_e x$; $p < .001$).

Isolation. Isolation factors (Table 8-2) were significantly correlated with numbers of forest interior species on the islands ($r = .44$, $p < .05$). The manner of selection of islands strongly influenced this correlation, and warrants discussion. Small and medium-sized islands (1.1-14 ha) were chosen (with a single exception) that were well isolated from mainland forest. Thus, the subset of islands in these size categories formed a relatively tight cluster; each of these islands had a high isolation value. On the other hand, the large islands occurred in one of two situations. In one, the extensive forest systems of the Beltsville area provided large (mainland) plots. In the second instance, the large islands bordered reservoirs along the Patuxent River. In each of these cases, the semicontinuous forest areas were divided into several subareas, each of which met the criteria for "isolation," but was in fact minimally isolated from adjacent subareas.

Only one of the total set of 30 islands did not fall into the categories described above. This was the island in the Laurel quadrangle (Table 8-2; lat. 39°01'00", long. 76°49'18") that formed the basis for the study of MacClintock et al. (1977). Eight forest interior species, each of which was common in the large nearby forest, were found in this fragment. MacClintock et al. (1977) concluded that this fragment, as a result of its minimal isolation from the large forest mainland, supported forest interior species not found on more isolated islands. Isolation is clearly one of the important factors in determination of breeding presence of forest interior bird species.

Species Associated with Small Islands. About 30 bird species were found to breed commonly in small to medium-sized (less than about 15 ha) forest islands of the Maryland Piedmont (Table 8-5, in part). Of these, only two species (Acadian flycatcher and scarlet tanager) were restricted to forest interior habitat. Four additional species (great crested flycatcher, eastern wood pewee, wood thrush, and red-eyed vireo) are neotropical migrants that utilize both forest interior and edge habitat. The remaining 24 species associated with small to medium-sized islands of deciduous forest are either permanent residents or edge species, or both. Some of the edge species are neotropical migrants (e.g., indigo bunting). Very small forest islands (less than 5 ha) were dominated by edge or field-edge species (e.g., mourning dove, common flicker, common crow, house wren, gray catbird, brown thrasher, American robin, starling, white-eyed vireo, common yellowthroat, yellow-breasted chat, northern oriole, common grackle, indigo bunting, and American goldfinch). Some of these species also occur (rarely) within the interior of extensive forest, particularly along floodplains and in other areas where canopy openings are common. Ranney (1977) indicates that small forest islands may lack the structural characteristics of true forest interior, and are entirely "edge." Because we did not measure the physiognomic characteristics of the islands we studied, we do not know the extent to which the absence of particular forest interior bird species from small forest islands may be related to subtle structural differences of the islands, as opposed to their area and isolation.

Urban vs. Agricultural Isolation. Twelve of the 20 small forest islands studied were surrounded by agricultural land, whereas eight were bounded by low-density suburban housing developments. The results of χ^2 analysis suggest that some bird species respond differently to these two modes of isolation. Acadian flycatcher, Kentucky warbler, and scarlet tanager were more abundant ($p < 0.05$) in forests within an agricultural context, whereas American robin was more common ($p < 0.05$) in suburban areas. The habitat associations of these species are also reflected in the Miniroute areas (Appendix C).

The Regional Species Pool

Composition. Contemporary species occurrence was determined by a combination of point survey, BBS, Breeding Bird Atlas, and, especially, Miniroute analyses (Appendix C) throughout the study region. About 94 species have been recorded, at one time or another, in forested habitat in our study area (Appendix E). One of the species (passenger pigeon) is now extinct. Most of these "forest birds" are not obligate inhabitants of forest interior, but may occur on the margins of forest islands, or in association with adjacent field habitat. In all, 73 of the species could be analyzed, to some extent, for the context of their occurrence in fragmented habitat. The remaining species, for various reasons (e.g., primary association with water, crepuscular habits, or status as raptors or game birds) could not be evaluated by any of our census or survey methods. Of the 73 species recorded in fragmented forest, only about 45 occur in forest interior and only 19 are obligately confined to forest interior (Table 8-6, Appendix E). Most of the remaining 28 species probably were not residents of the unbroken presettlement eastern forest, but resided instead in the savanna or prairie. These "edge" or "field-edge" species, which rarely (if ever) have territories within forest interior, cannot be properly evaluated in a study of forest islands alone.

Point Survey Estimates of Species Pool. The point survey method was used to determine the regional forest species pool of two extensively forested areas (Appendices B and D). Fifteen points in the interior of extensive forest at Beltsville and 18 points in the Seton Belt forest were selected for the surveys. In general, the points maximized the sampling of forest interior. These data showed that certain species that were rare or absent in small to medium-sized forest islands were common in extensive forest systems. These species included blue-gray gnatcatcher, yellow-throated vireo, black-and-white warbler, hooded warbler, Kentucky warbler, ovenbird, and Louisiana waterthrush. In the extensive forest of the Beltsville region all of these species were routinely encountered at suitable survey points (Appendix D; MacClintock et al. 1977).

In the Seton Belt forest system, although several hundred hectares of contiguous woodland are present, the forest is considerably more fragmented than at Beltsville (R. F. Whitcomb et al. 1977). Two of these fragments (North and South Tracts of the Seton Belt forest), composed of mature forest partially isolated from much larger but less mature tracts, have been censused by spot-mapping techniques (R. F. Whitcomb et al. 1977), whereas other portions of the forest system were analyzed by the point survey method (Appendix B). Possibly as a result of the relative isolation of the 14.6 ha South Tract, its avifauna has undergone a marked reduction since the site was first censused by Stewart and Robbins (1947) three decades ago. Worm-eating and black-and-white warblers have disappeared completely from the tract, and most of the remaining neotropical migrant species have declined in abundance (R. F. Whitcomb et al. 1977, L. MacClintock et al. 1978, N. MacClintock et al. 1978). With the exception of worm-eating warbler, the species that have declined in recent years at the South Tract or that are absent from forest islands (species intolerant to fragmentation) are still present in large populations in the extensive Beltsville forest. From such observations, we conclude that species that do not tolerate fragmentation have persisted in the contemporary regional species pool, but that their regional survival is largely dependent on the continued existence of extensive forests. This observation is crucial in assessing the significance of the decline of such species in smaller, more isolated tracts of forest.

Utilization of Forested Habitats

The coefficient of forest habitat utilization, F_i, expresses the ability of a species (i) to utilize forest of varying successional maturity, vegetational composition, and position on the moisture gradient. For statistical analyses, F_i included a correction for detectability, but values given in Table 8-5 and Appendix D were not so corrected. Uncorrected coefficients (F) for the four "field-edge" species were predictably low ($\bar{F}$ = 0.330, SD = 0.238), whereas F's for "forest-edge generalists" were high ($\bar{F}$ = 0.607; SD = 0.278, n = 18). Similarly, forest interior species had coefficients of intermediate value ($\bar{F}$ = 0.461, SD = 0.271, n = 10). Many of the forest interior species were, however, able to use a wide range of forest types. For example, the virtual disappearance of black-and-white warbler from the Seton Belt forests over the past 30 years (R. F. Whitcomb et al. 1977) occurred despite the ability of the species to utilize a wide range of forested subhabitats (F = 0.55). In fact, the actual F values for this species would have been considerably higher than the value computed, had the observations been confined to the Beltsville forests. High F values also were found (Table 8-5) for several other forest interior species (e.g., ovenbird, Kentucky warbler, hooded warbler). Therefore, it is highly unlikely that the regional decline of these species can be attributed to unusually narrow habitat requirements.

Breeding Bird Atlas Data: Evidence for Regional Extirpations

Breeding Bird Atlas data were examined retrospectively for evidence of regional extirpation of species intolerant to fragmentation. The data showed that a number of neotropical migrant species, particularly those associated with forest interior habitats, are absent from entire 25 km^2 blocks in the study region (last column of Table 8-5). The tendency for species absence is greatest in blocks that have undergone extensive deforestation, and where the remaining forest is highly fragmented. Since most of the missing species utilize a broad range of forest types (Table 8-5), it is reasonable to presume that before deforestation they were widely distributed throughout the area surveyed for the Atlas project. Some neotropical migrant species (e.g., ruby-throated humming-bird, yellow-throated vireo, and northern parula warbler) that are able to utilize both open habitats and forest interior habitats also show evidence of regional declines that ultimately may result in regional extirpation.

Only three permanent resident species associated with forest interior (pileated woodpecker, hairy woodpecker, and white-breasted nuthatch) are distributed patchily among the Breeding Bird Atlas blocks. All three have relatively large territories, and require fairly mature forest. However, all three species tolerate a considerable degree of fragmentation of their forested habitat, as long as the aggregate area of forest is sufficient. Thus, these three species probably occupy a large fraction of suitable habitat. Long-term census data (Kendeigh 1981) show that white-breasted nuthatch and hairy woodpecker regularly recolonized an Illinois forest island after extirpation, and their broad geographic ranges include parts of the central United States where forest is naturally discontinuous. In addition, both species (and, to a lesser extent, pileated woodpecker) commonly use open habitat during the winter. We conclude that these species are less sensitive to fragmentation per se than to reduction in total habitat area.

Some Atlas blocks lacked certain forest interior bird species, despite the presence of forest tracts of 50 ha or more. In 1975, to confirm these findings, we reexamined an archipelago of forest islands (12, 16, 32, 61, and 121 ha) surrounded by agricultural land in Howard County, and separated from extensive nearly continuous forest by a distance of 800-1000 m. All five islands lacked worm-eating warbler, northern parula warbler, and hooded warbler. Black-and-white warbler and American redstart were missing from four of the five, whereas veery nested only on the two largest islands. At the other extreme, Acadian flycatcher, blue-gray gnatcatcher, and Kentucky warbler were found on all five of the islands. A similar survey of two tracts (45 and 93 ha) separated by no more than 300 m from wooded corridors connected to the extensive Patuxent floodplain forest, similarly revealed the absence of a number of forest-interior bird species (yellow-throated vireo, black-and-white warbler, worm-eating warbler, northern parula warbler, ovenbird, and hooded warbler). We could not discern any structural or vegetational peculiarities that would account for these absences.

The Edge-Scrub Assemblage of Neotropical Migrants

Because of their ability to utilize nonforested habitat, "edge-scrub" and "field-edge" species are unlikely to be subject to complete isolation by forest fragmentation. Nevertheless, suitable habitat for certain of these species may be both scarce and patchily distributed. Breeding Bird Atlas data suggest that neotropical migrant species associated with edge and open habitats (e.g., prairie warbler, yellow-breasted chat, orchard oriole, and blue grosbeak) resemble, to some extent, forest-dwelling neotropical migrants in their patchy distributions among survey blocks (Fig. 8-7b). These species tend to occur in loose "colonies" that appear to be localized in especially suitable patches of habitat. The critical habitat dimensions for edge species have not been thoroughly studied, but are evidently more complex than those of forest interior species (Roth 1976). Although we have no reliable means of measuring habitat isolation for these species, much ostensibly suitable habitat appears to be unoccupied. Careful study of the habitat requirements of "edge" and "field-edge" species may well reveal that for this ecological grouping, as for forest-interior species, a high sensitivity to habitat fragmentation is associated with a highly migratory life-history pattern.

χ^2 Analysis: Taxonomic, Ecological, and Migratory Classification

Species-specific isolation ($p < 0.0001$), migratory strategy ($p < 0.001$), major habitat association ($p < 0.005$), and taxon ($p < 0.01$) all were highly significantly associated with tolerance to fragmentation in our χ^2 analyses. These analyses involved the 32 species for which indices of fragmentation tolerance were available (Table 8-5). We then wished to study intercorrelations between life history features. To do this, we expanded the sample size to the 70 species in the pool of species that might appear in diurnal forest island surveys (Appendix E). Major habitat association ($p < 0.001$) was the variable most closely associated with isolation, followed by nest type and location ($p < 0.005$), migratory strategy ($p < 0.025$) and taxonomic affiliation ($p < 0.025$). Finally, neotropical migration was most closely associated with taxonomic affiliation ($p < 0.005$), but was not highly correlated with major habitat association ($p > 0.05$). Similar independence ($p > 0.05$) was found between migratory and nest strategies.

These results, which led (in part) to formulation of the path diagram are portrayed diagrammatically in Fig. 8-7. Habitat and migration strategy provide a useful binary classification for bird families. The dimensions are sufficiently broad so that no family occupied even a major fraction of the two-dimensional area (Fig. 8-7). Analysis of life history features, tolerance to fragmentation, and tendency for regional extirpation in terms of this binary classification revealed important differences between members of the nine classes. In particular, forest interior neotropical migrants were shown to have lower reproductive rates, lower tolerance to fragmentation, and a greater tendency to become regionally extirpated (Fig. 8-7). These results suggested that other statistical methods might be appropriate to study the intercorrelation between tolerance to fragmentation and life history features. Correlation and stepwise regression were chosen for that purpose.

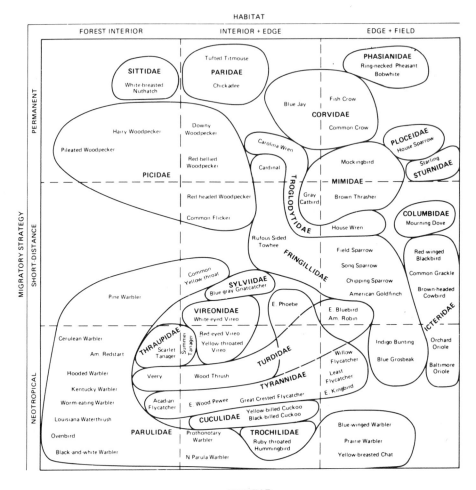

HABITAT

		FOREST INTERIOR		INTERIOR + EDGE		EDGE + FIELD	
MIGRATORY STRATEGY	**PERMANENT**	RE = 4.10 RE(L) = 836 ATLAS BLOCKS = .79 TF(MD) = .50	(n=3) (n=3) (n=3) (n=3)	RE = 6.20 RE(L) = 776 ATLAS BLOCKS = 1.00 TF(MD) = 1.19	(n=7) (n=7) (n=7) (n=7)	RE = 9.91 RE(L) = 1553 ATLAS BLOCKS = 1.00 TF(MD) = 2.16	(n=3) (n=3) (n=4) (n=2)
	SHORT-DISTANCE	RE = 3.80 RE(L) = 422 ATLAS BLOCKS = .11 TF(MD) = .50	(n=1) (n=1) (n=1) (n=1)	RE = 7.71 RE(L) = 979 ATLAS BLOCKS = .94 TF(MD) = 1.50	(n=7) (n=7) (n=7) (n=5)	RE = 8.18 RE(L) = 958 ATLAS BLOCKS = .98 TF(MD) = 1.39	(n=11) (n=11) (n=11) (n=4)
	NEOTROPICAL	RE = 4.57 RE(L) = 789 ATLAS BLOCKS = .59 TF(MD) = .19	(n=11) (n=11) (n=10) (n=8)	RE = 4.70 RE(L) = 753 ATLAS BLOCKS = .86 TF(MD) = .66	(n=9) (n=9) (n=8) (n=6)	RE = 4.68 RE(L) = 839 ATLAS BLOCKS = .81 TF(MD) =	(n=9) (n=9) (n=8) (n=0)

Correlation Matrix

Significant Regressions. Examination of the matrix of correlation coefficients for life history features showed that 59 of 153 (39%) interactions among variables (excluding tolerance to fragmentation) were significant at the 5% level (Table 8-7). Some of these significant interactions resulted from the presence in the matrix of different estimates of the same life history feature (e.g., migratory strategy estimate either by simple classification or by the distance between breeding and wintering grounds). Even when this was taken into consideration, however, at least 19 of the interactions were statistically significant.

   Tolerance to Fragmentation. Seven variables were significantly correlated with tolerance to fragmentation. These were, in the order of their probabilities of significance: habitat utilization strategy ($r = -0.648$, $p = 0.001$, $n = 36$); reproductive effort ($r = 0.619$, $p = 0.001$, $n = 36$); bridge utilization ($r = 0.581$, $p = 0.003$, $n = 34$); Atlas data ($r = 0.560$, $p = 0.004$, $n = 36$); regional abundance ($r = 0.566$, $p = 0.006$, $n = 36$); migratory strategy ($r = -0.440$, $p = 0.007$, $n = 36$), and number of broods ($r = 0.373$, $p = 0.0250$, $n = 36$). Not all of these variables measure independent attributes. For example, number of broods and reproductive effort were closely correlated statistically ($r = 0.573$, $p = 0.0001$, $n = 36$), but measure, in part, the same feature. Similarly, bridge utilization, fractional habitat utilization, and regional abundance all reflect, at least in part, species-specific isolation. Actually, such isolation, in the context of a fragmented environment, is likely to be a result as well as a cause of intolerance to fragmentation. For example, general habitat utilization (B'_i) measures the ability of a species to utilize the disturbed habitat between forest islands, and therefore reflects the adaptation of the species to patchy habitat, as well as its ability to adapt to entirely new habitat types that have arisen as a result of human activities. However, the manner in which this variable was measured assures a certain tautology that makes a close correlation with tolerance to fragmentation predictable. Thus, B'_i measures (in part) habitat utilization strategy as well as isolation, as is shown by the extremely high cor-

---

Fig. 8-7. (a) Diagrammatic representation of a 2 X 2 classification of forest species by migratory strategy and habitat choice (residence). Only one family (Fringillidae) has members with both permanent residents and neotropical migrants. Similar, but not quite so restrictive partitioning exists between residence in open habitats and forest interior. Since migratory and residence strategies have many consequences, there is a strong element of taxal idiosyncrasy in analyses of avian life history strategies. (b) Mean reproductive effort (RE), mean reproductive effort multiplied by maximum longevity [RE(L)], mean tendency for regional extinction, as measured by fraction of Breeding Bird Atlas blocks present (ATLAS BLOCKS), and mean tolerance to fragmentation in the Maryland study area [TF(MD)] classified according to residence and migratory strategy. Data were not available for ring-necked pheasant, black-billed cuckoo, red-headed woodpecker, fish crow, willow and least flycatchers, house sparrow, and red-winged blackbird. Bobwhite was not included in calculations of reproductive effort because its annual effort was much higher than any of the passerine species. Prothonotary warbler, blue-winged warbler and cerulean warbler were not included in the Atlas data because their ranges do not extend through the study area.

Table 8-7. Correlation Coefficients for Life History Features and Tolerance to Fragmentation of Maryland Forest Birds[a]

	Migratory strategy classification	Migratory strategy	Habitat utilization (class)	Habitat utilization (ordination)	Nest type	Nest height	Clutch size
Migratory strategy	.81*						
Habitat utilization (class)	.19	.26*					
Habitat utilization (ordination)	-.03	.12	.65*				
Nest type	.46*	.38*	-.06	-.29			
Nest height	-.24	-.01	.07	.22	-.23		
Clutch size	-.36*	-.34*	-.11	.21	-.21	-.14	
Number of broods	-.24	-.28*	-.34*	-.34*	.20	-.34*	-.07
Reproductive effort	-.42*	-.40*	-.34*	-.26	.00	-.31*	.75*
Longevity	.28*	.20	.21	-.04	-.08	.11	-.20
Territory size	.07	.06	-.07	-.04	.23	-.20	-.15
Body weight	-.42*	-.32*	-.20	-.14	-.12	.37*	.15
Bridge utilization (B_i)	-.33*	-.36*	-.76*	-.37*	.09	.01	.17
General habitat utilization (B_i')	-.31*	-.35*	-.78*	-.43*	.13	.04	.11
Atlas blocks	-.48*	-.35*	-.44*	-.17	-.12	.03	.16
Regional abundance	-.20	-.27	-.73*	-.43*	.03	.06	.12
Forest habitat utilization (F_i)	.06	.04	.06	.14	-.01	-.09	.32*
Tolerance to fragmentation (Maryland)	-.44*	-.42*	-.65*	-.33	-.27	-.03	.24

[a] Significance at 5% level is indicated by *.

relation ($r = -0.841, p = 0.0001, n = 39$) between them. Finally, tolerance to fragmentation measured in Maryland was significantly correlated with tolerance measured in Wisconsin (Bond 1957) ($r = 0.602, p = 0.0383$), even though only 12 species present in both species pools could be analyzed.

Species-Specific Isolation. Five variables of Table 8-7 are potential measures of species-specific isolation. Fractional habitat utilization and bridge utilization are closely related variants of a single measure, and accordingly, show the highest correlation in the table ($r = 0.975, p = 0.0001, n = 39$). Atlas data give a somewhat similar measure, but stress the tendency of a species to be absent from entire sections of the study area. Because all such sections probably contained some suitable habitats for the species, Atlas data stressed the tendency for species to become regionally extirpated. Finally, a rough approximation of species-specific isolation could be obtained from simple Miniroute totals (regional abundances from Miniroutes in Howard and Prince George's counties). All four of these measures were significantly correlated with tolerance to fragmentation (Table 8-8) and had roughly equivalent r values.

A fifth measure, forest habitat utilization (F_i) was, however, not correlated with tolerance to fragmentation ($r = -0.085, p = 0.634, n = 34$). Thus, the habitat breadth

Number of broods	Reproductive effort	Longevity	Territory size	Body weight	Bridge utilization (B_i)	General habitat utilization (B'_i)	Atlas blocks	Regional abundance	Forest habitat utilization (F_i)
.57*									
-.40*	-.41*								
.43*	.12	-.36*							
.02	.16	.04	-.27*						
.35*	.44*	-.32*	.27	.31*					
.41*	.46*	-.32*	.33*	.34*	.97*				
.42*	.37*	-.36*	.10	.21	.46*	.43*			
.23	.37*	-.18	.53*	.23	.77*	.77*	.31		
-.21	-.06	.14	-.23	-.08	.31	.25	-.10	-.13	—
.37*	.62*	-.33	.14	.07	.52*	.51*	.56*	.57*	-.24

of species within forest (as opposed to within all available habitat) has no bearing on ability of bird species to adapt to fragmentation.

Correlations among Life History Features. Although consideration of each regression is beyond the scope of this paper, certain patterns are relevant to the species' response to fragmentation. For example, of the variables entered in the correlation study, migratory strategy and reproductive effort were significantly correlated with a number of other variables. Thus, it is reasonable to examine the role of these variables in organizing the life history strategies of avian species.

Choice of Parameters for Stepwise Regression. Although several forms of multivariate analysis are appropriate to life history data, we chose stepwise regression for the purposes of this chapter. In this procedure, we attempted to explain tolerance to fragmentation in terms of various sets of independent variables. Classification of habitat utilization strategy (major habitat association), nest type, nest height, reproductive effort, longevity, territory size, forest habitat utilization, and body weight were included in all approaches. Two other variables (migratory strategy and B'_i) were included in some analyses. Classification of migratory strategy yielded more significant regressions and was chosen as the representative variable for that attribute. On the other hand, B'_i was chosen, although it was approximately equivalent to B_i, because it seemed to be a conceptually superior estimate of species-specific isolation.

Table 8-8. Avifaunal Compositions of Some Extensive, Predominantly Deciduous Forests[a]

Species	Pennsylvania 1972	Virginia 1974	Indiana 1974	Tennessee 1972	Maryland 1975	Georgia 1969
Trunk foragers						
Common flicker	0[b]	0	0	4	12	0
Pileated woodpecker	3	0	5	4	0	17
Red-bellied woodpecker	0	10	25	4	23	0
Hairy woodpecker	0	0	0	0	1	17
Downy woodpecker	10	5	0	4	1	17
White-breasted nuthatch	3	5	15	4	1	0
Black-and-white warbler	0	0	0	19	23	101
Flycatchers						
Great crested flycatcher	13	10	10	0	12	34
Eastern wood pewee	0[c]	5	31	8	17	0
Acadian flycatcher	6[c]	0	31	12	17	151
American redstart	0	0	0	108	0	0
Canopy specialists						
Yellow-billed cuckoo	3	0	10	15	23	0
Blue-gray gnatcatcher	13	0	10	4	12	0
Yellow-throated vireo	0	0	5	19	12	0
Red-eyed vireo	38	19	41	81	105	185
Worm-eating warbler	0	15	20	12	0	84
Cerulean warbler	0	0	20	66	0	0
Pine warbler	0	0	0	0	23	8
Scarlet tanager	3	0	15	42	23	50
Summer tanager	0	0	5	0	0	0
Other	0	0	0	0	0	185
Canopy generalists						
Blue jay	0[d]	15	10	4	12	34
Carolina chickadee	3[d]	15	0	8	17	34
Tufted titmouse	10	15	25	8	23	84

Species						
Shrub species						
Hooded warbler	134	17	73	30	0	0
White-eyed vireo	0	0	12	0	0	0
Kentucky warbler	0	12	42	5	0	0
Carolina wren	34	17	15	10	15	0
Gray catbird	0	1	0	0	5	0
Cardinal	84	17	27	15	15	0
Other	34	0	0	0	0	0
Ground foragers						
Wood thrush	17	58	39	25	15	3
Ovenbird	34	279	12	46	19	6
Louisiana waterthrush	34	6	0	0	5	3
Rufous-sided towhee	0	23	12	10	0	0
Edge species						
Indigo bunting	0	0	23	1	0	6
Other	0	12	58	0	5	0
Field-edge species						
Mourning dove	0	12	0	10	0	2
Brown-headed cowbird	0	12	12	10	5	0
Other	0	0	0	0	5	0
Total density[e]	1285	830	734	473	199	144

[a] From the censuses of Simons and Simons (1972), Woodward and Woodward (1972), Crooke and Webster (1974), Yahner (1972), Whitcomb et al. (1975), and Mellinger (1969), respectively.

[b] Territorial males/km^2.

[c] Geographical replacement: least flycatcher.

[d] Geographical replacement: black-capped chickadee.

[e] Total density (territorial males/km^2) of all species including raptors, nocturnal, and crepuscular species not listed individually in the table.

Stepwise Regression

Results from stepwise regressions varied somewhat depending on the variables chosen for analysis. When migratory strategy was absent from the variables, parameters relating to major habitat association tended to be selected. For example, in an analysis performed by a program that terminated when variables failed to meet the 0.50 significance level for entry into the model, a three variable model was obtained, with reproductive effort ($F = 6.66, p > F = 0.0159$), nest type ($F = 5.28, p > F = 0.0299$), and general habitat utilization (B_i') ($F = 3.04, p > F = 0.0929$) being chosen.

On the other hand, when the square of migratory strategy was entered as a variable, a two-variable model was chosen with B_i' ($F = 7.15, p > F = 0.0159$) and nest height ($F = 3.79, p > F = 0.0625$) being chosen. Results of six other analyses are given in Appendix F. We recognize the inherent difficulties in use of stepwise regression with "independent" variables that may be correlated. However, the effects we observed were extremely obvious and were confirmed at high levels of statistical significance by every method by which they were examined. For example, the stepwise regression analyses confirmed the χ^2 analysis in the importance of migratory strategy and major habitat association. In addition, nest height and location were indicated as additional parameters of importance, and reproductive effort was identified as a parameter closely correlated with migratory strategy that might alternatively explain tolerance to fragmentation.

Discussion

The Presettlement Forest

A full interpretation of the changes in bird communities that have resulted from deforestation and fragmentation would require an accurate knowledge of the presettlement forest community. Unfortunately, early records concerning the vegetation and avifauna of the primordial eastern deciduous forest are virtually nonexistent. Remaining fragments of virgin forest (if any) in our study area are much too small to support a full complement of forest bird species (i.e., regional species pool). Moreover, some bird species have become extinct in historical times, and others may have undergone genetic changes as a result of dwindling population size and increasing isolation from potential sources of genetic variation. Finally, the forest itself has been profoundly altered by the near extinction of dominant trees (e.g., American chestnut), and by introductions of ecologically important plant species (e.g., Japanese honeysuckle, kudzu). Tracts of upland forest that presently exist on the central Coastal Plain and adjacent Piedmont section of Maryland are composed almost entirely of second or third growth, and the associated bird communities have been severely affected both by deforestation (i.e., reduction in total area of forest) and fragmentation (insularization of remaining forest). Existing tracts, even if mature forest, are invariably so small that their breeding bird populations are of insufficient size to be self-sustaining. The persistence of individual bird species must therefore depend on their regional status and the biogeographic setting of the individual islands. For example, if a patch of mature forest (whether an island, a fragment or part of a large tract) is surrounded by younger or more disturbed forest, the older forest may serve as a "sink" (B. L. Whitcomb et al. 1977) in which the density of breeding birds may be abnormally high. The artificially high bird populations sometimes observed in such patches may not be an accurate indication of the primordial avifaunal composition or density. Despite this potential drawback, bird censuses from patches of old growth forest, taken together with the composition of more extensive, but less mature forest, provide the best available contemporary estimates of primordial bird communities. Such communities certainly were composed predominantly of neotropical migrants (Tables 8-8 and 8-9).

Geographic Consistency of the Island Assemblage

Large forest tracts tend to contain most or all of the regional species pool of forest-interior species (Table 8-8). As a result, neotropical migrant forest interior bird species are found primarily in large forests (Table 8-9). Such conclusions, of course, are subject to demonstrated altitudinal (Able and Noon 1976) and latitudinal (Rabenold 1978, 1979) gradients. Declines in species richness after fragmentation recall the "relaxation" of supersaturated land-bridge oceanic islands to lower equilibria (Diamond 1972, Terborgh 1973, Wilcox 1978). Table 8-10 documents the avifaunal composition of a forest fragment in Maryland which has not yet reached the typical island equilibrium, and of other islands of eastern forest in Illinois, Delaware, Michigan, and Ohio that are more or less equilibrial. Although the complex chain of causality that leads to the decline of forest island bird communities is incompletely understood, the pattern of decline is surprisingly consistent over wide geographic areas (Lynch and Whitcomb 1978). Particular consistency is seen in the equilibrial composition of island avifauna.

Table 8-9. Large Forest Requirement of Migratory Forest Bird Species: Evidence from Miniroute Data

Species	Extensive forest[a] (n=84 points)	All other habitats[b] (n=157 points)
Blue-gray gnatcatcher[c]	8	0***
Yellow-throated vireo	8	5*
Black-and-white warbler	14	0***
Worm-eating warbler	2	0
Prothonotary warbler	4	0**
Northern parula warbler	25	2***
Pine warbler[c]	15	0***
Ovenbird	30	3***
Louisiana waterthrush	5	0**
Kentucky warbler	14	2***
Hooded warbler	12	1***
American redstart	13	0***

[a] Number of points at which the species was recorded. (61 points were entirely within forest; 23 points had extensive forest on one side of the road.)

[b] Asterisks denote significant differences in frequencies at 0.05, 0.01, and 0.001 probability levels as determined by χ^2 analysis.

[c] Short-distance migrants; other species are neotropical migrants.

Table 8-10. Avifaunal Composition of Eastern Deciduous Forest Islands[a]

	Mature forest (Maryland) 16.2 ha		Trelease Woods (Illinois) 22.3 ha		Urban woodlots (Delaware)[b] 1966-1969	Oakland Co. (Michigan) 14.2 ha 1976	Sandusky Co. (Ohio) 12.9 ha 1974
	1947	1976	1934	1976			
Trunk foragers							
Common flicker	0	7	4	27	60	44	12
Red-bellied woodpecker	62	62	0	22	40	0	0
Red-headed woodpecker	0	0	18	13	2	14	6
Hairy woodpecker	7	7	4	9	6	5	0
Downy woodpecker	34	41	22	13	18	23	0
White-breasted nuthatch	14	17	4	4	9	5	0
Black-and-white warbler	27	0	0	0	0	0	0
Flycatchers							
Great crested flycatcher	21	0	22	22	17	9	62
Acadian flycatcher	82	48	0	0	4	0	62
Eastern wood pewee	48	41	13	31	31	5	87
American redstart	0	0	0	0	3	0	0
Canopy specialists							
Yellow-billed cuckoo	7	1	4	18	2	0	12
Blue-gray gnatcatcher	0	0	0	0	0	0	1
Yellow-throated vireo	27	7	0	0	0	0	12
Red-eyed vireo	247	209	36	27	63	14	62
Worm-eating warbler	7	0	0	0	0	0	0
Scarlet tanager	64	27	0	0	9	5	0
Canopy generalists							
Blue jay	7	21	0	58	62	14	0
Carolina chickadee	7	14	0	0	28	0	0
Tufted titmouse	24	31	22	0	38	5	25
Shrub species							
Carolina wren	21	31	4	9	15	0	0
House wren	0	0	13	31	7	9	12

Gray catbird	21	21	4	13	88	9	25
White-eyed vireo	21	0	0	0	2	0	0
Kentucky warbler	41	21	0	0	18	0	0
Hooded warbler	21	7	0	0	0	0	0
Cardinal	48	55	13	31	80	30	62
Ground foragers							
Wood thrush	100	144	13	27	176	9	12
Ovenbird	130	48	0	0	24	9	0
Rufous-sided towhee	21	21	4	9	90	5	37
Edge specialists							
Common yellowthroat	27	27	9	13	35	5	0
Indigo bunting	27	48	99	72	30	5	87
Field-edge species							
Mourning dove	0	7	0	27	6	0	37
Common crow	1	1	9	1	10	14	0
American robin	21	1	0	45	105	21	37
Starling	0	21	36	314	60	65	0
Common grackle	0	0	0	1	51	1	0
Brown-headed cowbird	7	21	0	1	18	0	25

[a] Data are numbers of territorial males/km^2. From census data of Stewart and Robbins (1947), R. F. Whitcomb et al. (1977), Kendeigh (1948), Kendeigh and Edgington (1977), Linehan et al. (1967), Jones (1969), Challis (1977), and Tolle (1974).

[b] Mean density for eight urban woodlots.

Illinois. Trelease Woods in Illinois (Kendeigh 1981) is an example of a forest island that has been highly isolated for such a long time that its avifauna has reached an approximate equilibrium. Superficial comparison of recent censuses (Kendeigh and Edgington 1977) with those taken on this 22.3-ha tract 40 or more years ago (Kendeigh 1944, 1948) might appear to suggest that few changes have occurred during that interval, except for an increase in the abundance of a few common species (e.g., mourning dove, blue jay, American robin, and starling). However, detailed examination of yearly censuses reveals a very high rate of extinction and colonization of bird species, especially during the period in the early 1950s when the forest was modified by the extirpation of a codominant canopy species (*Ulmus americana*) by Dutch Elm disease. The significance of such high turnover was discussed by Lynch and Whitcomb (1978), who also pointed out (1) that only nine bird species (of a pool consisting of 62 species) have bred in Trelease Woods each year since it was first censused in the early 1930s, and (2) that almost all species intolerant of fragmentation have disappeared. Clearly, for all of the nonpersistent breeding species, and probably even for the nine persistent breeders as well, Trelease Woods is insufficient as a reserve. Removal of external sources of colonists would lead inexorably to a species richness even lower than the present depauperate level.

Delaware. In a study of birds that inhabit urban woodlots in Delaware, Linehan et al. (1967) emphasized the value of small forest islands as avifaunal reserves. Such conclusions are less pessimistic than our own. Indeed, the censuses of Delaware forest islands did show a high density of breeding birds but the species involved were the same small subset that inhabits forest islands in Maryland and elsewhere in the eastern United States. In their studies, as in our own, sensitive forest interior species (particularly neotropical migrants) were rare or absent.

Wisconsin. In a study of the ecological factors determining bird distributions in deciduous forests of southern Wisconsin, Bond (1957) considered forest area to be one of many potentially important variables. Fortuitously, two of the size classes that Bond used to categorize forest islands correspond closely to those in our study, and the relative tolerances to fragmentation he calculated (using a method somewhat different from that employed here) are basically in accord with our own (Table 8-11). The computed tolerances to fragmentation from the two geographic areas are significantly correlated ($r = 0.602, p = 0.0383$).

Another important observation from the Wisconsin study was the importance to some birds of the position of the forest on the moisture gradient. Bond observed that mesic forest patches often contain more than the expected number of species intolerant to fragmentation: "It is as if the birds were responding simultaneously to 'largeness' and to 'mesic conditions' with largeness compensating in some manner for the lack of mesic conditions in the more xeric stands." Data for North Hemmer Woods in southern Indiana (Webster and Adams 1971) provide additional evidence that even isolated patches of mesic forest may support a high bird density, including some neotropical migrants. These findings are reinforced by our own observations (B. L. Whitcomb et al. 1977) that particularly rich forest islands may serve as "sinks" for local bird populations.

Table 8-11. Comparison of the Tolerance to Fragmentation of Bird Species in Deciduous Forest in Central Maryland and Southern Wisconsin

Species	Central Maryland	Southern Wisconsin (from Bond 1957)
Permanent residents		
Red-bellied woodpecker	1.00[a]	.53[b]
Hairy woodpecker	.50[c]	1.42
Downy woodpecker	1.25	2.46
Blue jay	1.00	1.99
Common crow	1.00	—
Carolina chickadee	1.22	1.25[d]
Tufted titmouse	.87	—
White-breasted nuthatch	1.00[c]	1.00
Carolina wren	1.78	—
Starling	3.33	—
Cardinal	1.18	—
Short-distance migrants		
Mourning dove	.89	—
Common flicker	1.83	—
Gray catbird	2.33	—
American robin	1.67	—
Blue-gray gnatcatcher	.00	.34
Common grackle	2.00	—
Brown-headed cowbird	—	1.02
Neotropical migrants		
Yellow-billed cuckoo	.14	1.43
Great crested flycatcher	1.67	1.86
Acadian flycatcher	.44	.11
Eastern wood pewee	1.20	1.14
Wood thrush	.61	.14
Yellow-throated vireo	.00[c]	.67
Red-eyed vireo	.35	.95
Black-and-white warbler	.00	—
Worm-eating warbler	.00	—
Cerulean warbler	—	.17
Ovenbird	.10	.69
Louisiana waterthrush	.00[c]	—
Kentucky warbler	.33	—
Hooded warbler	.25	—
American redstart	—	.31
Scarlet tanager	.43	.75

[a] Number of territories/point in forest islands 6-14 ha divided by number in forests 70 ha or larger.
[b] Relative frequency of birds in forest islands 6-14 ha divided by number in forests 32 ha or larger.
[c] Small sample size.
[d] Black-capped chickadee.

New Jersey. One of the most substantial attempts to quantify the relationship between forest area and bird community composition has been the research of Forman and colleagues in the Piedmont of New Jersey (Forman et al. 1976, Galli et al. 1976). These workers concluded that only 18 of a pool of 44 bird species were sensitive to forest area. Of the remaining 26 species, 17 were classed as area-independent, and nine were too infrequently encountered to be evaluated. Because these results appear to contrast sharply with ours, we will consider the apparent discrepancies in some detail.

One problem in interpreting the results of the New Jersey study is that multiple transect lines were used to assess the occurrence and abundance of bird species. Because more transect lines were employed in large tracts than in smaller ones, and because the data were not corrected for sampling intensity, numbers of almost all species appeared to be higher on large islands. In addition, records from as many as three survey visits to a single island were given as simple totals with the result that an individual bird seen on three visits was accorded the same status as three different individuals seen once each. These doubly confounded data do not provide a reliable basis for comparison of densities on large vs. small islands, nor is it possible to determine the fraction of islands of a given size on which a given bird species occurred.

A second problem is that Forman et al. (1976) and Galli et al. (1976) assessed area dependency on the basis of occurrences "over the entire range of forest sizes greater than 0.01 ha." Most of the "size-independent" species listed in the study of New Jersey woodlots are in fact edge and field edge species (e.g., house wren, starling, song sparrow, and red-winged blackbird). Such species occur, both as casual visitors and breeders, in many habitats (Appendices C and E).

For the most part, species classified as "size-dependent" in the New Jersey study are those that can use forest edge as well as forest interior (11 species). Only five species are restricted to forest interior habitat, an indication that a substantial number of characteristic forest interior species were not even encountered in the course of the New Jersey study. Conspicuous absences from the New Jersey species pool include *Empidonax* flycatchers, blue-gray gnatcatcher, yellow-throated vireo, worm-eating warbler, and hooded warbler. Some of these may be present in moist bottomland forest not studied by Forman and colleagues, but the total absence of these birds from the forest fragments surveyed in the New Jersey Piedmont suggests that the area has already experienced the disastrous cascade of local extirpations that is currently underway in our Maryland study area.

The minimum forest area reported as suitable for most species in the New Jersey study is only slightly larger than the size of a single territory (Table 8-12). In a few instances (e.g., tufted titmouse), the minimum area is even smaller than the usual territory size. Because the observers did not separate breeding individuals from transients, such findings must be viewed with caution. Situations in which species are unable to persist on islands much larger than their minimum territorial size are more difficult to interpret. In the New Jersey study, ovenbird and black-and-white warbler are especially good examples of this type of area-sensitivity (Table 8-12). Interestingly, both species were found to be highly intolerant of fragmentation in our study area (Table 8-5).

When the foregoing considerations are taken into account, most apparent discrepancies between the Maryland and New Jersey studies disappear. The conclusion of Galli et al. (1976) and Forman et al. (1976) that area sensitivity is a function of territorial size results from the fact that these workers studied only small to medium-sized forest islands. The paucity of forest interior neotropical migrants encountered in the New Jersey study (three species, in comparison with 11 in the Maryland study), and of neotropical migrants generally (12 species in New Jersey vs. 33 in Maryland) is most likely to reflect the historical effects of deforestation and fragmentation. Thus, it appears that the New Jersey workers inadvertently focused their attention on a

Table 8-12. Comparison of Minimal Island Size Required for Presence of Bird Species in New Jersey Piedmont with Minimum Territorial Requirements for the Species[a]

Species	Defended territory Maryland[b] (ha)	Computed territory Maryland[c] (ha)	Minimum size New Jersey (ha)
Yellow-billed cuckoo	2.19	2.38	4.0
Red-bellied woodpecker	.89	1.40	3.0
Downy woodpecker	1.21	1.93	1.2
Eastern wood pewee	1.13	1.69	2.0
Black-capped chickadee[d]	1.46	1.04	2.0
Tufted titmouse	1.58	.72	2.0[e]
Wood thrush	.40	.32	.8
Red-eyed vireo	.32	.29	.8
Black-and-white warbler	1.34	1.50	7.5
Ovenbird	.32	.35	4.0
Scarlet tanager	1.82	1.50	3.0
Cardinal	.81	.42	.8

[a] Data from Galli et al. (1976).
[b] From Maryland census maps.
[c] Computed from density values in Maryland forests.
[d] Data for Carolina chickadee in Maryland.
[e] Minimum size for tufted titmouse may have been 2.0 ha, since a sharp increase in number of islands occupied was observed at that size. Occupancy of smaller islands may have been by nonterritorial birds.

reduced species pool that was predominantly composed of ecologically generalized species that have survived, and possibly even profited from, insularization of forested habitats.

Our critique of the New Jersey study is intended to be constructive. Indeed, interpreted properly, the data of Forman and colleagues contribute importantly to our knowledge of forest island bird communities. Our main concern is that the results of that research not be misconstrued as evidence that meaningful preservation of forest birds can be accomplished by a system of small forest fragments such as that existing on the New Jersey Piedmont. On the contrary, forests much larger (by several orders of magnitude) than the 24 ha maximum considered in the New Jersey study are certainly required for the preservation of many forest-inhabiting bird species.

Our concern that data from minuscule islands might be used in conservation decisions is underscored by a recent study of small islands in a Minnesota lake (Rusterholz and Howe 1979), in which island sizes essentially adequate for single territories of birds were identified. On this basis, the authors concluded that the . . . "small island effect is important for conservation . . ." and urged that the "critical range . . . be identified" so that "land use planners can strive to maintain optimally sized habitat patches." Morse (1977) concluded from a similar study of small islands in Maine that dispersion from mainland sources was adequate to insure saturation of the islands. Such results, and those from our studies, make it clear that such small islands are nearly useless for avifaunal preservation, and that land use planners must turn instead to the extensive surrounding forests such as those from which bird inhabitants of the Burntside and Maine islands are no doubt recruited.

The conclusions of Galli et al. (1976), Forman et al. (1976) and Rusterholz and Howe (1979) are in some ways analogous to those reached by Simberloff and Abele

(1976b). These authors interpreted the pattern of arthropod diversity observed on a series of fragmented mangrove islets as evidence that numerous small faunal preserves might in certain instances be preferable to one large preserve. In both studies, only a fraction of the regional species pool actually was encountered. More importantly, in both instances that fraction was clearly a nonrandom subset of the regional species pool. The arthropod fauna of tiny mangrove islands resembles the avifauna of small New Jersey woodlots in the dominance of generalized "weedy" species that are not highly sensitive to the size or isolation of habitat patches. Unfortunately, this is the fraction of the regional species pool that is least in need of protection, and conservation policy should not be based on the ecology of such species (Whitcomb et al. 1976).

Causes for Decline: Habitat Association

Although our results are concerned primarily with a single biome (Eastern Deciduous Forest), the regional species pool of forest birds may have been derived from three or possibly four biomes (prairie, savanna, and deciduous and southeastern coniferous forests). Fonaroff (1974) stated that garden birds on Trinidad came from the savanna rather than the forest assemblage. The common inhabitants of suburban and open rural habitats (edge or field-edge species) in our area also probably have this origin. Such observations have significance in considering the limited sample of bird families represented in the forest avifauna. Most taxa are restricted in their ability to inhabit different biomes. This is demonstrated in Fig. 8-7 in which bird families we studied are portrayed in a binary classification according to their migration and habitat strategies.

Causes for Decline: Relevant Habitat Dimensions

Ecological variables that may influence the area requirements of species on oceanic islands have been discussed by Diamond (1976b). Several general descriptors are of special interest in a mainland biogeographic context:

Quantity of Habitat (= Island Size). A self-sustaining colony of birds requires enough habitat to support an adequate breeding population. Although "adequacy" varies among species and circumstances, we envision a population as being adequately large when there is a relatively low probability per unit time of becoming extinct as a result of stochastic fluctuations. For species that exploit unpredictable resources (e.g., warblers that capitalize on periodic local outbreaks of spruce budworm) regional persistence may require the preservation of either several large discontinuous patches of habitat, or an enormous contiguous area containing numerous habitat patches.

Isolation of Habitat. Isolation may function on the mainland in many of the same ways it functions on oceanic islands. If the isolation of a forest tract is sufficient to preclude regular recolonization, maintenance of a bird population on the tract will depend entirely on internal productivity. This point has been stressed by Terborgh (1974, 1975) in his arguments for very large faunal preserves. Little information exists concerning the minimal area above which postfledging dispersal of young tends to occur within, rather than among patches, but it seems certain that large tracts of the order of hundreds or even thousands of hectares are required for maintenance of stable population levels of even small bird species in the absence of regular recolonization from outside. An additional effect of fragmentation of large forests into islands is the openness of the island communities to competition and predation from the adjacent habitats (MacArthur and Wilson 1967).

Quality of Habitat. Diamond (1975a) envisioned habitat quality on oceanic islands in terms of isoclines of productivity. Habitat quality in deciduous forest must vary over a wide range, although this variability has rarely been studied. Potential nest areas exhibit a gradational range in the degree of protection they afford from predators, nest parasites, weather, and physical disturbance. Other gradients must also exist in the amount and quality of food resources among sites. Because the regional carrying capacity for a species is determined by the mean carrying capacity per unit of habitat (i.e., habitat quality) as well as by the total number of habitat units (i.e., habitat quantity), factors related to habitat quality may either offset or augment the purely geometric effects of habitat area and isolation. An interesting, but poorly studied aspect of habitat quality is the possible diminution of the resource base in island habitats. If the species richness of most taxa is a function of area, the net result from the point of view of a predator should be a less diverse prey assemblage. In fact, a decreased diversity of arthropods on islands has been noted (Mühlenberg et al. 1977a), but may differ among arthropod taxa (Mühlenberg et al. 1977b). Decreased densities of insects have also been observed on small oceanic islands (Jaenike 1978). If arthropod crops of deciduous forest islands were significantly lower than those of mainland forests, the avifauna could obviously be affected. Faeth and Kane (1978) found that species richness of Diptera and Coleoptera in urban parks was a function of park areas. Decreases in insect diversity and/or abundance may explain, in part, why urban parks fail, disastrously, to preserve forest bird species (Lynch and Whitcomb 1978).

Heterogeneity of Habitat. Even an apparently uniform expanse of forest is, at some level of discrimination, a mosaic of habitat patches. Although many bird species appear to be associated with a particular patch type (e.g., floodplain forest), many species utilize two or more major patch types. This may involve contemporaneous utilization of different habitats (e.g., use of open feeding areas by species that require forest edge for nesting) or seasonal habitat shifts such as those documented by Root (1967) in his classic study of foraging patterns of the blue-gray gnatcatcher. This species nested in bottomland forest, but family groups moved into drier upland habitats after the fledging period. Many neotropical migrants utilize very different types of foraging habitat in tropical or subtropical latitudes (Keast and Morton 1980). Habitat shifts that are results of migratory strategy are relatively well known, but the more subtle short-distance movements that occur after fledging but before migration are poorly documented for most species. This omission is critical, because postfledging dispersal may be instrumental in determining both survivorship and ultimate spatial distribution of young birds.

Interrelatedness of Forest Patches. Censuses and surveys within the extensive Beltsville forest (Whitcomb et al. 1975, MacClintock et al. 1977, Appendix B) document the complex nature of the relationship between spatiotemporal habitat heterogeneity and bird distribution and diversity. In this relatively xeric oak-hickory upland forest, there are comparatively few breeding bird species. Except for ovenbird and black-and-white warbler, most species that do breed in the Beltsville oak-hickory forest are less abundant than in nearby mesic or bottomland forest. In dry forests of this type, bird diversity and abundance is greatest in the vicinity of local patches of mesic vegetation which develop around small streams and topographic depressions. These patches not only harbor larger numbers of the usual species associated with xeric oak-hickory forest, but also provide breeding habitat for certain species (e.g., Acadian flycatcher, Louisiana waterthrush, Kentucky warbler, and hooded warbler) that do not occur elsewhere in dry forest (Whitcomb et al. 1975). The same mesic patches are utilized by blue-gray gnatcatcher during the breeding season, but family groups of this species forage extensively in the adjacent xeric oak-hickory forest after the fledging period.

Secondary plant succession is also an important source of habitat heterogeneity within forested areas. The results of our point surveys within tracts of different successional maturity (Whitcomb and Robbins, unpublished) generally agree with the findings of other workers who have studied the responses of bird communities to secondary plant succession (e.g., Johnston and Odum 1956, Oelke 1966). Even such area-sensitive species as black-and-white warbler and ovenbird are capable of occupying very early successional forest (e.g., pine-scrub stage) if a sufficiently large area of such habitat is available. As deciduous trees gradually supplant conifers in the regenerating forest, additional forest interior bird species colonize the habitat, but such colonization generally occurs well before the forest would be considered "mature" on vegetational or physiognomic grounds. Thus, it would appear that for many (perhaps most) conservation purposes, extensive tracts of early-middle growth forest may be preferable to small areas of old growth forest.

Not only do forest interior birds appear at a surprisingly early stage of secondary succession, but "pioneering" bird species may persist long after their preferred habitat has ceased to dominate the successional sequence. For example, pine warbler, a species that nests only in pines, colonizes regenerating forest in the late pine-scrub stage, and reaches its maximum abundance in pine-dominated forests. As succession continues, reproduction by pine dwindles or ceases, and the oldest pines begin to senesce and die. As this process occurs, pine warbler territories are increasingly organized around remaining groups of pines (Whitcomb et al. 1975), but the species may persist in the forest as long as a few pines survive. At this relatively advanced successional stage, pine warbler presumably shifts much of its foraging activity from pines to deciduous trees. Similarly, prairie warbler colonizes successional forests in the early pine-scrub stage, and may persist in pine stands of moderate maturity that would be unsuitable for establishment of a new colony. Such relationships underscore the importance of historical factors in the composition of local bird populations. Willmot (1980) recently reported that numbers of woody species in English hedgerows increased with hedgerow age. This has a counterpart in the avifauna of such hedges (N. E. Moore, personal communication).

Within bottomland forest, considerable heterogeneity arises from differences in the size of streams. The community of birds associated with mesic forest becomes more diverse as one proceeds from narrow headwater streams to broad lowland floodplains. In our study area, prothonotary warbler is restricted to broad floodplains, whereas

American redstart, although abundant in broad floodplains, utilizes narrower riparian strips as well. Northern parula warbler utilizes these habitats, but also occurs along even smaller water courses.

We have discussed these somewhat obvious sources of forest heterogeneity in order to emphasize the complex nature of the continuum between "xeric" and "mesic" vegetation and between "young" and "mature" forest; such complexity, of course, and the amenities it may provide to bird species, cannot exist in small fragments of the entire forest system.

Habitat Quality and Social Organization. Some migratory species (e.g., ovenbird, hooded warbler) tend to occur in loose "colonies" that appear to be associated with particular types of habitat. In such colonies, territories near the colony center approach minimal size. For example, ovenbird territories may be as small as 0.25 ha at the center of a colony, or as large as 1 ha at the periphery (Hann 1937, Whitcomb et al. 1975). Territory size in this species has been shown to be inversely proportional to density of food organisms (Stenger 1958, Stenger and Falls 1959, Zach and Falls 1975). Other species have been less intensively studied, but similar relationships presumably hold for other small insectivorous species. The carrying capacity of a given area is increased by the ability of birds to "track" resource availability by adjusting territory size. This in turn should lower the probability of stochastic extirpation of local populations.

A second potentially important result of the loose colonial habits of species such as ovenbird is that existing colonies may actively attract new colonists, and that previously unoccupied areas may be avoided. Such a behavioral pattern would lower the rate of colonization of unoccupied habitat. Given a regionally declining population of a colonial species, we might expect to find partitioning of remaining individuals among a few discrete, isolated colonies. Territorial densities within these remnant colonies might appear to be normal, despite the existence of vast expanses of unoccupied but suitable areas between colonies. Mayfield (1973) has, in fact, noted such partitioning in populations of Kirtland's warbler.

Causes for Decline: Migration Strategy

The importance of migratory strategy as an organizing force in the eastern forest biome was recognized by MacArthur (1959). In general, neotropical migration appears to organize many life history features (Table 8-7). Neotropical migrants in the eastern forest biome are, with few exceptions, open nesters; they tend to nest low or on the ground ($r = -0.239$, $p = 0.059$, $n = 63$) (also see Preston and Norris 1947); they have small clutch sizes; and they tend to be single-brooded and therefore have low annual reproductive rates. However, the relative longevity of neotropical migrants, particularly when body weight is taken into account, is greater than that of species with other migratory strategies, so that lifetime productivity of similarly sized birds of the two migratory groups may not differ greatly (Fig. 8-7). Such observations recall the theory of r- and K-selection put forth by MacArthur and Wilson (1967). Although recent critiques (Stearns 1977) have modified or extended the concepts set forth by

MacArthur and Wilson, sacrifice of reproductive effort is often noted in life history observations, and has generated many theories to explain its origin. Finally, neotropical migrants are small, and judging by their inability to utilize patchy or disturbed habitat, may be considered to be more "behaviorally rigid" than short distance migrants, or especially, permanent residents. Thus, we interpret neotropical migration as a powerful organizing force, perhaps the most important of all the life history features.

Other Impacts Associated with Insularization of Forest

In addition to the constraints imposed by habitat association and migration strategy, bird species must contend with additional difficulties if they inhabit forest islands. Many of these difficulties are similar to those discussed by Diamond (1975a, 1976b) for oceanic islands. Others are specific to mainland habitat islands.

Psychological Avoidance of Small Tracts. Forest fragments that are smaller than the minimum territory size for a species cannot, of course, be inhabited by that species (Moore and Hooper 1975, Forman et al. 1976). However, as noted earlier, forest islands many times larger than the territorial requirements of small birds also may be avoided by certain species. We suggest that the "gestalt" (*sensu* James 1971) of tracts below a certain size may cause behavioral rejection of the tracts as suitable breeding sites. For some species this pattern may reflect a psychological need for a substantial forested buffer zone around the smaller area that is actually utilized. Examples of such species include ovenbird, hooded warbler, and Kentucky warbler, all of which tend to locate their territories toward the center of small forest tracts, and tend to be scarce or absent in the vicinity of forest margins (Fig. 8-5a). Interestingly, the same pattern appears to hold for bird species that inhabit grassland islands surrounded by forest. Such birds tend to choose grassland patches with the lowest available perimeter/area ratio (P. A. Allaire, personal communication).

Brood Parasitism and Nest Predation. Associated with fragmentation and other disturbance of forested habitat is an influx of large numbers of birds usually associated with more open habitats. Some of the highly detrimental effects that these intruders may have on the forest interior bird community include brood parasitism by brown-headed cowbird (Mayfield 1977), or nest predation by blue jay (Fretwell 1972), common crow (Bent 1946), and common grackle (Bent 1958). Jays may be especially prevalent in city parks (Fretwell 1972), whereas crows are more important in agricultural areas.

Diffuse Competition with Permanent Residents. The loss of neotropical migrant species from forest islands is often accompanied by increases in the populations of permanent resident species (Lynch and Whitcomb 1978). Such observations recall the extensive theories on competition (Cody 1974) that have been built to explain avian community structure. It is possible that such increases in permanent residents are in part attributable to competitive release, as suggested by Yeaton and Cody (1974) in their study of birds on the California Channel Islands. Once numbers of permanent residents have increased to unusually high levels, for whatever reasons, such species may have the potential to impede recolonization and population growth of neotropical migrant species with similar feeding or nesting requirements.

Mortality during the Nonbreeding Season. Fretwell (1972) has argued that many bird populations in temperate zones may be regulated by mortality during the winter, because resources during the breeding season are unlikely to be in short supply. The limitation, of course, is apt to differ among species (Morse 1980). Whether or not this hypothesis is valid, high mortality of neotropical migrants on wintering grounds or migratory routes would be expected to affect populations of large forest areas as well as smaller ones, so winter mortality per se is not the explanation for selective deterioration of bird communities on small forest islands. The latter point is crucial, for if the observed trend toward depletion of numbers and disappearance of species were characteristic of large continuous areas of forest habitat as well as small islands, we must look for explanations that possibly are unrelated to island biogeography. For example, it might appear that accelerating destruction of forest in the neotropical area might be the key to the decline of neotropical migrants. Although these factors may play a role, they cannot be the entire explanation for the decline of neotropical migrants, for many forest interior warblers and vireos shift to successional and edge habitats on their wintering grounds. Moreover, Breeding Bird Survey data compiled over the last decade show statistically significant increases in the eastern populations of some of the warblers intolerant of fragmentation and no significant changes in population indices of most remaining sensitive species (C. S. Robbins, unpublished data). However, species that require forest interior habitat in all seasons (e.g., black-and-white warbler, worm-eating warbler, ovenbird, Louisiana waterthrush, and hooded warbler) are certain to be affected by the continued and accelerating destruction of tropical forest (Keast and Morton 1980). Observations (Johnston 1974) that DDT residues in migratory passerines are decreasing are more encouraging.

We do recognize that models based on losses on the wintering grounds could be constructed to explain losses of neotropical migrant species from forest islands. Such models would have to assume that declining populations selectively abandoned forest islands as territories became available in presumably more suitable habitat in larger tracts. According to such models, densities in larger tracts might appear to remain stable, while marginal populations decreased and individual birds abandoned the less suitable island habitats.

Colonization Biology. There are taxonomic differences in the sensitivity of bird species to fragmentation. Tyrannids such as great crested flycatcher and eastern wood pewee inhabit many fragments in the Maryland Piedmont, and were ubiquitously distributed at Miniroute stations (Appendix C). They are also known to breed in isolated urban habitats, such as cemeteries (Lussenhop 1977). Other species, including many parulids, may occur in high densities in large forest tracts, but are almost never observed in small forest fragments. On the basis of these observations, we predicted that the dispersion biology of the Tyrannidae should differ from that of other taxa. Analysis of data supplied by the Bird Banding Laboratory showed that North American tyrant flycatchers, banded as nestlings and recovered in subsequent years during the breeding season, dispersed more widely than young of most other passerine families (Fig. 8-8).

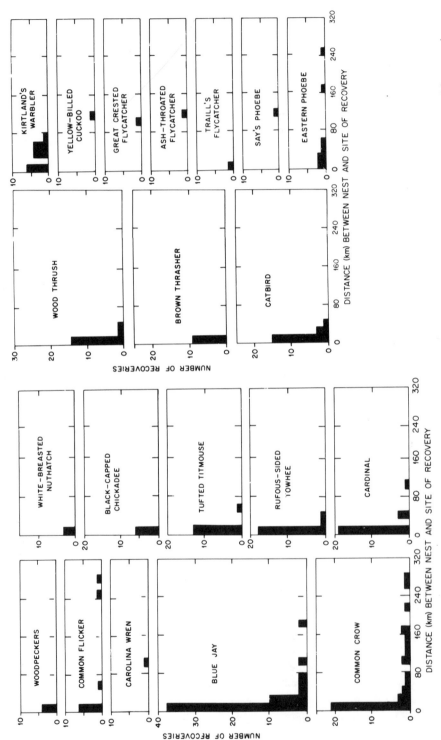

Fig. 8-8. Banding recoveries of birds banded as nestlings and recovered in subsequent years during the breeding season. Most species are recovered within 48 km of the site of fledging, but limited data suggest that Tyrannidae may have a more dispersive strategy than other families.

Thus, it appears that for species of this group, the disadvantages of a highly migratory life history strategy are compensated for by unusual colonizing abilities.

Other aspects of tyrannid biology deserve some comment. The family is predominantly associated with the neotropical savanna and forest breaks (Fitzpatrick 1980); thus the adaptation of new world flycatchers to fragmented environment may reflect their historical origin. In terms of the r-K continuum, one might expect that this strategy would lean toward r-selection. Indeed, flycatchers are exceptional in having high lifetime reproductive efforts (Table 8-4). A final observation concerns the tendency of some tyrannids such as eastern wood pewee to remain on territory throughout the late summer months. The possibility that progeny locate territory for subsequent years before migration should be examined. Löhrl (1959) hand-reared Old World flycatchers (Muscicapidae), and discovered that only after a certain period had elapsed, did progeny develop site fidelity that permitted their return after migration. Summer dispersal of some bird species, especially European parids, has been relatively well studied. For example, postfledging mortality of juvenile great tits was found to be very high in an area where forest islands were relatively small and isolated (Dhont 1979). Study of such aspects of the dispersion biology of migratory passerine New World birds would greatly enhance our ability to interpret the causal sequence of events that underlies avifaunal change in the eastern deciduous forest.

Causes for Decline: Path Diagram

The results of our χ^2 correlation and stepwise regression analysis emphasize the importance of migratory strategy, habitat utilization, nest site and type, and reproductive strategy (Fig. 8-9). We suggest that the genetic composition of existing bird families in eastern North America determines to a major degree (a) their migration strategy and (b) their adaptation for forest or savanna-type habitat. The large number of significant correlations between migration strategy and other life history attributes (Table 8-7) suggests that this feature, in particular, "organizes" life history strategies. Habitat adaptation, in this analysis, showed fewer correlations with other attributes. From the combination of these two basic organizing life history features are derived the "choices" of reproductive effort, dispersal strategies, behavior, habitat breadth, nest type, and nest height. Actually, these "choices" are governed by the restrictive laws that are described in the regressions we have computed. Such laws are an intrinsic part of the processes of natural selection, and are as important, if not as visible, as the physiognomic features of the forest within which evolution has proceeded. Decline of a given species may result from the unsuitability of any one or any combination of the life history features of bird species exposed to the stress of a newly fragmented environment. It has been useful to us to think of the organization of life history traits of forest bird species in terms of r- and K- selection (MacArthur and Wilson 1967). During the course of our analyses, Brewer and Swander (1977) independently noted the tendency of bird species of the deciduous forest to budget lower relative amounts of reproductive effort, in contrast to species associated with grassland, marsh, or forest edge. Our studies are in complete accord with their analysis; in Table 8-13 we have phrased possible discrimination between life history strategies in fragmented environments in terms of r- and K- selection.

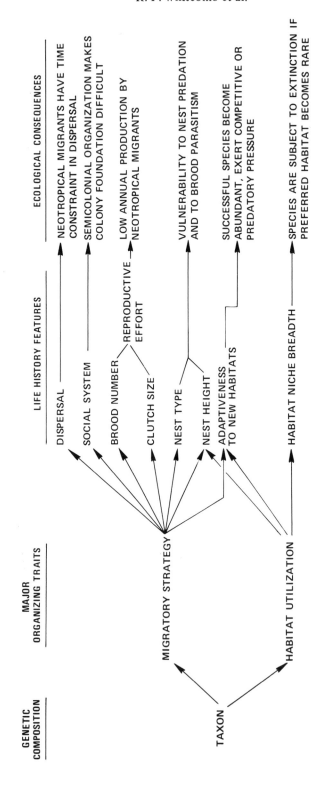

Fig. 8-9. Path diagram showing interrelationships between bird taxa, migratory strategy, residence and various life history features. These features have important consequences in a newly fragmented environment (see also Table 8-4).

Table 8-13. Possible Mechanisms for Discrimination between r- and K-Selected Bird Species in a Newly Fragmented Environment

	Responses	
Process	r-Strategist	K-Strategist
Fragmentation of primitive habitat	Mean distance between habitat patches decreases; dispersal becomes easier	Mean distance between suitable areas of habitat increases, leading to isolation and dispersal losses
Creation of new disturbed habitats	Favored, because of greater adaptability; "new" niches may resemble their primitive niches	Primitive niche was homogeneous forest; behavioral rigidity impedes utilization of disturbed habitat
Small size of new fragments	If too small, may be insufficient for a single territory; mean territorial sizes are larger than those of specialists of same body weight	Forest interior specialists may avoid small forest islands
Disturbance of migratory pathways	Permanent residents or short-distance migrants not seriously affected	New hazards may reduce numbers
Disturbance of winter habitat	Permanent residents or short-distance migrants can disperse over short distances to locate adequate food supply	Not definitely known to affect neotropical migrants, but may affect forest interior specialists
Spillover from adjacent heterogeneous habitats	May be pressure from competition for nesting holes. Open nesters may be subject to nest predation or brood parasitism	Almost all species are open-nesters. Combination of nest parasitism and predation of known importance
Local extinctions of other species	Local extinction of specialists leads to competitive release	May be some competitive release for surviving canopy species
Increased mortality (from all causes)	High reproductive effort buffers losses	Low reproductive effort makes replacement difficult
Trends toward warmer, drier climate	Favored by climatic changes that tend to open up forest	Not favored by disruption of mesic forest
Oscillations in population size	May have smaller amplitude	Trans-gulf migrants or species adapted to xeric niches may have large amplitude

Two examples suffice to illustrate this point. We discovered significant correlations between nest height and reproductive effort ($r = -0.3133, p = 0.013, n = 63$), and nest height and body weight ($r = 0.3713, p = 0.003, n = 63$). In the case of reproductive effort, it is possible that predation of nests is significantly higher for species that nest on or near the ground than for canopy species. Therefore, only those species able to maintain a relatively high reproductive effort may have been able to adopt low-nesting strategies. Among the species that became extinct on Barro Colorado Island (Willis 1974) after its creation by canal construction were a large group of low-nesting species, whose reproductive effort failed to replenish losses. Why should any species, then, be a low nester? If small body weight is correlated with low-nesting, it may be that the energetics of feeding young dictates that nests should be located as close as possible to resources that are being harvested by the parent birds. A further factor in nest location in savanna or deciduous forest habitats is that a significantly large fraction of potential nest locations are on the ground, or particularly, in the shrub layer.

A detailed analysis of such life history patterns is beyond the scope of this chapter, and could well occupy a monograph in its own right. The important aspect that we must note here is that the strategy of neotropical migration apparently has involved several important, obligate life history features that are unsuited to a newly fragmented environment. This view is in accord with other studies that conclude that response to disturbance in a community is best explained by life-history characteristics of individual species (Sousa 1980). A qualitative multifactorial model for avifaunal change is presented in Table 8-14.

Causes for Decline: A Dispersal Model

Given the observation that upland forest islands are inhabited by a characteristically depauperate subset of the regional forest avifauna, and noting that many neotropical migrants are intolerant of fragmentation, the question arises: Is there a causal link between intolerance to fragmentation and migratory patterns? In this section we propose a model for such a link, and discuss evidence from a "natural experiment" that appears to support the model.

We begin with the obvious fact that extinction and recolonization of neotropical migrant bird species within their breeding range occur every year when individuals of these species fly south for the winter and return for spring breeding. Banding data (Hann 1937, Robbins 1969) reveal that recolonizing passerines fall into two categories. Territorial birds show a strong tendency to return to the same territories they have established in the previous breeding season. However, as a consequence of mortality during migration and on the wintering ground, 40-60% of these birds fail to return, and their former territories are eventually occupied by young birds breeding for the first time. On the other hand, banding data also show that first year birds rarely return to the same plot where they were reared (Hann 1937, Nolan 1979; C. S. Robbins, unpublished data). A necessary consequence of this pattern is that most of the birds that replace adults lost in the nonbreeding season must have fledged elsewhere. These "new" territorial individuals comprise a second class of annual migratory colonizers of breeding habitat.

A simple equation expresses these relationships:

$$T_i = kT_{i-1} + C \tag{5}$$

where T_i, the number of territorial birds in year i, is the product of the number of territorial individuals in the previous year (T_{i-1}) times the annual survivorship rate (k), plus the number of "new" colonists (C). At equilibrium, the number of territories will be approximately constant from year to year, and the number of "new" birds will balance losses resulting from winter mortality. Thus:

$$C = (1 - k)T_{i-1} \tag{6}$$

Substituting, we can rewrite Eq. (5) as follows:

$$T_i = \frac{kC}{1-k} + C = \frac{C}{1-k} \tag{7}$$

This relationship, which has been expanded and presented formally by Robert M. May in Chapter 9, predicts that the apparent equilibrium carrying capacity of a forest island is strongly dependent on the colonization rate by first-year birds fledged elsewhere, as well as the mortality rate of previously resident individuals. This model suggests that the colonization rate is mainly determined by events that occur outside the island (i.e., in the region from which colonists might be recruited). In consequence, the number of highly migratory birds that breed on a forest island may be largely regulated by events that occur elsewhere, and observed declines could occur even in the absence of significant ecological changes within the forest islands themselves. Data gathered for the present study and previously published data, some of which were summarized by Lynch and Whitcomb (1978) clearly indicate that bird communities of many eastern forest islands are currently undergoing nonequilibrial decay. This decline is a regional phenomenon, and is being experienced by a wide variety of bird species, most of which are neotropical migrants associated with forest interior habitat (Lynch and Whitcomb 1978).

An example of avifaunal deterioration that may have been induced by external, as opposed to internal, habitat changes is provided by long-term census data (Criswell and Gauthey 1980) for Cabin John Island, a small (7.3 ha) forested island in the Potomac River near Washington, D.C. Annual censuses were initiated in 1947 and have continued, with the exception of a few years, to the present time. Before construction of the George Washington Parkway in 1958, the island had been a functional part of an extensive continuous strip of rich riparian forest that supported large breeding populations of most bird species characteristic of riparian forest. After construction of the new parkway, Cabin John Island, which is separated from the riverbank by only a few meters, became a true habitat island. Over the next four years several neotropical migratory species (e.g., red-eyed vireo, northern parula warbler, and American redstart), whose populations had oscillated about a fairly stable mean for the previous 11 years, declined to substantially lower, but relatively stable, levels (Fig. 8-10a, b). This decline was preceded by a temporary population increase the year after disturbance, and featured a "bottoming-out" before attainment of a new lower population level. We attribute the initial increase to an influx of displaced birds. A possible interpretation of the bottoming-out in terms of age structure of the population is presented in Chapter 9. Some common species (e.g., Kentucky warbler and wood thrush) disappeared entirely from the island after 1958, despite the fact that the habitat on the island remained essentially unchanged. The decrease in the abundance of neotropical migrants contrasts with the stability of populations of such permanent resident species as cardinal and woodpeckers (Fig. 8-10c), or the sharp increase in numbers of starling (Fig. 8-10d)

Table 8-14. Multifactorial Model for Susceptibility of Bird Species to Forest Fragmentation, a Qualitative Evaluation of Potential Susceptibility of Forest Species to Stresses Associated with Forest Fragmentation[a]

| | | | | | Stress | | | | |
Species	Large territorial size	Open nester	Ground or low nester	Narrow habitat selection	Forest interior specialist	Low reproductive rate	Diffuse competition	Poor colonist	Mean score[b]
Permanent residents									
Pileated woodpecker	+	0	0	0	+	+	0	0	3
Red-bellied woodpecker	0	0	0	0	0	0	+	0	1
Hairy woodpecker	+	0	0	+	+	±	0	0	3.5
Downy woodpecker	0	0	0	0	0	±	0	0	0.5
Blue jay	0	+	0	0	0	0	0	0	1
Common crow	±	±	0	0	0	0	+	0	1
Carolina chickadee	0	0	0	0	0	0	+	0	1
Tufted titmouse	0	0	0	0	0	0	0	0	1
White-breasted nuthatch	±	0	0	+	+	0	+	0	2.5
Carolina wren	0	+	+	0	0	0	+	0	3
Starling	0	+	0	0	0	0	0	0	0
Cardinal	0	+	+	0	0	0	0	0	2
Neotropical migrants									
Yellow-billed cuckoo	0	+	0	0	+	+	0	0	3
Great crested flycatcher	0	0	0	0	±	+	+	0	2.5
Acadian flycatcher	0	+	0	+	+	+	0	0	4
Eastern wood pewee	0	+	0	0	0	+	0	0	2
Wood thrush	0	+	0	0	0	0	+	+	3
Blue-gray gnatcatcher	0	+	0	+	±	+	+	0	4.5
Yellow-throated vireo	0	+	0	+	±	+	+	+	5.5
Red-eyed vireo	0	+	0	0	0	+	+	+	4
Black-and-white warbler	0	+	+	+	+	+	±	+	6.5
Worm-eating warbler	0	+	+	+	+	+	+	+	7
Northern parula warbler	0	+	0	0	0	+	+	+	5
Ovenbird	0	+	+	0	+	+	+	+	6
Kentucky warbler	0	+	+	+	+	+	+	+	7

									[b]
Hooded warbler	0	+	+	+	+	+	+	+	7
American redstart	0	+	0	+	+	+	0	+	5
Scarlet tanager	0	+	0	0	0	+	+	+	4

[a] +, present; 0, absent.
[b] Sum of + values.

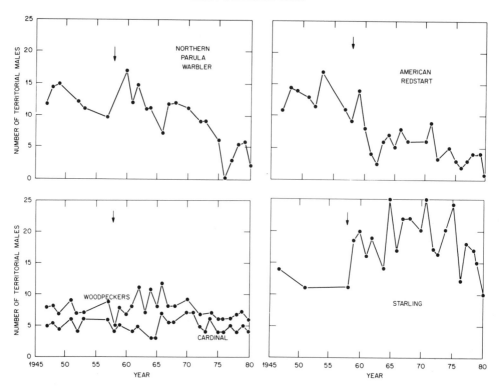

Fig. 8-10. Avifaunal populations on Cabin John Island before and after destruction of adjacent habitat (↓). Populations of northern parula warbler increased somewhat after first impact, presumably as a result of colonization by displaced territorial birds from the destroyed habitat. Within the next several years, however, the population decreased to an all-time low, followed by a temporary partial recovery. Populations of American redstart increased slightly at first, but then decayed to a low in 1962. In subsequent years, there was a slight recovery from this low but the population appeared to be at a new, lower equilibrium level governed by the new biogeographical circumstance. Woodpecker and cardinal populations were not affected by the disruption. Starlings have increased, presumably as a result of an increase in nearby open foraging areas.

after highway construction. The richness of the Cabin John avifauna has continued to decline slowly, so a new equilibrium has not yet been attained. Some of the declines in recent years may reflect cyclical habitat change. In particular, opening of the forest after the 1972 floods associated with hurricane Agnes has resulted in loss of forest interior habitat (J. Criswell, personal communication). Because there is no reason to assume that the intrinsic carrying capacity of the island has declined markedly over the past 20 years, it seems reasonable to attribute the observed decline in the avifauna to a reduced rate of colonization by "outside" birds. This reduction could be the result of at least two factors: (1) removal of extensive areas of nearby breeding habitat that may have been the main source of colonists, and (2) the reduction of the "psychological" acceptability of Cabin John Island, by virtue of its increased isolation from extensive tracts of similar habitat and disturbances related to the nearby parkway.

Our model implies that islands tend to "leak" progeny at a greater rate than new colonists are accumulated if their basic attractiveness as habitat is equal to that of competing forest islands of larger size. This implies the existence of "tradeoffs" between size, isolation, and habitat quality, some of which are diagrammatically illustrated in Fig. 8-11.

Summarizing, we emphasize two consequences of the simple model we have presented: (1) both the species composition and the overall abundance of birds on forest islands may be altered drastically by disruption of habitat outside the islands themselves, and (2) for birds, and presumably other organisms as well, the realized productivity of small, isolated habitat islands may be substantially less than the upper limit imposed by their intrinsic carrying capacity. Such constraints on carrying capacity were predicted theoretically by Gadgil (1971).

Niche Shifts and Density Compensation

There are many empty niches in depauperate communities characteristic of oceanic islands, and niche shifts are commonly observed (Diamond 1970). Forest island communities also lack many species that could contribute to the efficient utilization of all available resources. Niche shifts are also observed on forest islands. For example, house sparrow colonized Trelease Woods during an episode following an epidemic of Dutch Elm disease. Also, indigo bunting apparently inhabits the forest interior of this island (Kendeigh 1981). This is not to say that the abundance of certain bird species on forest islands might not be as high or higher than in an equivalent area of large forest. Data for Trelease Woods show that this is in fact commonly observed; the house wren population is a particularly good example (Kendeigh 1981). Also, wood thrush populations on forest islands may temporarily explode (B. Whitcomb et al. 1977). Mainland populations are regulated by predation, territorial spacing, or other factors not directly related to local resource levels (MacArthur et al. 1972). The persistence of all bird species able to inhabit small forest islands must, for the reasons noted, depend on subsidization by outside colonists. The subsidization, or "rescue effect" was discussed by Brown and Kodric-Brown (1977) in their study of local colonization and extinction of insect populations associated with vegetation patches. These authors postulated that the rate of extinction of island populations close to a mainland source of propagules is less than that predicted by simple MacArthur-Wilson dynamics, by virtue of regular replenishment from outside.

Subsidization of Local Populations

Recognition that the dynamic processes of subsidization may operate even within large forest systems has important consequences in conservation applications. For example, of 85 bird species known to breed in a particularly well-studied 1130-ha segment (BFMA:Beltsville Federal Masterplan Area) of the Beltsville forest area, we estimate that at least 35 (41%) are represented by marginal populations whose centers of productivity are elsewhere. The status of these populations is precarious even if the BFMA forest were to remain intact. For example, if the rich bottomland forest along the nearby Patuxent River were destroyed or greatly reduced in area, we would predict

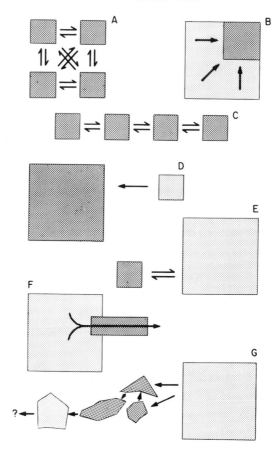

Fig. 8-11. Possibilities for leakage of progeny from forest islands. (A) An archipelago of mesic islands of equal size could be theoretically at equilibrium. (B) Even within large forest, mesic areas might serve as "sinks." (C) More centrally located islands might accumulate colonists at the expense of peripheral islands. (D) Xeric islands would be expected to lose progeny to more attractive mesic areas, but progeny from a larger xeric habitat (E) might choose a smaller but otherwise more suitable habitat. (F) An attractive habitat corridor might funnel progeny away from undesirable habitat. (G) The confusion of small fragments could result in loss of progeny from an entire region.

that many of these 35 species would decline, possibly to extinction. Extirpations from xeric forests such as the BFMA tract might reflect population oscillations whose amplitude exceeds that in mesic forests (Fig. 8-12). Differences between xeric and mesic forests could arise as the size of the regional population changes, or could also reflect greater resource predictability in mesic forests. Although this model is speculative, we believe it deserves thoughtful testing.

An example will illustrate the potential importance of outside subsidization of bird populations. In any given year, 4-5 pairs of red-shouldered hawks breed in the BFMA. These birds are outliers of a much larger population that is centered along the Patuxent

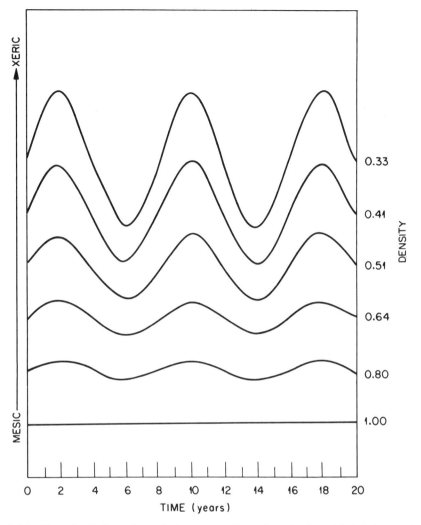

Fig. 8-12. Hypothetical gradient between mesic and xeric forest habitats in which increases in amplitude are associated with decreasing density of hypothetical species. Stronger oscillations in mean density and productivity might be expected in more xeric forest.

River. This population has been intensively studied by Henny et al. (1973). From such studies it can be surmised that marginal habitat such as that found in the BFMA is unsuitable for optimal reproduction. It is reasonable to predict that extirpation of the Patuxent population of red-shouldered hawks would be followed by extinction of the BFMA population, even if the existing habitat at Beltsville were preserved. However, the status of wide-ranging raptorial species such as red-shouldered hawk may depend on events occurring over a wide region (Snyder and Snyder 1975).

Documentation of subsidization patterns would require long-term study of a regional population. Such studies are understandably rare, but there is evidence that the situation described for red-shouldered hawk is not unique. For example, Thompson and Nolan (1973) found that an Indiana yellow-breasted chat colony was not self-sustaining, but depended on subsidization by recruits from more productive sites.

Figure 8-13 illustrates a hypothetical distribution of recruitment rates for the set of species in the BFMA, based on our intensive, but qualitative, long-term observations. Quantitative data of this type would permit us to predict which species are likely to be extirpated in the event of major habitat disruption outside the BFMA. Even species that appear to be well established could face local extirpation if the forest should become isolated from sources of potential colonists. In addition to the "leakage" of dispersing juveniles after fledging, isolated avian colonies face increased risk of extinction from such factors as ordinary stochastic fluctuations in population size, increased competition with "edge" species not normally found within forest, increased numbers of nest predators and nest parasites (e.g., grackles, crows, jays, and cowbirds) which tend to be most abundant in disturbed areas, and increased human intrusion and disturbance (Lynch and Whitcomb 1978).

Conservation and Island Biogeography

The status of bird populations on any forest fragment must be interpreted in a regional context. Although the area, vegetational composition, physiognomic characteristics, and successional maturity of a forest island all may affect the numbers and kinds of birds that breed there, the spatial context of the island relative to other wooded areas and the total amount of such habitat in the region is probably crucial. Some, and possibly all, of the progeny produced by breeding birds on a small forest island are lost

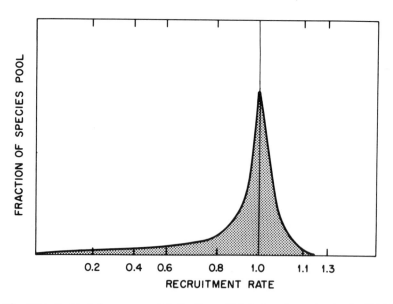

Fig. 8-13. Hypothetical frequency distribution of bird recruitment rates in 1000 ha of upland forest, edge, and field. Species with rates less than 1.0 would be locally extirpated if surrounding habitat were destroyed.

to the combined effects of mortality and postfledging dispersal. These "missing" young birds may make an important contribution to the regional balance between immigration and emigration, but they are largely irrelevant to the maintenance of a viable long-term breeding population in any particular forest island.

Our multifactorial model suggests that typical neotropical migrant forest interior bird species are poorly adapted for survival on forest fragments. The regional preservation of such species requires the sequestering of extensive tracts of forested habitat. Just how large such tracts must be is an unresolved question with different answers for different species. Minimum reserve size also may depend on the overall productivity of the particular type of forest considered. One approach would be the use of long-term data on fluctuations of local populations for estimation of time-specific probabilities of stochastic extinctions. The ultimate decision on an "acceptable" length for the expected life of a population would still be arbitrary, but such an approach would at least provide an objective basis for comparison. Of course, any conservation actions taken on the basis of such computations would have to take into account the possibly detrimental effects of unanticipated destruction of nearby habitat as well as increases in the amount of human disturbance (Geis 1974, Liddle 1975).

One can envision a continuum of island sizes that vary in their value as avifaunal reserves. At one extreme are tracts smaller than the minimum territorial size of many bird species. By definition, such forest tracts have negligible value for avifaunal conservation. However, many bird species avoid, or cannot utilize forest islands that are considerably larger than the size of their territories. Any habitat islands smaller than the behaviorally programmed minimum acceptable size are also virtually useless in a conservation context. Somewhat larger forest islands may be colonized by several breeding pairs, but are not large enough to support a self-sustaining population. Forest fragments of this size are commonly set aside as nature preserves and parks, and, if numerous, may have some utility on a regional level. However, documented failures of such parks as reserves (Lynch and Whitcomb 1978, Butcher et al. 1980, Kendeigh 1981) warn us that they are much too small. The final class of reserves comprises tracts of forest that are large enough to maintain sizable breeding populations of many bird species and to prevent excessive postfledging "leakage" of young birds. Only preserves of the latter size can function without continual subsidization by outside colonists (Terborgh 1975, Diamond 1976a).

Although many preservation schemes are directed toward individual rare or endangered species, ecologists and other conservationists presumably aim to set aside entire functioning ecosystems. Although there may be no universally accepted definition of an "entire" ecosystem, a simple rule of thumb in the establishment of faunal preserves might focus on the status of those animals at the highest trophic levels. A preserve large enough to maintain a viable population of timber wolves, for example, would be likely to preserve primary producers, herbivores, and insectivores as well. Conversely, an area too small to permit long-term survival of large carnivores runs the risk of preserving a misleading caricature of an entire ecosystem.

In conclusion, a faunal reserve should be large enough to preserve not only a large fraction of the entire regional complement of forest bird species, but a complete complement of other taxa as well. Such megapreserves would have areas comparable to those of our National Forests and larger National Parks.

Conclusions

1. Deciduous forest covers about 22% of the portion of central Maryland considered (Fig. 8-1) in the present study (Table 8-1). The existing forest comprises a complex archipelago of dendritic strips and islands (Figs. 8-2 and 8-3) that vary in area, shape, successional maturity, vegetational composition, level of disturbance, and surrounding land use.

2. A point survey method (Fig. 8-4) was used to survey forest island avifauna. Small forest islands are inhabited by a very limited subset of common forest-dwelling bird species, together with a large number of species that prefer brushy or edge habitats (Table 8-5). No bird species is restricted to small islands.

3. The ratio of the mean number of territories observed per sample point on forest islands of intermediate size (6-14 ha) to the mean number observed on larger (70+ ha) islands provided a useful index of tolerance to fragmentation (Table 8-5).

4. Neotropical migrant bird species that are restricted as breeders to forest interior (Fig. 8-5a) are rare on small (1-5 ha) forest islands, somewhat more common on fragments of intermediate (6-14 ha) size, and are most abundant on large (70+ ha) forested tracts (Table 8-5, Fig. 8-6). Species capable of utilizing forest-edge or patchy habitats as well as forest-interior (Fig. 8-5b) are equally abundant on forest islands of all sizes (Fig. 8-6).

5. An isolation factor was computed for each studied island (Table 8-2) by dividing the area of each nearby (0.1-3 km) island at least 25 ha in size by the squared distance from the studied island and taking the reciprocal of the quotient. Factors calculated in this way were significantly correlated with numbers of breeding forest interior bird species. The one island in the two smaller subsets that was situated close to large mainland forests supported certain forest interior species that were not found on any other island.

6. Neotropical migrant bird species typically account for 80-90% of breeding individuals in extensive tracts of eastern deciduous forest (Table 8-8), but are much less common on small tracts, where they generally constitute less than half the breeding individuals. Small forest islands in Maryland, Illinois, Ohio, Delaware, Wisconsin, New Jersey, and Michigan all have similarly depauperate avifaunas (Table 8-10).

7. Most permanent residents are not restricted to forest interior. Only three such species (pileated woodpecker, hairy woodpecker, and white-breasted nuthatch) breed primarily in forest interior habitat in the central Maryland study area. These species have large territories and require relatively mature forest, but remain common in the study area.

8. An extensive forest system near Beltsville, Maryland, was found to support typical densities (Appendix D) of all small bird species that have declined in or disappeared from small tracts in the study region.

9. Point surveys in a mature tulip-tree-oak tract and in surrounding immature forest (Appendix B) indicated that only yellow-throated vireo and white-breasted nuthatch were restricted to the mature forest. Large tracts of early successional forest served as important population reservoirs for such forest-interior species as Acadian flycatcher, northern parula warbler, ovenbird, Louisiana waterthrush, Kentucky warbler, hooded warbler, and American redstart.

10. Forest patches surrounded by fields or other nonforested habitat can be considered true habitat "islands" for about 19 of the 93 species that comprise the regional pool of forest species (Appendix E). Of the remaining species, some (e.g., certain raptors and woodpeckers) have very large home ranges and can combine several isolated tracts within a single territory; others typically inhabit fields, hedgerows, residential areas, and other nonforested situations, but occasionally "spill over" into forest interior, particularly in small woodlots. Many species are able to utilize a broad spectrum of habitats. Such habitat generalists, many of which are permanent residents or short-distance migrants, tend to utilize small parks, suburbs, and other disturbed areas that maintain some tree cover. These species, many of which are probably derived from the primordial savanna assemblage, do not depend on large tracts of undisturbed forest for their regional survival.

11. Absence of bird species that are intolerant to fragmentation, such as worm-eating, hooded, and black-and-white warblers, from entire 25-km^2 blocks surveyed for Breeding Bird Atlas studies indicates that these species are subject to regional as well as local extirpation (Table 8-5).

12. Statistical analyses (χ^2 and stepwise multiple regression) revealed a highly significant correlation between tolerance to fragmentation and migration strategy, degree of habitat specialization, nest type, and nest height (Table 8-7, Appendix F). Species that are intolerant of fragmentation tend to be highly migratory, to be specialized for forest interior habitat, to build open nests, and to nest on the ground. Life history features (Table 8-4) such as migratory strategy and habitat association were strongly correlated with taxonomic position (Table 8-7).

13. Long-term census data for Cabin John Island in the Potomac River near Washington, D.C., illustrate the reequilibration of an island bird community following an increase in the degree of isolation. Six species of neotropical migrants responded similarly to the destruction of forest adjacent to the island. These species increased slightly in the year following the habitat destruction, presumably as a result of an influx of birds displaced by deforestation. Subsequently, however, they declined markedly in abundance, and some disappeared altogether from the census tract. A new, lower equilibrium species diversity and total population level is apparently being approached (Fig. 8-10a, b).

14. A preliminary model attempts to explain the dispersal dynamics involved in differential tolerance to fragmentation as a result of the combined effects of progeny "leakage" following fledging and the difficulties of recolonizing forest tracts each year following a long-distance migratory flight (Fig. 8-11).

15. Annual reproductive effort (estimated by multiplying mean clutch size by mean annual brood number) tends to be low in forest interior species and in neotropical migrants generally (Fig. 8-7b). Neotropical migrants that are also forest interior specialists therefore have a relatively low reproductive effort, whereas permanent residents and short-distance migrants associated with open habitats have the highest reproductive effort (Fig. 8-7b). On the other hand, neotropical migrants tend to be longer lived than other birds, so their lifetime reproductive effort may be similar to that of nonmigratory species of similar size (Table 8-4, Fig. 8-7b). We suggest that these two ecological groupings fall toward opposite poles of the r-K selective continuum.

16. Avifaunal change in fragmented forest systems probably results from unsuit-

ability of several life history features common to forest interior neotropical migrants. These include low annual reproductive effort, dispersal strategy, nest type and location associated with vulnerability to nest predation (Tables 8-4 and 8-14, Figs. 8-8 and 8-9). The tendency of forest interior neotropical migrants to utilize forest interior while wintering in the tropics ultimately could reduce continental populations as tropical deforestation proceeds, but should not affect island populations differentially.

17. Forest tracts that are sufficiently large to preserve populations of some organisms (e.g., most plant species) may nevertheless be inadequate as reserves for bird species. The maintenance of forest bird communities requires wooded tracts of hundreds or even thousands of hectares. Management of such preserves should be aimed at minimizing disturbance of the forest interior.

Acknowledgments. We are indebted to many members of the Maryland Ornithological Society who participated in the Breeding Bird Atlas and Miniroute surveys within our study region, and in particular to D. Holmes, L. MacClintock, N. MacClintock, R. Patterson, and J. Solem. We thank Bernard Wineland of the USDA-SEA data processing unit for his advice and expertise in running computer analyses. J. M. Diamond, M. Erwin, P. Geissler, M. Howe, and B. K. Williams made many constructive comments that improved the manuscript.

Appendix A

Computation of Detectability Coefficients

To compare habitat utilization by different bird species, it was necessary to calculate coefficients of detectability (D_i). These were derived from a set of 86 point surveys taken throughout the study area, in which the 60 min of observation at each point were recorded in 5-min intervals. From these observations, it was possible to construct discovery curves for each species. The species could be separated into two classes, those with low detectability, and those with high detectability. The fraction of 5-min intervals in which such species were recorded ($P_{5\ min}$) for the latter class was calculated directly from the data, and a detectability coefficient (in terms of the 6-min miniroute protocol) was computed:

$$D_i = \frac{1}{1 - (1 - P_{5\ min})^{6/5}}$$

Computation of D_i for species of low detectability was accomplished by recording the total number of intervals (out of 12) that the species were recorded. Then, the results from the 86 points could be arranged in a series, $N_1, N_2, \ldots, N_{12}$, where N_i is the number of 5-min intervals (out of the sample of 86 points) in which the species was recorded with a frequency of $i/12$ intervals. N_0, an unknown number of points in which the species could be estimated to have been present but undetected (and which was confounded with points in which the species was not present at all) was estimated by fitting the series, $N_1, N_2, \ldots, N_{12}$ to a Poisson distribution (Snedecor 1956). Then, the detectability coefficient was computed:

$$D_i = \frac{1}{1 - \left\{ 1 - \left[\dfrac{N_1 + N_2 + \ldots + N_{12}}{N_0 + N_1 + \ldots + N_{12}} \right]^{6/5} \right\}}$$

Appendix B

Estimation of Local Forest Interior Bird Species Pool for the Seton Belt Mature Tulip-Tree-Oak Forest[a]

Species	North tract[b] Spot-mapping census	North tract[b] Point survey	South tract[b] Spot-mapping census	South tract[b] Point survey	Mature forest Wind-damaged[c]	Mature forest Penin-sula[d]	Mature forest Seep-age[e]	Middle succession Mixed[f]	Early succession Xeric[g]	Early succession Mesic[h]	Late old-field Contigu-ous[i]	Late old-field Island[j]
Bobwhite	0	0.00	0	0.0	+	0	0.00	1.0	0	0.0	0.5	0
Mourning dove	14	0.67	0	0.0	0.0	1	0.43	0.5	0	0.0	0.0	0
Yellow-billed cuckoo	14	1.00	7	1.0	0.5	0	1.00	0.5	0	0.0	0.5	1
Ruby-throated hummingbird	0	0.00	0	0.0	0.0	0	0.00	0.0	0	0.0	1.0	0
Common flicker	14	1.00	+	0.5	0.5	0	0.14	1.0	0	0.0	0.0	0
Pileated woodpecker	+	0.00	+	0.0	0.0	0	0.14	1.0	0	0.0	0.0	0
Red-bellied woodpecker	56	2.00	48	1.0	1.0	1	1.00	1.0	1	0.5	0.5	1
Hairy woodpecker	7	0.33	7	0.5	0.5	1	0.00	0.5	0	0.0	0.0	0
Downy woodpecker	42	1.00	34	1.0	1.0	1	0.71	1.0	0	0.5	0.0	1
Great crested flycatcher	14	0.33	0	0.0	0.5	0	0.00	0.0	0	0.0	0.0	0
Acadian flycatcher	46	1.67	41	1.0	0.5	2	2.00	1.0	1	2.0	1.0	0
Eastern wood pewee	53	1.00	58	1.5	0.5	1	1.00	0.5	1	0.0	0.0	0
Blue jay	14	+	13	+	+	0	+	+	+	+	+	+
Common crow	+	+	+	+	+	+	+	+	+	0.0	0.0	0
Carolina chickadee	32	0.67	24	0.5	0.5	1	1.00	1.0	1	1.0	1.0	1
Tufted titmouse	25	1.00	38	1.5	1.0	1	1.29	1.0	1	1.0	0.5	1
White-breasted nuthatch	21	1.00	17	1.0	0.5	0	0.00	0.0	0	0.0	0.0	0
Carolina wren	39	1.67	31	2.0	1.0	1	1.43	1.5	2	1.0	2.5	1
Gray catbird	7	0.00	0	0.0	0.0	0	0.00	0.5	0	0.5	0.5	0
Brown thrasher	+	0.00	0	0.0	0.0	0	0.00	0.0	0	0.0	0.0	0

Species	North Tract[a]	[b]	South Tract[a]	[c]	[d]	[e]	[f]	[g]	[h]	[i]	[j]
American robin	0	0.00	0	0.0	0.0	0.00	0.0	0	+	0.0	0
Wood thrush	162	3.33	261	4.5	1.5	1.86	1.0	2	1.0	1.0	2
Eastern bluebird	0	0.00	0	0.0	0.0	0.00	0.0	0	0.0	0.5	0
Blue-gray gnatcatcher	7	0.00	+	0.0	0.5	0.58	1.0	0	0.0	0.0	0
Starling	+	0.00	7	0.0	0.0	0.00	0.0	0	0.0	0.5	0
White-eyed vireo	25	0.00	0	0.0	1.0	1.00	0.5	0	0.0	2.0	1
Yellow-throated vireo	18	0.67	3	0.0	0.5	0.00	0.0	0	0.0	0.0	0
Red-eyed vireo	158	1.67	247	2.5	2.5	2.29	2.5	2	2.5	1.5	1
Black-and-white warbler	0	0.00	0	0.0	0.5	0.00	0.0	0	0.0	0.0	0
Northern parula warbler	0	0.00	+	0.0	0.0	0.71	1.0	1	1.0	1.0	0
Ovenbird	11	0.33	65	1.5	0.5	2.57	1.5	1	3.0	0.0	0
Louisiana waterthrush	0	0.00	0	0.0	0.0	0.29	1.0	1	0.0	0.0	0
Kentucky warbler	49	2.33	27	1.5	1.5	1.14	1.0	2	1.0	1.0	1
Common yellowthroat	14	0.33	0	0.0	1.0	0.29	1.0	1	0.0	0.5	0
Yellow-breasted chat	0	0.00	0	0.0	1.5	0.00	0.0	0	0.0	1.0	0
Hooded warbler	77	1.67	0	0.0	1.5	1.14	0.0	0	0.5	0.0	0
Northern oriole	7	0.33	0	0.0	0.0	0.00	0.0	0	0.0	0.0	0
Brown-headed cowbird	21	0.33	14	0.5	0.5	0.86	0.5	0	1.0	0.5	0
Scarlet tanager	42	1.00	34	1.5	1.0	0.86	0.5	1	1.0	1.0	1
Cardinal	92	2.33	65	1.0	1.5	1.86	2.0	2	2.0	2.5	1
Indigo bunting	39	1.00	0	0.0	0.5	0.14	1.0	0	0.5	0.5	1
Rufous-sided towhee	49	1.33	14	0.5	1.0	0.85	1.0	1	1.5	1.0	1

[a] North and South Tract Spot-Mapping Census data are territorial males/km²; all other data are average numbers of territories recorded per point for the n points surveyed. Plus (+) indicates presence only.

[b] Fragments were described by B. L. Whitcomb et al. (1977) and R. F. Whitcomb et al. (1977).

[c] Forest similar to North and South tracts, but with abundant treefalls. Represented by two point surveys (n = 2).

[d] Peninsular forest similar to South tract and contiguous with it (n = 10).

[e] Seepage forest dominated by pin oak, sweetgum, red maple, and tulip-tree (n = 7).

[f] Disturbed forest similar to seepage forest but with many canopy breaks and heterogeneous patches (n = 2).

[g] Early successional forest dominated by tulip-tree (n = 1).

[h] Early successional forest, dominated by red maple, with closed canopy (n = 2).

[i] Late old-field dominated by black locust, Japanese honeysuckle, and poison ivy (n = 2).

[j] Similar to footnote i, but isolated as a 1.0 ha island surrounded by plowed fields (n = 1).

Appendix C1

Habitat Utilization by Forest Interior Birds[a]

Species	Forest		Residential		Rural	Open	
	Bottomland	Upland	Suburban	Subdivision		Field	Industrial
	(26)	(35)	(8)	(36)	(32)	(38)	(10)
Permanent residents							
Pileated woodpecker	.08	.00	.00	.00	.00	.00	.00
Hairy woodpecker	.04	.00	.00	.00	.03	.00	.00
White-breasted nuthatch	.00	.00	.00	.03	.00	.00	.00
Short-distance migrants							
Pine warbler	.00	.31	.00	.00	.00	.00	.00
Neotropical migrants							
Acadian flycatcher	.85	.20	.13	.06	.00	.00	.00
Black-and-white warbler	.23	.17	.00	.00	.00	.00	.00
Worm-eating warbler	.08	.00	.00	.00	.00	.00	.00
Northern parula warbler	.81	.05	.00	.00	.03	.00	.00
Cerulean warbler	.04	.00	.00	.00	.03	.00	.00
Ovenbird	.27	.40	.00	.03	.03	.00	.00
Louisiana waterthrush	.15	.00	.00	.00	.00	.00	.00
Kentucky warbler	.35	.03	.00	.00	.00	.05	.00
Hooded warbler	.15	.14	.00	.00	.00	.00	.00
American redstart	.35	.03	.00	.00	.00	.00	.00
Scarlet tanager	.42	.29	.13	.03	.06	.08	.00

[a] Numbers are fractions of Miniroute points of the given habitat type where the species was observed. Numbers in parentheses are number of Miniroute points in the indicated land use type.

Appendix C2

Habitat Utilization by Forest Interior and Edge Bird Species[a]

Species	Forest		Residential		Open		
	Bottomland	Upland	Suburban	Subdivision	Rural	Field	Industrial
	(26)	(35)	(8)	(36)	(32)	(38)	(10)
Permanent residents							
Red-bellied woodpecker	.35	.23	.25	.06	.22	.16	.10
Downy woodpecker	.16	.11	.13	.06	.09	.08	.00
Blue jay	.08	.17	.38	.17	.19	.13	.30
Carolina chickadee	.38	.29	.50	.08	.25	.11	.30
Tufted titmouse	.62	.40	.63	.14	.41	.11	.50
Carolina wren	.92	.37	.75	.19	.50	.53	.60
Cardinal	.88	.46	.75	.17	.78	.61	.60
Short-distance migrants							
Common flicker	.04	.06	.13	.06	.06	.03	.10
Eastern phoebe	.27	.00	.13	.00	.03	.03	.10
Gray catbird	.38	.14	.50	.19	.50	.37	.40
Blue-gray gnatcatcher	.19	.03	.00	.00	.00	.00	.00
White-eyed vireo	.46	.06	.00	.03	.03	.16	.00
Common yellowthroat	.54	.17	.00	.06	.31	.45	.50
American goldfinch	.16	.11	.00	.03	.03	.18	.00
Rufous-sided towhee	.23	.34	.13	.11	.31	.29	.20
Neotropical migrants							
Yellow-billed cuckoo	.08	.03	.00	.00	.13	.05	.00
Ruby-throated hummingbird	.04	.03	.00	.00	.00	.03	.00
Great crested flycatcher	.08	.06	.25	.03	.13	.00	.10
Eastern wood pewee	.27	.11	.13	.14	.31	.21	.20
Wood thrush	.85	.51	.75	.08	.31	.08	.40
Yellow-throated vireo	.16	.03	.00	.03	.06	.03	.00
Red-eyed vireo	.88	.37	.25	.08	.47	.37	.30
Prothonotary warbler	.16	.00	.00	.00	.00	.00	.00
Northern parula warbler	.81	.06	.00	.00	.03	.00	.00

[a] Numbers are fractions of Miniroute points of the given habitat type where the species was observed. Numbers in parentheses are number of Miniroute points in the indicated land use type.

Appendix C3

Habitat Utilization by Edge and Scrub plus Field-Edge Bird Species[a]

Species	Forest		Residential		Open		
	Bottomland	Upland	Suburban	Subdivision	Rural	Field	Industrial
	(26)	(35)	(8)	(36)	(32)	(38)	(10)
Permanent residents							
Bobwhite	.38	.31	.00	.03	.59	.68	.50
Edge and scrub species							
Short-distance migrants							
House wren	.08	.03	.88	.19	.31	.13	.20
Brown thrasher	.08	.03	.25	.03	.13	.24	.00
Field sparrow	.12	.05	.13	.00	.41	.53	.50
Neotropical migrants							
Prairie warbler	.04	.03	.00	.00	.00	.08	.00
Yellow-breasted chat	.08	.03	.00	.00	.09	.16	.00
Orchard oriole	.00	.05	.00	.03	.09	.08	.00
Northern oriole	.04	.00	.13	.00	.03	.00	.00
Blue grosbeak	.00	.00	.00	.00	.00	.26	.00
Indigo bunting	.42	.20	.13	.08	.47	.87	.40
Field-edge species							
Permanent residents							
Common crow	.65	.40	.50	.14	.78	.76	.60
Mockingbird	.16	.05	.75	.28	.78	.74	.40
Starling	.16	.11	.50	.22	.81	.63	.80
House sparrow	.00	.00	.75	.17	.53	.16	.20
Short-distance migrants							
Mourning dove	.27	.23	.88	.25	.72	.58	.40
American robin	.12	.14	.88	.31	.84	.71	.30
Eastern bluebird	.08	.00	.00	.00	.06	.18	.00

Red-winged blackbird	.27	.03	.00	.00	.19	.47	.30
Common grackle	.19	.11	.63	.22	.63	.63	.80
Brown-headed cowbird	.16	.09	.13	.03	.19	.16	.10
Chipping sparrow	.04	.00	.13	.14	.44	.24	.30
Song sparrow	.16	.05	.88	.25	.72	.66	.60
Neotropical migrants							
Eastern kingbird	.04	.00	.00	.03	.13	.24	.00

^{a} Numbers are fractions of Miniroute points of the given habitat type where the species was observed. Numbers in parentheses are number of Miniroute points in the indicated land use type.

Appendix D

Bird Species Habitat Utilization in Extensive Beltsville Forest according to Successional Stage, Dominant Overstory Species, and Position in the Elevational Gradient[a]

| Species | Early | | | Middle | | | Mature | | | | Utilization (uncorrected for detectability) |
| | Pine | Deciduous | | Pine | Deciduous | | Pine | Deciduous | | | |
	Xeric (n=3)	Xeric (n=1)	Mesic (n=2)	Xeric (n=3)	Xeric (n=4)	Mesic (n=3)	Xeric (n=3)	Xeric (n=5)	Mesic (n=2)	Floodplain (n=4)	
Permanent residents											
Pileated woodpecker	0.00	0.0	0.0	0.00	0.00	0.33	0.33	0.20	0.5	0.00	0.136
Red-bellied woodpecker	0.00	1.0	0.5	0.33	1.00	1.00	0.67	1.00	1.0	1.00	0.750
Hairy woodpecker	0.00	0.0	0.0	0.67	0.50	1.00	0.33	0.00	0.5	0.50	0.350
Downy woodpecker	0.33	0.0	0.5	0.00	0.50	0.33	0.00	0.80	1.0	0.50	0.396
Blue jay	1.00	1.0	0.0	1.00	1.00	1.00	1.00	1.00	1.0	1.00	0.875
Common crow	1.00	0.0	0.5	1.00	0.50	0.00	1.00	0.80	0.5	0.75	0.625
Carolina chickadee	1.00	1.0	1.0	1.00	0.75	0.67	1.00	1.00	1.0	1.00	0.772
Tufted titmouse	1.00	1.0	1.0	1.00	1.00	1.00	1.00	1.00	1.0	1.00	1.000
White-breasted nuthatch	0.00	0.0	0.0	0.00	0.00	0.00	0.00	0.75	1.0	0.25	0.200
Carolina wren	0.00	1.0	1.0	0.67	0.75	1.00	0.67	1.00	1.0	1.00	0.809
Starling	0.00	0.0	0.0	0.00	0.00	0.00	0.33	0.20	0.0	0.00	0.053
Cardinal	0.00	1.0	1.0	1.00	0.00	1.00	1.00	1.00	1.0	1.00	0.800
Short-distance migrants											
Mourning dove	0.00	0.0	0.0	0.33	0.75	0.67	0.33	0.80	0.0	0.50	0.338
Common flicker	0.33	0.0	0.0	0.00	0.25	0.33	0.33	0.60	0.5	0.25	0.259
Gray catbird	1.00	0.0	0.5	0.33	0.25	0.33	0.00	0.40	0.0	0.50	0.331
Brown thrasher	1.00	0.0	0.0	0.67	0.00	0.00	0.33	0.20	0.0	0.00	0.220
American robin	0.67	0.0	0.5	0.67	0.25	0.33	0.33	0.20	0.0	0.50	0.345
Eastern bluebird	0.00	0.0	0.0	0.00	0.25	0.00	0.00	0.00	0.0	0.00	0.025
White-eyed vireo	0.33	1.0	0.0	0.00	0.00	0.00	0.00	0.20	0.0	0.75	0.228
Pine warbler	0.33	0.0	0.0	1.00	1.00	0.33	1.00	0.00	0.0	0.00	0.366
Yellow-breasted chat	0.67	0.0	0.0	0.00	0.00	0.00	0.00	0.00	0.0	0.25	0.092
Common yellowthroat	1.00	1.0	0.0	1.00	0.00	0.00	0.00	0.00	0.0	0.25	0.378

Species											
Common grackle	0.33	0.0	0.0	1.00	0.00	0.33	0.00	0.20	0.0	0.00	0.286
Brown-headed cowbird	0.33	0.0	1.0	0.33	0.25	0.33	1.00	1.00	0.5	1.00	0.441
Rufous-sided towhee	1.00	1.0	1.0	1.00	0.00	1.00	1.00	0.80	0.5	0.50	0.780
Neotropical migrants											
Yellow-billed cuckoo	0.75	0.0	0.0	0.33	0.50	0.67	0.67	0.80	1.0	0.25	0.497
Great crested flycatcher	0.75	0.0	0.0	0.67	0.75	0.67	1.00	0.40	0.0	0.75	0.499
Acadian flycatcher	0.00	1.0	1.0	0.67	0.00	0.33	0.00	0.80	1.0	1.00	0.580
Eastern wood pewee	0.75	1.0	1.0	0.33	0.75	0.67	0.33	1.00	1.0	0.75	0.658
Wood thrush	0.33	1.0	1.0	1.00	1.00	1.00	1.00	1.00	1.0	0.75	0.908
Blue-gray gnatcatcher	0.00	0.0	0.0	0.00	0.00	1.00	0.00	0.80	0.0	1.00	0.280
Yellow-throated vireo	0.00	0.0	0.0	0.00	0.00	1.00	0.00	0.20	0.5	0.50	0.220
Red-eyed vireo	0.33	1.0	1.0	1.00	1.00	1.00	1.00	1.00	1.0	1.00	0.930
Black-and-white warbler	0.33	0.0	0.0	0.67	0.75	1.00	1.00	1.00	0.0	0.75	0.550
Worm-eating warbler	0.00	0.0	0.0	0.00	0.00	0.00	0.00	1.00	0.0	0.00	0.100
Northern parula warbler	0.00	1.0	1.0	0.00	0.00	0.33	0.00	0.00	0.0	0.75	0.308
Prairie warbler	0.67	0.0	0.0	0.00	0.00	0.00	0.00	0.00	0.0	0.00	0.067
Ovenbird	0.67	1.0	1.0	1.00	1.00	1.00	1.00	1.00	0.0	0.75	0.842
Louisiana waterthrush	0.00	0.0	0.0	0.00	0.00	0.00	0.00	0.40	0.0	0.25	0.065
Kentucky warbler	0.00	1.0	1.0	0.00	0.00	0.67	0.00	0.60	1.0	1.00	0.527
Hooded warbler	0.00	0.0	0.5	0.00	0.75	1.00	1.00	0.40	0.0	0.75	0.440
American redstart	0.00	0.0	0.0	0.00	0.00	0.00	0.00	0.00	0.0	0.50	0.050
Northern oriole	0.00	0.0	0.0	0.00	0.00	0.00	1.00	0.00	0.0	0.00	0.125
Scarlet tanager	0.33	1.0	1.0	1.00	1.00	1.00	0.67	1.00	1.0	0.75	0.875
Indigo bunting	0.33	0.0	0.5	0.33	0.25	0.33	0.00	0.00	0.0	0.25	0.199

[a] Data are mean numbers of territories per point at the n points surveyed.

Appendix E

The Regional Species Pool

Forest Interior Species

Permanent residents. Only three of the permanent resident species are primarily forest interior inhabitants and all three are trunk foragers. These are pileated woodpecker (*Dryocopus pileatus*), currently enjoying an expansion of its local range, hairy woodpecker (*Picoides villosus*), and white-breasted nuthatch (*Sitta carolinensis*). The territories of all three of these species are much larger than those of any passerine neotropical migrant.

Short-distance migrants. Pine warbler (*Dendroica pinus*), which is dependent on pine forest for breeding, is the only forest interior short-distance migrant in this region.

Neotropical migrants. Wood warblers of the family Parulidae are dominant within the forest interior. Five of the forest warblers in this region are ground nesters, but only two, ovenbird (*Seiurus aurocapillus*) and Louisiana waterthrush (*Seiurus motacilla*), also forage on the ground. Black-and-white warbler (*Mniotilta varia*) is largely a trunk forager, while worm-eating warbler (*Helmitheros vermivorus*) forages in deciduous understory, especially in xeric upland patches. Kentucky warbler (*Oporornis formosus*) forages in low shrub and understory, where it may compete with Carolina wren (*Thryothorus ludovicianus*). Hooded warbler (*Wilsonia citrina*) nests and forages in dense understory and may therefore profit from modest forest disturbances that lead to an increase in density of the shrub layer. American redstart (*Setophaga ruticilla*), an understory nester, and cerulean warbler (*Dendroica cerulea*), a canopy nester, are characteristic floodplain species in central Maryland, and therefore do not appear in surveys or censuses of upland forest. In addition, cerulean warbler is at the easternmost edge of its present range in the western portion of the study area. Both of these floodplain species occur in upland forest in Maryland's western mountains. Yellow-throated warbler (*Dendroica dominica*) occurs locally in Coastal Plain habitats, both in sycamore (*Platanus occidentalis*) associations in floodplain, and in loblolly pine. Scarlet tanager (*Piranga olivacea*) is a canopy nester and forager throughout upland and bottomland forests, but Acadian flycatcher (*Empidonax virescens*) reaches xeric forest only in the vicinity of thriving populations in nearby mesic forest, which it prefers. Veery (*Catharus fuscescens*) has become established locally within our study area in the last three decades.

Forest Interior and Edge Species

Permanent residents. Seven permanent resident species utilize a broad range of habitat types, and therefore occupy most residential areas as well as forest interior. Carolina wren (*Thryothorus ludovicianus*), a semihardy species, may survive unusually harsh winters principally in residential areas and other localized favorable habitats. From these focal areas, the population may rebuild to maximal levels. The cardinal (*Cardinalis cardinalis*) also profits from agricultural habitats and urban areas that provide a thermal shield and rich winter food resources, including those at feeding stations. Red-bellied woodpecker (*Melanerpes carolinus*), downy woodpecker (*Picoides pubescens*), black-capped chickadee (*Parus atricapillus*), Carolina chickadee (*P. carolinensis*), tufted titmouse (*P. bicolor*), and blue jay (*Cyanocitta cristata*) also profit from such factors, and occur ubiquitously in forest, rural residential, edge, and successional habitats.

Short-distance migrants. Common flicker (*Colaptes auratus*) regularly occurs in forest interior as well as open habitats. Eastern phoebe (*Sayornis phoebe*) is found both in open country and along streams, where bridges are often used for nest support. A few nest under overhanging stream banks. Brown creeper (*Certhia familiaris*) has recently colonized local floodplains, or at least has greatly increased in abundance there. If the species was present in the study area prior to 1964, it was overlooked, possibly due to its inconspicuous song. Gray catbird (*Dumetella carolinensis*) and common yellowthroat (*Geothlypis trichas*), while preferring shrubby, edge, or early successional habitats, also occur in forest interior, especially floodplain, when openings are present. Rufous-sided towhee (*Pipilo erythrophthalmus*) occurs in both forest interior and edge. Two other species, white-eyed vireo (*Vireo griseus*), and blue-gray gnatcatcher (*Polioptila caerulea*), must (by our classification) be placed in this category because they are able to survive the winter in the southern United States. In both cases, however, the majority of the population winters in tropical areas. Prothonotary warbler (*Protonotaria citrea*), the area's only hole-nesting parulid, is a species of major floodplains.

Neotropical migrants. Only five common neotropical migrants, red-eyed vireo (*Vireo olivaceus*), Eastern wood pewee (*Contopus virens*), great crested flycatcher (*Myiarchus crinitus*), yellow-billed cuckoo (*Coccyzus americanus*), and wood thrush (*Hylocichla mustelina*) utilize both forest interior and edge. Use of edge and marginal habitat by wood thrush and red-eyed vireo may possibly be a reflection of full occupancy of nearby forest interior. At the Seton Belt tract, for example, a red-eyed vireo sang from a hedgerow, but four other territorial males in the nearby mature forest were audible from the hedgerow. Two tyrannids, great crested flycatcher and eastern wood pewee, breed in open habitats with limited canopy as well as in forest interior. Eastern wood pewee, for example, was found in grazed pasture where only mature oaks remained of the original forest. It also nests in suburbs, where it occasionally follows lawnmowers. Two uncommon species, ruby-throated hummingbird (*Archilochus colubris*) and northern parula warbler (*Parula americana*) require rich forests. Yellow-throated vireo (*Vireo flavifrons*) shows considerable tolerance for edge. This species originally may have inhabited stream edges in the forest; certainly, most territories we encountered were either centered along streams or in rich mesic woodlands. Also, mature trees appear to be important to this vireo. The black-billed cuckoo (*Coccyzus erythropthalmus*) and summer tanager (*Piranga rubra*) are rare summer residents in our area; the former is at the southern edge of its range, the latter at the northern.

Forest Edge and Scrub Species

Permanent residents. Bobwhite (*Colinus virginianus*) is a common species in hedgerow-wood margin habitat.

Short-distance migrants. Brown thrasher (*Toxostoma rufum*), American goldfinch (*Carduelis tristis*), and field sparrow (*Spizella pusilla*) are typical dominants of hedgerow, old field, and wood margins. Red-headed woodpecker (*Melanerpes erythrocephalus*) once nested commonly in our area, but is now rare. House wren (*Troglodytes aedon*) rarely nests in mature forests of our study area, but may achieve dominant status locally, as in maple-dominated floodplain with abundant nestholes. This wren is mainly associated with residential areas. Cedar waxwing (*Bombycilla cedrorum*) nests very locally in hedgerows and old orchards.

Neotropical migrants. Least flycatcher (*Empidonax minimus*) breeds rarely in the northwestern section of the study area, where it inhabits wood margins. Three parulids occur entirely outside the forest interior. Blue-winged warbler (*Vermivora pinus*) nests in late old-field habitats with trees 3-6 m in height, but is at the southern limit of its range in the northwest portion of the study area. Prairie warbler (*Dendroica discolor*) packs to high density in early successional Virginia pine, but occurs in deciduous scrub habitats, especially sweet gum (*Liquidambar styraciflua*), as well. Yellow-breasted chat (*Icteria virens*) uses second growth and other shrubby or edge habitat. Orchard oriole (*Icterus spurius*) and northern oriole (*I. galbula*) nest throughout the study area in rural or suburban habitats with isolated trees. Northern oriole also breeds locally along streams and in canopy breaks in the forest. Indigo bunting (*Passerina cyanea*) occurs abundantly in hedgerows and along wood margins. It is also able to utilize large clearings within forest, such as treefalls. Blue grosbeak (*Guiraca caerulea*), on the other hand, occurs only locally in similar edge and hedgerow habitats, usually in association with large fields.

Field-Edge Species

Permanent residents. Three ubiquitous permanent residents of open habitats, common crow (*Corvus brachyrhynchos*), starling (*Sturnus vulgaris*), and house sparrow (*Passer domesticus*) require no comment. Fish crow (*Corvus ossifragus*) has expanded its range inland from coastal areas and is now a fairly common resident in our area. Mockingbird (*Mimus polyglottos*), like Carolina wren and cardinal, is currently undergoing a northward expansion of its breeding range, chiefly in suburban and agricultural habitats, where buffering of conditions makes overwintering feasible.

Short-distance migrants. Mourning dove (*Zenaida macroura*), American robin (*Turdus migratorius*), and common grackle (*Quiscalus quiscula*) feed primarily in field or residential habitats. Red-winged blackbird (*Agelaius phoeniceus*), although listed by some authors as a "forest island" species, is well adapted to breed in treeless areas and requires open habitat for foraging. Brown-headed cowbird (*Molothrus ater*) forages largely or entirely in field habitats. Chipping (*Spizella passerina*) and song (*Melospiza* *melodia*) sparrows are dominant species in residential as well as agricultural areas. All of the species in the category are common, with the exception of the eastern bluebird (*Sialia sialis*). The limitation of the bluebird is apparently imposed by scarcity of, and competition for, nest holes, as the species is capable of achieving regional abundance if nest boxes are provided and competitors are discouraged (Zeleny 1976).

Neotropical migrants. Eastern kingbird (*Tyrannus tyrannus*) nests in isolated trees but forages almost entirely in open fields.

Species Not Analyzed

Water species. Several species recorded in our censuses occupy niches along wooded streams. These species, such as green heron (*Butorides striatus*), wood duck (*Aix sponsa*), belted kingfisher (*Megaceryle alcyon*), and rough-winged swallow (*Stelgidopteryx ruficollis*), occupy forest habitats secondarily and will not be discussed in the context of forest islands.

Raptors. Territorial sizes of raptorial birds are much too large for fine-grained analyses of forest area. Some raptors that are commonly associated with extensive forest, notably sharp-shinned hawk (*Accipiter striatus*) and Cooper's hawk (*A. cooperi*), have been extirpated from this region. Others, such as turkey vulture (*Cathartes aura*), black vulture (*Coragyps atratus*), red-tailed hawk (*Buteo jamaicensis*), red-shouldered hawk (*B. lineatus*), and broad-winged hawk (*B. platypterus*), remain, but nest only locally. In general, all but the American kestrel (*Falco sparverius*) are associated with large forests. Owls, because of their nocturnal habits, were not censused or surveyed by our methods. However, general observations indicate that barred owl (*Strix varia*) and great horned owl (*Bubo virginianus*) are associated with extensive wooded tracts, while barn (*Tyto alba*) and screech (*Otus asio*) owls regularly utilize open country and wood margins.

Other species presenting census problems. Late migration poses a significant difficulty in June censuses and surveys. Yellow-billed cuckoo, blue jay, and American redstart are the most troublesome species, since some migration often continues long after nest building or even fledging of summer resident populations. Even in spot-mapping censuses, the exact status of jays and cuckoos may be difficult to define. Migrating American redstarts, however, usually leave central Maryland by June 5, and most of the birds on territory after that date are breeding individuals.

Diurnal survey methods are, of course, inadequate for all nocturnal and crepuscular species; therefore American woodcock (*Philohela minor*) and whip-poor-will (*Caprimulgus vociferus*) were poorly sampled in our surveys. Distributional records indicate that whip-poor-will occurs in association with large forest tracts, while American woodcock uses brushy fields and wood margins for nesting.

Ruby-throated hummingbirds also are undersampled by normal census techniques. This species is associated primarily with large rich, productive forest tracts.

Finally, the common brown-headed cowbird poses exasperating problems. There is no standard means of estimating breeding numbers of this nonterritorial species, even in spot-mapping census work.

Extinct, extirpated, and introduced game birds. The passenger pigeon (*Ectopistes migratorius*) was apparently originally resident in Maryland forests, but is now extinct throughout its range. Two other game birds, turkey (*Meleagris gallopavo*) and ruffed grouse (*Bonasa umbellus*), characteristic of extensive forested areas, have been largely extirpated from our study area. On the other hand, ring-necked pheasant (*Phasianus colchicus*) has been introduced successfully, particularly in the upper Piedmont, where it occupies open field habitats.

Appendix F

Stepwise Regression Analyses for Tolerance to Fragmentation with Various Avian Life History Attributes as Independent Variables[a]

Variable	Analysis[b]					
	1	2	3	4	5	6
Migratory strategy	7.53[c]	N[d]	$-$[e]	—	—	—
Square of migratory strategy	—	—	—	—	10.82	N
Reproductive effort	N	8.41	8.07	18.41	N	18.41
Major habitat association	14.01	—	5.80	—	11.94	—
General habitat utilization (B'_i)	—	N	—	N	—	N
Nest type	N	6.99	6.18	6.99	N	6.99
Nest height	3.95	N	N	N	5.12	N
Square root of body weight	N	N	N	N	N	N
Longevity	N	N	N	N	N	N
Territorial size[f]	N	N	N	N	N	N
Forest habitat utilization (F_i)	N	N	N	N	N	N

[a] Computed by SAS stepwise procedure; see text.
[b] Six analyses were performed in this run, using different transformations of variables.
[c] F value.
[d] N, failed to meet 0.50 significance level for entry into the model.
[e] Not entered as independent variable in this analysis.
[f] Computed from territorial densities (Table 8-4).

9. Modeling Recolonization by Neotropical Migrants in Habitats with Changing Patch Structure, with Notes on the Age Structure of Populations

ROBERT M. MAY
DEPARTMENT OF BIOLOGY
PRINCETON UNIVERSITY

This chapter elaborates on the discussion in Chapter 8 concerning the "island biogeography" of neotropical migratory birds in temperate forest fragments, and presents two models that capture the behavior of these migrants. First, we outline a simple model that describes the change in the equilibrium number of individuals in a particular forest island in response to changes in the overall average rate of production of first-order colonists in the larger temperate region from which the patch draws its recruits. Second, a more complicated model is presented that includes the effects of age structure so that a richer range of dynamical behavior can be exhibited.

As discussed by Whitcomb et al. (Chapter 8), the island biogeography of neotropical migrants offers some interesting special features. For a given species within a given patch, all breeding individuals go "extinct" each year as they migrate south. Those breeding individuals that survive their winter excursion return to the same patch to breed again the following year. The losses from mortality are balanced by new, first-year colonists, drawn from a pool of individuals from the surrounding region. Such individuals have a low order of site fidelity to the patch in which they were fledged. Thus, the equilibrium number of breeding individuals in a patch depends on two factors that have essentially nothing to do with the patch under consideration, namely, the year-to-year survival probability and the overall rate of production of new colonists within the wider area from which the island draws its first-year colonists.

A Simple Model without Age Structure

Let T_t be the number of territorial birds breeding in year t, and C_t the number of "new" colonists in year t. The birds that breed in a given patch in year t are assumed either to survive and return the next year, with probability k, or to die. Therefore, we can write

$$T_t = kT_{t-1} + C_t \tag{1}$$

If the number of new colonists is steady at some equilibrium value C, the system will settle to have an equilibrium number of territorial birds T^*

$$T^* = C/(1-k) \tag{2}$$

This formula makes explicit the way in which the equilibrium breeding density depends on the factors C and k, both of which are exogenous to the temperate forest island under study.

Suppose now that C_t has been steady at the value C_b (b for "before") for many years, but that as a result of deforestation and fragmentation it changes to the lower value C_a (a for "after") for $t \geqslant 1$. The original equilibrium number of breeding birds

$$T_b^* = C_b/(1-k) \tag{3}$$

will move to a new, lower equilibrium value

$$T_a^* = C_a/(1-k) \tag{4}$$

The dynamics of the process are described by Eq. (1). We find, in successive years starting from $T=T_b^*$ in year $t=0$ and using Eq. (4) to rewrite C_a

$$T_1 = T_a^* + k(T_b^* - T_a^*) \tag{5}$$

and

$$T_2 = T_a^* + k^2(T_b^* - T_a^*) \tag{6}$$

and in general

$$T_t = T_a^* + k^t(T_b^* - T_a^*) \tag{7}$$

That is, T relaxes steadily and monotonically (at a rate dependent on the magnitude of k) from its original value, T_b^*, to the new, lower value, T_a^* (Fig. 9-1a, c).

Another possible change can come from a decrease in k, the overwintering survival probability. Destruction of tropical habitat for the species is one possible source of such diminishing k values. Specifically, suppose k has the value k_b for the pristine system, and that it suddenly decreases to the lower value k_a for $t \geqslant 1$. The analysis is similar to that above, and leads to the conclusion that the number of breeding birds in the temperate forest patch will decrease from

$$T_b^* = C/(1-k_b) \tag{8}$$

to the lower equilibrium value

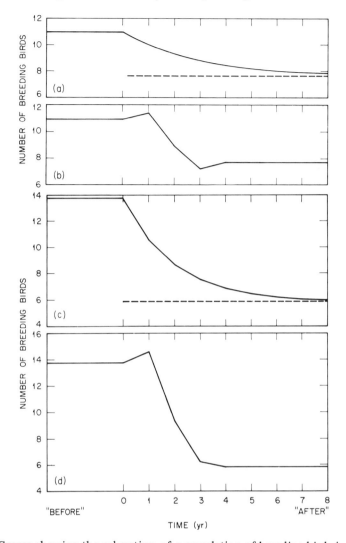

Fig. 9-1. Curves showing the relaxation of a population of breeding birds in a temperate forest island to new equilibria, following a reduction in the annual number of new colonists (see text for details). (a) The system here is described by the simple model with pure Type II survivorship and no age structure. Specifically, the parameters are chosen to be C_b = 5, C_a = 3.5, and k = 0.7, when $T_b^* \simeq 11.0$ and $T_a^* \simeq 7.7$. (b) As above, except now the system is described by the more complicated model, with age structure and with extra colonization by displaced breeders in the first year following disturbance. The parameters are as in (a), but in addition, we have D = 2.5. This curve illustrates the initial overshoot and final undershoot that can arise for particular combinations of parameter values. Also notice that the final equilibrium is reached after four time steps, in contrast with the asymptotic approach to equilibrium in the simple model with pure Type II survivorship. (c) As in (a), except here C_b = 7, C_a = 3, and k = 0.6. Consequently, $T_b^* \simeq 13.7$ and $T_a^* \simeq 5.9$. (d) This curve bears a relation to (c) similar to that of (b) to (a). The parameters are as in (c), but in addition, D = 4. Here there is still an initial upswing produced by "displaced" birds, but the subsequent decline to the new equilibrium is smooth (with no undershoot).

$$T_a^* = C/(1-k_a) \tag{9}$$

Note that this temperate-zone phenomenon is here possibly caused by tropical-zone events. The dynamics of the process can be shown to be similar to those analyzed above; for $t \geqslant 1$ we have

$$T_t = T_a^* + k_a^t(T_b^* - T_a^*) \tag{10}$$

Again, the decline is monotonic and at a rate that is most rapid when k_a is small.

Because decreases in survival probabilities, k, and in numbers of new colonists, C, would have similar effects on T_t, we cannot distinguish the roles of survival and colonist numbers in the declines of neotropical migratory species described in Chapter 8. Perhaps a combination of both effects is actually occurring.

A Model with Age Structure

As long as age structure is ignored, and survival described by the pure "type II" mortality factor k, changes are bound to occur smoothly. We now sketch a more complicated model in which the patterns of change can manifest "overshoot" and "undershoot."

Specifically, consider a neotropical migrant species that lives up to about four years. That is, we assume: the probability for a first year breeder to survive and return to breed a second year is k; likewise, the probability for a second year breeder to return for a third year is k; but after this, all birds die. In any given year, t, we then have three age classes of breeding birds labeled $T_t(1)$, $T_t(2)$, and $T_t(3)$, respectively. The extension to more general age-specific survivorship assumptions is obvious; our specific assumptions (corresponding roughly to the American redstart, for example) help keep the analysis more concrete.

As before, we assume that originally a constant number of new colonists, C_b, enter our forest island each year. In the initial, undisturbed system, these are all first-year breeders (last year's fledglings). In later years, they will be territorial birds returning to the previous year's territory. Thus, equations can be written for each age class separately.

$$T_t(1) = C_b \tag{11}$$

$$T_t(2) = kT_{t-1}(1) \tag{12}$$

and

$$T_t(3) = kT_{t-1}(2) \tag{13}$$

The total number of breeding birds is

$$T_t = T_t(1) + T_t(2) + T_t(3) \tag{14}$$

The equilibrium in the original system is therefore

$$T_b^* = (1+k+k^2)C_b \tag{15}$$

Of these, the proportions of first, second, and third year birds are $1:k:k^2$. [The earlier result, Eq. (3), is recovered by remembering that birds simply survived from year to year with probability k, regardless of their age: the result would be a sequence $1+k+k^2+k^3+\dots$ in Eq. (15). This infinite series sums to $1/(1-k)$, thus producing Eq. (3).]

Now suppose that a significant amount of deforestation and fragmentation takes place after year $t=0$. There are two effects. First, the total number of fledglings produced in year $t=1$ will be smaller, and hence the pool of new colonists will decrease; for $t > 1$ we assume a steady smaller value for the number of new colonists, C_a. Second, year $t=1$ will be special in that many second and third year breeders (as well as the newly arriving first year breeders) will find their territories gone, and will be forced to look elsewhere. We assume, consequently, that there is an extra pool of new colonists, over and above the "usual" C_b first year birds, arriving in our patch in year $t=1$. We arbitrarily assume that this pool of new colonists produced by disturbance of, and displacement from, other patches provides first, second, and third year breeders in the ratio $1:k:k^2$. This assumption allows for the second and third year breeders (whose numerical abundance is less than the $1:k:k^2$ ratio) to compensate for their lower numerical abundance by their greater "experience." We take the total number of new colonists to be $(1+k+k^2)D$, corresponding to the assumption that in year $t=1$ the number of new colonists in the three age classes are

$$C_1(1) = D \tag{16}$$

$$C_1(2) = kD \tag{17}$$

and

$$C_1(3) = k^2D \tag{18}$$

The total number of breeding birds in our patch, enumerated by age class, are

$$T_1(1) = C_1(1) \qquad = D \tag{19}$$

$$T_1(2) = kC_b + C_1(2) = k(C_b + D) \tag{20}$$

and

$$T_1(3) = k^2C_b + C_1(3) = k^2(C_b + D) \tag{21}$$

Here we have used the fact that $T(1)=C_b$, $T(2)=kC_b$ and $T(3)=k^2C_b$ for all times $t \leqslant 0$. For $t > 1$, the only new colonists are the C_a first year birds; all the effects of displacement are assumed to be sorted out in the first year. We can then see the disturbance work its way through the age structure of the population:

$$T_2(1) = C_a \tag{22}$$

$$T_2(2) = kT_1(1) = kD \tag{23}$$

$$T_2(3) = kT_1(2) = k^2(C_b + D) \tag{24}$$

and

$$T_3(1) = C_a \tag{25}$$

$$T_3(2) = kT_2(1) = kC_a \tag{26}$$

$$T_3(3) = kT_2(2) = k^2 D \tag{27}$$

Finally,

$$T_4(1) = C_a \tag{28}$$

$$T_4(2) = kT_3(1) = kC_a \tag{29}$$

$$T_4(3) = kT_3(2) = k^2 C_a \tag{30}$$

The population is now, four times steps later, at its new equilibrium, appropriate to the new, lower rate of influx of new colonists ($C_a < C_b$).

Of main concern is the *total* number of breeding birds. Using the general expression (14) on Eqs. (19)-(30), we have

$$T_1 = (k+k^2)C_b + (1+k+k^2)D \tag{31}$$

$$T_2 = C_a + k^2 C_b + (k+k^2)D \tag{32}$$

and

$$T_3 = (1+k)C_a + k^2 D \tag{33}$$

For $t \geqslant 4$, the system is in the new equilibrium,

$$T_a^* = (1+k+k^2)C_a \tag{34}$$

which is in contrast with the old equilibrium value of Eq. (15).

The situation described above can show, with appropriate values for D (in relation to C_b, C_a, and k), an initial *rise* in the number of breeding birds ($T_1 > T_b^*$), followed by a fall to levels that go *below* the ultimate equilibrium ($T_3 < T_a^*$). To be explicit, there will be a rise in year $t=1$ if $(1+k+k^2)D > C_b$, and a dip below the final equilibrium level if $C_a > D$. Since $C_b > C_a$, and $k < 1$, there is only a rather narrow window of D values for which both overshoot and undershoot are exhibited (see Fig. 9-1b, d). The cause of the initial rise is the large number of new colonists produced by displacement from disturbed or destroyed habitats. The dip, enroute to equilibrium, is less intuitively explicable, but is basically due to the effects of age structure (whereby the upsurge of displaced second and third year breeders in year $t=1$ can produce an "over-aged" population, and a dip in year $t=3$).

The presentation of this age structured model has been horribly heavy-handed. It has the virtue of illustrating the complicated events that can occur once age structure is accounted for. The simpler model that was discussed first could describe an initial upswing in breeding birds (due to displacement effects), but could not explain an "undershoot" (because such effects are explicitly related to age structure).

The simple model exemplifies some of the general points about island biogeography and neotropical migration that were made by Whitcomb et al. (Chapter 8). The age structured model, although less transparent, shows how quite complicated dynamical responses to changing environmental circumstances can be described by a few basic and sensible assumptions.

Acknowledgments. I am indebted to J. F. Lynch and R. F. Whitcomb for their guidance in this work, and to the NSF for their partial support under grant DEB 77-01565.

10. Modeling Seed Dispersal and Forest Island Dynamics

W. Carter Johnson
Department of Biology
Virginia Polytechnic Institute and
State University

David M. Sharpe
Department of Geography
Southern Illinois University

Donald L. DeAngelis
Environmental Sciences Division
Oak Ridge National Laboratory

David E. Fields
Health and Safety Research
Division
Oak Ridge National Laboratory

Richard J. Olson
Environmental Sciences Division
Oak Ridge National Laboratory

Structural modification of natural landscapes is a major consequence of settlement. In much of the Eastern Deciduous Forest Biome, habitat fragmentation has progressed to the point where remnant forest patches exist as more or less isolated islands surrounded by agricultural and urban land. These forest islands are of varying size, shape, degree of isolation, and pattern. The surrounding area, or landscape matrix, often contains structural corridors such as fencerows that serve to connect isolates, but the connections between patches differ widely among dissected landscapes. In addition to fields, fencerows, and settlement, the landscape matrix includes patches of seminatural vegetation (e.g., old fields) that may in time be sites of new islands. Other matrix refuges for forest species include roadside verge, untilled corners of fields, and suburban lots. Clearly, modification of landscape organization of the magnitude realized in the eastern deciduous forest suggests concomitant change in the pattern and process of landscape ecosystems.

The ecological consequences of altering landscape structure have just begun to be assessed, despite the fact that culturally fragmented ecosystems have long been a characteristic of both American and European landscapes. These principles, first explored for ocean islands (Preston 1962, MacArthur and Wilson 1963, 1967, Diamond 1972), generally have been found also to apply to landscapes (Elfstrom 1974, Forman and Elfstrom 1975, Scanlan 1975b, Tramer and Suhrweir 1975, Forman et al. 1976, Galli et al. 1976), but there are complicating factors. For example, terrestrial island shape is an important determinant of richness because it strongly influences microclimate (Levenson 1976, Wales 1972, Ranney 1978). As the perimeter/area ratio increases, the island interior becomes more xeric and floristic richness declines. Also, for circular islands there is a threshold size below which a mesic interior does not develop and mesic species are thus excluded. A landscape of islands all below the threshold size would be expected to contain fewer total species than a landscape of fewer larger islands of the same total area. Total floristic richness has also been shown to be negatively correlated with the degree of isolation, suggesting that dispersal of some species is inadequate to bridge interisland distances.

Other general studies indicate that a major consequence of anthropogenic dissection of forests is alteration of seed dispersal patterns which ultimately affect composition, structure, and successional development of remnant patches and their aggregate, the regional ecosystem. Auclair and Cottam (1971) found that some oak islands in south-central Wisconsin were being invaded by *Prunus serotina* rather than by *Acer saccharum,* the climax dominant in the region. They attributed this to a number of factors, especially the isolation and disturbance of the woodlots and the high dispersibility of the bird-carried *Prunus* seeds relative to the heavy, wind-dispersed seeds of *Acer.* Gomez-Pompa et al. (1972) noted a similar problem in the tropics of Latin America where regeneration of certain tall upper-canopy tree species was thought to be declining because of their inability to recolonize large areas once opened to intensive agriculture. Curtis (1956) theorized that forest patch isolation caused local species extinctions that would not have occurred in presettlement landscapes.

The effects of spatial arrangement on interisland seed dispersal and seed pools, and the impacts of seed dispersal and patch seed budgets (Kellman 1974) on the dynamics of forest islands are ecologically significant in rapidly changing landscapes. A tenable hypothesis is that the exogenous component of the seed budget of a forest island is, in part, a function of its spatial relationship with respect to other forest patches and the quality of the intervening matrix. As a result, species replacement patterns and subsequent steady-state composition should differ between isolated and contiguous patches, as suggested by the results of Auclair and Cottam (1971). Matrix quality affects the composition, distribution, numbers, and movement pattern of animal dispersers and hence the deposition pattern of seeds. Also, different landscape configurations may have differential effects on the relative success of wind- or animal-dispersed species. For example, heavy, wind-dispersed species may be at a competitive disadvantage relative to light, wind-dispersed or bird-dispersed species. If matrix features constitute barriers to dispersal, species that are preadapted for continuous disturbance will be favored (Harper et al. 1970). Therefore, there appears to exist a complex feedback relationship between landscape pattern, matrix quality, and forest island structure, composition and dynamics as mediated by seed dispersal.

In this chapter, we report on the development, modification, and use of several mathematical models for the analysis of dispersal and vegetation dynamics in landscapes of forest islands. We consider two basic components: (1) the effect of island pattern and matrix quality on the amount and quality of exogenous seed dispersed to island ecosystems, and (2) the overall effect of altered seed supply, both in numbers and composition, on ecosystem attributes. For the analysis of component (1), we have developed two new models of seed dispersal, one for seeds dispersed primarily by wind and the other for seeds dispersed by animals. Briefly, the wind dispersal model (SEDFAL) (Fields and Sharpe 1980), computes the pattern of seed rain around a seed source based on specific meteorologic conditions, seed supply, and seed fall velocity. The model is used to show the annual pattern of dispersal of *Acer saccharum* seeds around a source tree for meteorologic conditions at Minneapolis, Minnesota.

The animal dispersal model traces the movement of individual animals assumed to be carrying seeds from source trees. An animal takes one spatial "step" or "wing-beat" at a time, the direction being selected randomly but biased by the animal's behavioral interactions with the environment. At each step there is a certain probability that a seed or seeds will be deposited or buried. The utility of the model was demonstrated by simulating the dispersal pattern of acorns by jays in a fabricated landscape with islands differing in their attractiveness as caching sites. For the analysis of component (2), we have modified an existing stand-level forest simulator (FOREST, Ek and Monserud 1974, Monserud 1975) to examine the effect of variable seed rain on the successional development of vegetation. Thus, FOREST served as a processor of the seed supply information produced by the dispersal models. Simulations show the rates of invasion of *Acer saccharum* into a pure stand of *Prunus serotina* based on three levels of seed supply of *Acer saccharum* generated from the wind dispersal model.

The Wind Dispersal Model

Introduction

Past investigation of seed dispersal by wind has had three dominant features. First, it has focused on conifers, the major wind dispersed species of commercial interest. Second, the issues have centered on the area over which seeds can be dispersed to provide satisfactory regeneration after timber harvest. Third, conclusions have been derived primarily from seed trap studies at particular sites.

In remnant forest island stands, isolation and disturbance together may lead to ecological effects that differ from effects in contiguous stands. One issue is the impact of the spatial arrangement of seed sources in the landscape on particular forest islands in which exogenous seed supplies are the sine qua non for continued species replacement. Here it is difficult to measure seed inputs directly, or to observe the long-term effect of isolation. Hence, the model SEDFAL was developed to compute total annual seed-fall at various points surrounding an isolated seed source, e.g., a tree or forest island, in order to evaluate the seed input to outlying natural patches.

Developing a Concept of Seed Dispersal by Wind

Seed dispersal by wind has been studied primarily through observations of numbers of seed falling into seed traps at various distances and directions from seed sources (Alexander 1969; Bjorkbam 1971, Dobbs 1976, Randall 1974, Roy 1960). Weekly or daily measures have identified the phenology of seed fall for some species (Harris 1969, MacKinney and Korstian 1938, Gashwiler 1969, Zasada and Viereck 1970). The cumulative seasonal seed fall has been used to describe how seed fall changes annually with distance from seed source.

Yet, the observed pattern of dispersal is the cumulative effect of numerous individual incidents of release of seeds from a plant and subsequent flight in response to gravity and wind. Each seed presumably could have a different time of release and move in response to different conditions of atmospheric motion. SEDFAL captures the essence of these processes as they control the distribution of dispersed seeds.

Dispersal Distance. One major finding of seed dispersal studies is that the distances that seeds are wind-borne vary by several orders of magnitude. Most seeds appear to fall within a short distance of their source. The conclusion that seeds are dispersed in adequate numbers for forest regeneration to a distance equivalent to only a few times the height of seed trees is firmly established by seed trap studies (Boyer 1958, Gashwiler 1969, Jemison and Korstian 1944). By contrast, incidents of long distance dispersal (on the order of kilometers) fix upper bounds. For example, storms have transported *Populus* seeds 30 km; birch, 1.6 km; and ash, 0.5 km. Scotch pine seeds have been transported 2 km, while maple has been carried 4 km (van der Pijl 1972).

One reason for these widely diverging reports of proximal and long distance dispersal is the markedly variable wind speeds to which seeds are exposed, from near calm to gale forces. Second, the rate of fall of seeds of a given species varies, even in calm air. McEwen (1971) reports that the average fall velocity of a sample of black spruce (*Picea mariana*) seeds was 61 cm/sec, with relatively small variation (10%) between individual seeds. Siggins (1933), in a classic paper, discussed the results of timing seeds in a 48.8 m fall down the elevator shaft of the campanile at the University of California at Berkeley. He, too, found that most seeds of a given species have a characteristic fall velocity (V_f), but that some fell either faster or slower. Results for one sample of loblolly pine (*Pinus taeda*) showed an average V_f of 124 cm/sec, with a range from 74 cm/sec to over 800 cm/sec. Furthermore, he found that the average V_f of seeds from seven different trees from the same locality in Virginia varied from 116 cm/sec to 149 cm/sec, and averaged 128 cm/sec. These, in turn, were all slower than the 162 cm/sec V_f of a lot of loblolly pine seeds from Texas. Seeds of different species, too, have different characteristic V_f's. Of the twelve species of pine studied, *Pinus contorta* had the lowest, 82 cm/sec, while *Pinus lambertiana* had the highest, 265 cm/sec. Characteristic fall rates of 13 other conifer species and of yellow birch (155 cm/sec) and tulip poplar (189 cm/sec) were also reported. McCutchen (1977) gave average values of V_f for tulip poplar (156 cm/sec), Norway maple (107 cm/sec), and *Ailanthus* (122 cm/sec), but no data on the variability of V_f.

The results of variable V_f have been studied by Isaac (1930) for a number of western U.S. conifers, and by Mair (1973) for Sitka spruce growing in Scotland. Both found that seedfall increased downwind to some distance, then declined beyond that point. Seedfall of Sitka spruce released from a height of 15 m peaked at about 9 m in a wind speed of 0.8 m/sec; at about 19 m at a wind speed of 1.9 m/sec; and had a double peak, one at 13 m, the other at 26 m at a wind speed of 3.4 m/sec (Mair 1973).

Greater release heights result in disproportionately greater dispersal distances (Isaac 1930). Seeds released at a wind speed of 3.1 m/sec from a height of 61 m were first recorded 61 m from the release point, peaked at a distance of over 300 m from the release point, and were still present to distances over 975 m. Seeds released under similar wind conditions but at a height of 30.5 m peaked at 122 m distance and were not recorded beyond 244 m. In each case, if all seeds had the same V_f, they would have landed en masse.

Cremer (1971) points out that all seeds falling in steadily moving air will travel a distance determined by time of flight and the mean wind speed along the seed's trajectory. For morphologically similar seeds (seeds from the same species of tree), an equation relating distance traveled to other important parameters is

$$D = V_w \times H/V_f$$

where D = distance of dispersal, V_w = mean horizontal wind velocity between ground and release height, H = release height, and V_f = fall velocity. The distribution of fall velocities is thus transformed into a distribution of dispersal distances.

Yet, the distances of seed dispersal measured by Isaac (1930) and Mair (1973) are considerably shorter than those noted as examples of long-range dispersal. Two conditions may account for this: long-range dispersal by occasional high winds that are not represented in the suite of conditions studied; and atmospheric turbulence that transports falling seeds to higher altitudes from which they can glide. The former can be analyzed by considering the full spectrum of wind speeds to which seeds may be exposed. The latter has been treated theoretically. Rombakis (1947) computed the probable flight path of seeds, taking into account their V_f, wind speed, and turbulence. Consideration of turbulence had the effect of prolonging seed flight time as seeds were first carried above their release height, then glided downward from this vantage point. Both Mair (1973) and Siggins (1933) observed seeds being carried upward by turbulence during the course of their experiments.

Although it would be desirable to include the effects of turbulence on seed dispersal, we conclude from a review of the literature on wind in forests that too little is quantitatively known about the phenomenon to draw generalities useful in this context (Moen 1974, Reifsnyder 1955, Bergen 1975).

Timing of Seed Release. A number of investigators have pointed out that the timing of seed release is not solely a function of the maturation date of the seed crop. Rather, seed release may extend for weeks or months, and occur at different times in successive years. Wind speed and/or drying conditions for the seeds or cones are major contributors to seed release timing. For example, Cayford (1964) noted that red pine seedfall in southeastern Manitoba had an unusual timing in 1959. Rather than the normal September-October pattern, only 18% of the seeds had fallen by the end of October; most fell the following spring. Cayford related his observations to heavy precipitation, low sunshine, and low temperatures during September and October, 1959.

Mair (1973) reported that seed release by Sitka spruce, western hemlock, and Douglas fir in Scotland was stimulated by dry easterly winds. Ruth and Bengtsen (1955) found that the first day of seedfall came with the first easterly winds after seed maturation, which usually caused a heavy seedfall. Godman (1953) and Harris (1969) found that seed dispersal in hemlock-spruce stands in southeastern Alaska was closely associated with atmospheric moisture, and that most seed is released during a few days of dry weather. Similar reports were given for loblolly pine by Jemison and Korstian (1944), and MacKinney and Korstian (1938). Allen and Trousdell (1961) noted that "after initial opening, loblolly pine cones tend to close in wet weather, restricting seedfall until they dry out and reopen." That is, pine cones exhibit xerochasy (van der Pijl 1972). Geiger (1966) draws upon a study by Kohlermann (1950) to relate seed release by pines to relative humidity. Few seeds were released when relative humidity was above 75%. Most were released during the frequent periods of 65-75% relative humidity, although significant numbers were released at lower humidity levels. However, as Siggins (1933) points out, cone scales in pines open from the base toward the apex as they dry out, so seeds are gradually dispersed from each cone rather than being dispersed simultaneously.

All of the above observations of humidity effects are confounded by unknown effects of wind speed on seed release. The reported periods of heavy seedfall came during periods of dry *and* strong winds (Jemison and Korstian 1944, Godman 1953). Sheldon and Burrows (1973) report that seed release of some of the Compositae does not occur until a threshold wind speed has been attained, although the degree of ripeness affects the threshold value. They further note that dispersal may be severely hampered by the lack of suitable winds at the time seeds mature.

Consequently, the direction and distance of seed dispersal need not be a function of prevailing winds. Quite the contrary has been reported when prevailing winds are associated with high humidity and/or precipitation (Benzie 1959, Ruth and Bengtsen 1955, Shearer 1959). However, except for Kohlermann's work reported by Geiger (1966), no specific relations have been drawn that are useful for simulating seedfall. We know only that seedfall is inversely related to humidity (e.g., precipitation, vapor pressure, or relative humidity) and directly related to wind speed.

Summary/Synthesis. Seed dispersal by wind thus involves a complex of processes whose outcome is difficult to predict. Seeds released from the upper crowns of emergents may be carried upward by local wind turbulence, carried tens or hundreds of meters into the air, and then allowed to glide laterally along a trajectory dictated by the turbulence encountered, seedfall velocities, and horizontal wind speed. By contrast, seeds caught in a downdraft, released in stable conditions and light winds, or released from portions of the canopy sheltered from the wind will be dispersed only a short distance. If these patterns occur in an extensive stand homogeneous in structure, short and long dispersal distances from various source regions will balance to create a uniform seedfall throughout.

A clearing in such a forest will be a sink for windblown seeds from the surrounding stand. The pattern and quantity of deposition within the open patch are functions of the quantity of seeds that are carried above the canopy, their trajectories and the wind field above the clearing, and the seeds swept through the canopy and to the subcanopy space of the stand along the border. While most seeds may originate from trees surrounding the clearing, it is a mistake to assume that all seeds do. The cumulative effect of combinations of seed release/transport conditions for short periods throughout the season of seed dispersal, together with the size of the seed crop, determine the seed input to the clearing. If light winds prevailed during a good seed crop year, the dispersal to a locale removed from the seed source might not exceed that brought about by a year with a moderate or poor seed crop, but exceptionally high winds. Likewise, the sequence of humidity/wind conditions during the period after seed maturation will control dispersal distances; i.e., the fortuitous combination of drying and transport conditions dictate distances that seeds are dispersed.

The quantity and distribution of seeds dispersed from discrete forest islands are the issues that provide the impetus for this modeling effort. The meteorologic and aerodynamic complexity of an extended forest with a clearing, especially the complexity of the wind field over the canopy, makes the problem currently intractable. Direct observation is the only recourse until we have a better understanding of forest meteorology. Likewise, the turbulence set up by larger forest islands currently poses great problems in predicting seed dispersal patterns. Our strategy, then, is to consider the case of the smallest forest island, a single tree, which can be assumed to have zero effect on the wind field. By so doing, the studies of Siggins (1933) and Isaac (1930) provide limited extant data for model validation.

A seed dispersal model should account for:

1. The variable fall velocity of windblown seeds, which leads to the variable dispersion of seeds under given conditions of wind speed and direction (Siggins 1933, Isaac 1930, Mair 1973).

2. The phenology of seed release, i.e., the month of initial seed release under normal conditions and the duration of the release period (Schopmeyer 1974).

3. The impact of short-term meteorologic events on seed release. These include the effect of humidity and/or temperature on the opening of cones and seed abscission (Siggins 1933, Kohlermann 1950, Chaney and Kozlowski 1969), and the effect of wind speed on seed release.

4. The coincidence of the conditions that bear on the presentation of seeds for wind dispersal, phenology and short-term meteorologic events, and the wind conditions that lead to dispersal. This is approximated by an array that shows the frequency of combinations of wind speed/wind direction/humidity during months of seed dispersal.

5. The wind speed profile between the ground and the height of seed release to obtain the mean wind velocity to which the seeds are subjected, and the effects of surface roughness on the wind speed profile.

Model Description

SEDFAL simulates seed dispersal about a point source of specified strength (i.e., number of seeds to be dispersed) [see Fields and Sharpe (1980) for a program description]. The driving force for seed flight to varying distances and directions is based on average wind rose data from NOAA. These give the proportion of hourly observations during each month having various combinations of wind speed and wind direction categories plus temperatures and relative humidity. The wind rose data are presented as an array whose axes are windspeed classes, wind direction classes, and vapor pressure deficit classes. The latter are computed from relative humidity and temperature by the Goff-Gratch saturation vapor pressure equation (Goff 1965) which is a definitive expression of drying conditions. Typically, 16 wind directions, 10 wind speed classes, and five vapor pressure deficit classes are considered.

Seed release is apportioned among the various windspeed-direction-humidity classes for the primary months of seed dispersal. The total number of seeds available for dispersal during each month is specified in the model, and seeds are allocated among the cells of the wind rose array in accord with the assumption that seed release is directly and linearly proportional to wind speed and vapor pressure deficit. Seed release is assumed to increase with increasing VPD at a given wind speed and with the increasing shear stress of higher wind speeds under given drying conditions. Highest seed release thus occurs during periods when highest VPD and highest wind speed prevail. Thus, we assume that postponement of seed dispersal during calm and/or wet periods occurs, but only within each month of normal seed dispersal. Also SEDFAL does not account for the sequencing of synoptic events, so that the effect previous weather events have on seed dispersal during subsequent events cannot be assessed. In this sense, SEDFAL is generalized to account for average conditions.

Both wind velocity at the specified height of seed release and the average wind velocity experienced by seeds during seed flight are computed from the mean of each of the 10 wind velocity classes on the assumption that the wind velocity profile is logarithmic. The mean of each wind velocity class is adjusted before its effect on seed release and dispersal is assessed.

The seed dispersal pattern for each month of significant seed dispersal (Schopmeyer 1974) is determined by mapping the histogram of seeds vs. fall velocity onto each of the 16 radial transects of the wind rose. This is done for each wind velocity class expressed in the wind rose data.

Model Simulation

A sample simulation run of SEDFAL shows sugar maple (*Acer saccharum*) seeds distributed under climatological conditions typical of the upper midwest (Fig. 10-1). The frequency distributions of wind speed, direction, and vapor pressure deficit were based on a 10-year record of hourly observations for the months of seed release (September, October, November) at Minneapolis, Minnesota. The seed supply was one million. The asymmetry of dispersal (Fig. 10-1) coincides with prevailing southwest winds during the dispersal season. Order-of-magnitude differences in seed densities occur at comparable distances but in different compass directions. To illustrate, density at 400 m in the northeast quadrant is approximately 1 seed/m^2 while only 0.1 in the southwest quadrant at the same distance. Seed rain density near the source (Fig. 10-1) is over 40 seeds/m^2 but declines to 1 seed/m^2 within several hundred meters. On the basis of this simulation, an adjacent forest island northeast of the source would receive approximately 40 *Acer* seeds/m^2, an island 350 m distant would receive an order-of-magnitude fewer seeds, and an island 550 m distant, two orders-of-magnitude fewer. These differences correspond to approximately 200 m and 300 m in the southwest quadrant. The effect of order-of-magnitude differences in seed supply on the successional development of vegetation is evaluated below.

Several conclusions emerge from the model simulation. First, the asymmetry of the dispersal pattern shows that the juxtaposition of forest islands with respect to prevailing winds can have a large effect on the quantity, and presumably the quality, of exogenous seed dispersed to other patches. Thus, it is plausible that abandoned agricultural patches would revegetate quite differently if positioned at different compass directions around a major source of seeds such as a forest island. Second, in the absence of updrafts, maple seed density is extremely low (1 seed/m^2) beyond 200–400 m from a

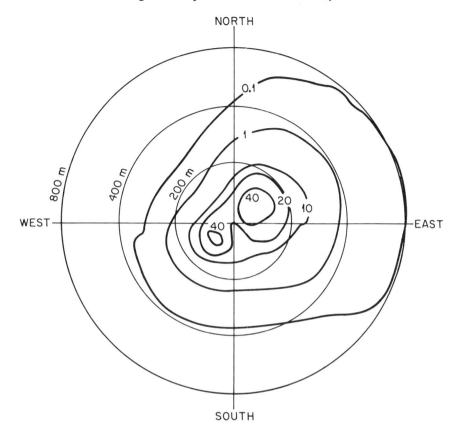

Fig. 10-1. Plot from the wind dispersal model, SEDFAL, showing the simulated dispersal pattern of maple seeds (*Acer saccharum*) for climatological conditions characteristic of Minneapolis, Minnesota. Isolines are seeds/m^2. The seed source, a single tree producing one million seeds, is located at the center.

seed source. Given the rather high probability that the 1 seed/m^2 is either inviable (empty) or is consumed, it is doubtful that sugar maple would make significant inroads into a nonmaple stand if the seed source were further than 200-400 m away. Thus, without updrafts, this could be considered a reasonable threshold distance for significant maple dispersal. Updrafts would certainly increase this threshold distance, but by how much is unknown. In a 184 km^2 area in southeastern Wisconsin, only 31 forest patches (ca. 10-15%) were within 250 m of their nearest neighbor (J. Tyburski, unpublished data).

Similar analyses of other wind dispersed species would enable us to assess the probable effects of the isolation of seed sources on their contribution of seeds to other forest islands in a landscape. We would probably find that equal interstand distances create dissimilarities in "functional" isolation between species. These may be expressed in relation to landscape pattern and successional pattern, and eventually in the ability of species to maintain themselves in the landscape.

The Animal Dispersal Model

General Description

This model simulates the dispersal activities of mammals and birds and their effects on seed dispersal. The direction of travel is a function of the animal's forward inertia, and the presence of attractors or repellers, including habitat boundaries or ecotones. At each step, there is a certain probability that a transported seed will be deposited or buried, and for some animals, this probability depends on the characteristics of the habitat or cover. Model output is a printed map of the seed deposition pattern or the animal movement pattern.

There are several advantages in a model of this type. First, it allows great flexibility in dealing with functional or behavioral differences among animal dispersers, and will deal as easily with a flying bird as a walking raccoon. Second, although spatial arrangements of land use or habitat types remain relatively stable over short periods of time, the interactions of different species with the same environment are often quite different. For example, a bird may fly across an open field but a tree squirrel may, in most cases, avoid such crossing because of the lack of cover and consequent greater exposure to predation. Also, the interactions of one species with a given environment may change over short time intervals, e.g., diurnally, and the flexibility to handle these is incorporated in the model. Third, the model provides output of either total seed distribution by all potential dispersal agents for one plant species (or individual), or the effects of one dispersal agent in the movement of seeds of many species of plants.

In an attempt to predict rates of seed transport among forest remnants and, ultimately, the patterns of seed distribution, one of the most important factors involves the behavioral differences in seed-dispersing animals encountering different habitat types. Environments differ in relative favorability (or risk) for different animal species. These differences influence the probabilities of animals entering or leaving a habitat type and the rates and direction of movement within it. The spatial distribution and geometric form of habitat patches may also influence the direction of movement within a patch.

To simulate the motion of animals through a complex environment, it is convenient to identify units of distance or to divide space into discrete parcels. Consequently, the landscape is represented by a grid of points. An animal then moves through the landscape by moving from point to point in the grid. The scaling of the grid (the distance between adjacent points) is arbitrary, but distances are usually from a few to several tens of meters.

Determination of Direction of Animal Movement. Consider a seed-carrying animal located at some point (i,j) in the grid of points (Fig. 10-2). The animal can move to one of eight adjacent points $(i+\delta, j+\eta)$, where δ and η can take the values -1, 0, and $+1$ (but both cannot be 0 simultaneously since the model is based on movement events, not time steps). Assume that the following factors influence the next location of the animal:

1. The tendency for the moving animal to continue in the general direction in which it is already headed. This is termed the "forward inertia."

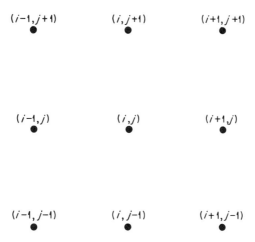

$(i-1,j+1)$　　　　$(i,j+1)$　　　　$(i+1,j+1)$
●　　　　　　　　　●　　　　　　　　●

$(i-1,j)$　　　　　(i,j)　　　　　$(i+1,j)$
●　　　　　　　　　●　　　　　　　　●

$(i-1,j-1)$　　　　$(i,j-1)$　　　　$(i+1,j-1)$
●　　　　　　　　　●　　　　　　　　●

Fig. 10-2. Grid of points in a model landscape used to simulate movement of animal seed dispersal agents.

2. The occurrence of attraction points, or "attractors," toward which the animal desires to move. For example, an animal may move preferentially through or toward forests.

3a. The occurrence of repulsion points, or "repellers," which the animal tends to avoid. For example, a seed-bearing animal may preferentially move away from the seed source.

3b. The boundaries or edges between habitats may also act as repellers. An animal that prefers one habitat over another may alter its path when it reaches a boundary.

These factors can be elucidated by examination of Fig. 10-3. Assume that the animal is located at point (i,j) and has just moved there from the point $(i-1,j+1)$. There is an attractor at point $(i+4,j-3)$ and a repeller at $(i-4,j+3)$. For simplicity, only two habitat types are pictured. The point (i,j) is in forest (shaded circles) and borders open field (open circles). The forest is an area that the animal prefers. It is likely that this animal would avoid the cultivated field. The most probable next movement of the animal might be to the point $(i+1,j-1)$, because the above factors positively influence such a move. A step to $(i,j-1)$ is also fairly probable. On the other hand, moves to any of the other six points are less likely, since these would be counter to the influence of one or more of the factors.

The identification of factors that can affect the direction of animal movement is convenient but quantification is still needed. There are no data on animal movements that allow unambiguous assignment of quantitative values to each factor. Routine field experiments do not fully distinguish their relative significance. But measurements using telemetry (e.g., Siniff and Jensen 1969), direct observation (e.g., Smith 1968), and use of radioactively tagged seeds (e.g., Tester 1963) provide some estimates of probability. Detailed mathematical characterization of the factors is given below.

Probability of Seed Deposition. Deposition probability is largely a function of the species of animal, seed or fruit (propagule) type, and habitat location. Types of depo-

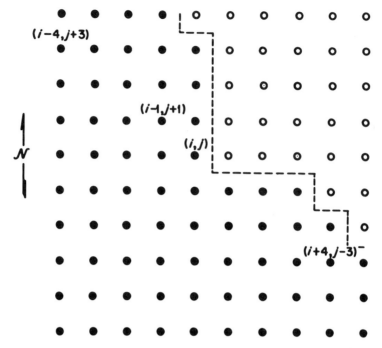

Fig. 10-3. Expanded grid overlain on two habitat types, forest (closed circles) and cultivated field (open circles), illustrating probabilities for directional movement of an animal constrained by attractor and repeller factors in the environment.

sition can be classed as passive (regurgitation, gut passage, attachment to fur), or active (scatter hoarding, caching). The deposition probabilities for each type require quite different estimation procedures. Passive deposition is largely a function of where a disperser spends time after feeding. Active dispersal, however, involves a disperser predisposed to approach a certain destination point after leaving the seed source area. Deposition in the latter case is complicated by choices regarding the habitat for burial or caching. A gray squirrel (*Sciurus carolinensis*), for example, has a greater probability of depositing a seed in a wooded area than in a cultivated field. In the present version of the model, it is assumed that, for a given habitat type, the probability of the animal dropping, burying, or caching a seed is the same for every unit of distance traveled. This assumption might be adequate for some types of seed dispersal, for example, seed attachment to fur, but may be poor for other types of dispersal, such as when seeds must pass through an animal's gut. In this case, the deposition location of seeds will be a function of the transmission time through the gut. Appropriate changes can be made in the model to accommodate almost any assumption concerning seed deposition probability.

Mathematical Description of the Model

Determination of Direction of Animal Movement. It is convenient to represent the probability of an animal moving one step from a point (i,j) to another point (k,m) in a two-dimensional grid as an element of a transition matrix, $p_{ij,km}$. Since the animal can only move from one grid point to an adjacent one in a single step, k and m are constrained as follows:

$$k = i + \delta \quad (\delta = -1, 0, +1) \tag{1a}$$

$$m = j + \eta \quad (\eta = -1, 0, +1) \tag{1b}$$

In all future discussions, k and m will be implicitly subject to these limitations.

The sum over all probabilities for direction of motion must equal unity:

$$\sum_{k=i-1}^{i+1} \sum_{m=j-1}^{j+1} p_{ij,km} = 1.0 \tag{2}$$

The model is event-oriented, where an event is a step in space. This means, given an animal initially at point (i,j), that the next moment of interest occurs only when the animal has moved to an adjacent grid point. Therefore, the probability of the animal being in its same position at the next time increment in the model is identically zero, or

$$p_{ij,ij} = 0.0 \tag{3}$$

All of the transition elements together define a transition matrix, P. Let $X(1)$ be the probability matrix for the position of the animal at a given moment. The elements of $X(1)$, which are $x_{ij}(1)$, represent the probabilities of the animal being located at any given point (i,j). The condition

$$\sum_{i=-\infty}^{+\infty} \sum_{j=-\infty}^{+\infty} x_{ij}(1) = 1.0 \tag{4}$$

must hold. Then

$$X(2) = \sum_{i=k-1}^{k+1} \sum_{j=m-1}^{m+1} x_{ij} p_{ij,km} = P \cdot X(1) \tag{5}$$

is the probability matrix for the position of the animal after its next movement to a new grid point.

If the movement of the animal from one grid point to the next is purely random (i.e., a "random walk"), then

$$p_{ij,km} = 1.0/8.0 = 0.125 \tag{6}$$

that is, there is an equal probability of 0.125 of the animal going to any of the eight adjacent points. However, as discussed above, the direction of animal movement is biased by its forward inertia, the presence of attractors and repellers, and the occurrence of habitat boundaries. Consider only the influence of forward inertia which introduces a directional bias on top of the random motion. The transition probability can then be written

$$p_{ij,km} = \left\{ 1.0 + I(k,m) \right\} / \xi \tag{7}$$

where ξ is the normalization factor

$$\xi = \sum_{i=k-1}^{k+1} \sum_{j=m-1}^{m+1} p'_{ij,km} \qquad (8)$$

and

$$p'_{ij,ij} = 0.0 \qquad (9a)$$

$$p'_{ij,km} = 1.0 + I(k,m) \quad (m \neq j, \text{ if } k = i) \qquad (9b)$$

The term $I(k,m)$, is a measure of the strength of forward inertia relative to random effects in determining the next grid point in the animal's course of movement. If $I(k,m) \ll 1.0$, then the random effects dominate the movement. On the other hand, if, say, $I(i+1,j+1) \gg 1.0$ and $I(i+1,j+1) \gg I(k,m)$ for all seven other pertinent values of k and m, then the animal is likely to move toward the "northeast" on its next step. The magnitude of $I(k,m)$ for particular values of k and m depends on the past motion of the animal. For this reason, P is not a Markov process matrix.

In a similar manner, the effects of attractors and repellers can be incorporated into this mathematical scheme. If $A_q(k,m)$ and $R_n(k,m)$ represent the strengths of the qth attractor and the nth repeller, respectively, then

$$p_{ij,km} = \left\{ 1.0 + I(k,m) + A_q(k,m) + R_n(k,m) \right\} / \xi \qquad (10)$$

where ξ is defined by Eq. (8), and now

$$p'_{ij,km} = 1.0 + I(k,m) + A_q(k,m) + R_n(k,m) \qquad (11)$$

In the model (as currently constructed), a maximum of one attractor and one repeller can affect an animal at a given moment. This constraint has been imposed for the sake of simplicity in the absence of knowledge on the possible interactions of numerous attractors and repellers.

The incorporation of the effects of boundaries between habitats requires a different approach. If $B(k,m)$ is a measure of the influence of habitat type on animal movement, then we assume that

$$p_{ij,km} = \left\{ 1.0 + I(k,m) + A_q(k,m) + R_n(k,m) \right\} B(k,m)/\xi \qquad (12)$$

where ξ is defined by (8), and now

$$p'_{ij,km} = \left\{ 1.0 + I(k,m) + A_q(k,m) + R_n(k,m) \right\} B(k,m) \qquad (13)$$

There is a reason for treating $B(k,m)$ differently from $I(k,m), A_q(k,m)$, and $R_n(k,m)$. If the animal is at some point (i,j) in forest and the points $(i+1,j)$ and $(i+1,j+1)$ lie in open field which the animal avoids (See Fig. 10-3), this can be simulated in the model by making

$$B(i+1,j) = B(i+1,j+1) = \beta_2 \qquad (14)$$

and also making

$$B(i-1,j-1) = B(i-1,j) = B(i,j-1) = B(i,j+1) = B(i-1,j+1) = B(i+1,j-1) = \beta_1 \qquad (15)$$

where $\beta_1 \gg \beta_2$. Since $p_{ij,km}$ for the six values of (k,m) in forest share the same factor, β_1, the relative probabilities of movement within forest depend on $I(k,m)$, $A_q(k,m)$, and $R_n(k,m)$. If $B(k,m)$ was added to the other factors as

$$1.0 + I(k,m) + A_q(k,m) + R_n(k,m) + B(k,m) \tag{16}$$

then $B(k,m)$ might tend to overshadow the other effects in forest, counter to our intentions. We want the animal to exhibit a strong tendency to stay on one side of the boundary, and we want its options on that side of the boundary to be governed by forward inertia and the presence of attractors and repellers.

It is now appropriate to discuss the detailed formulations of $I(k,m)$, $A_q(k,m)$, $R_n(k,m)$, and $B(k,m)$. These are developed in a simple and practical manner in the absence of definitive field measurements. Readers not interested in further mathematical details may wish to skip to the description of the model simulation below.

Inertia of Forward Movement, $I(k,m)$. Assume the animal is at point (i,j) and its preceding location was (i',j'), where

$$i = i' + \delta' \tag{17a}$$

$$j = j' + \eta' \tag{17b}$$

and where δ' and η' have the same ranges of values as δ and η [see Eqs. (1a) and (1b)]. Then $I(k,m)$, where k and m are given by Eqs. (1a) and (1b), is a conditional probability,

$$I(k,m) = \text{probability } (\delta,\eta \text{ given } \delta',\eta') \tag{18}$$

Where this probability is higher, the more positive the correlation becomes between (δ,η) and (δ',η'). In the model, a quantity, C, is defined, where

$$C = |\delta - \delta'| + |\eta - \eta'| \tag{19}$$

The bars represent absolute values of the enclosed differences. The quantity, C, can take on one of five different integer values, for each of which $I(k,m)$ is assigned a different value, e_i, as represented in Eq. (20),

$$I(k,m) = \begin{cases} e_1 & (C = 0) \\ e_2 & (C = 1) \\ e_3 & (C = 2) \\ e_4 & (C = 3) \\ e_5 & (C = 4) \end{cases} \tag{20}$$

where the constants e_i are chosen so that $e_1 > e_2 > e_3 > e_4 > e_5$. The model animal is likely to continue in the same direction because $I(k,m)$ is greatest when $\delta = \delta'$ and $\eta = \eta'$.

Attractor Terms, $A_q(k,m)$. Assume that the qth attractor is effective on the animal and is located at the point (i_q,j_q). The squares of the distances between the attractor and the eight possible succeeding points are computed. These are, respectively,

$$D_{ap} = (i_q - i)^2 + (j_q - j)^2 \tag{21a}$$

and

$$D_{af}(k,m) = (i_q - k)^2 + (j_q - m)^2 \tag{21b}$$

Then D_{ap} is compared with the eight values of $D_{af}(k,m)$ and the following equation employed to determine $A_q(k,m)$:

$$A_q(k,m) = \begin{cases} p_{a,q} > 0.0 & (D_{af}(k,m) < D_{ap}) \\ 0.0 & (D_{af}(k,m) > D_{ap}) \end{cases} \tag{22}$$

The quantitative value of the constant $p_{a,q}$ is assigned to reflect the strength of the attractor on the animal in question. To demonstrate how the attractor functions, consider that a squirrel has just crossed the edge of forest island 1 and has taken a step into the adjacent open field toward forest islands 2 and 3 which are connected by corridors (Fig. 10-4). In the model, any attractor positioned in the general direction of the animal's motion can become the effective attractor, and a new effective attractor is chosen every time the animal crosses a habitat boundary. If attractors are positioned in forest islands 2, 3 and corridor 6 each can differentially attract the squirrel. Points roughly in the centers of these regions are specified as the attractors. Only one of these will be the effective attractor at a given time. Which one actually functions as the attractor is decided with the help of a pseudo-random number generator.

Repeller Term, $R_n(k,m)$. Assume that a given repeller is located at the point (i_n, j_n). As in the case of the attractor, the distances D_{rp} and $D_{rf}(k,m)$ are computed and compared. Then $R_n(k,m)$ is given by

$$R_n(k,m) = \begin{cases} p_{r,n} > 0.0 & [D_{rf}(k,m) > D_{rp}] \\ 0.0 & [D_{rf}(k,m) < D_{rp}] \end{cases} \tag{23}$$

where

$$D_{rp} = (i_n - i)^2 + (j_n - j)^2$$

and

$$D_{rf}(k,m) = (i_n - k)^2 + (j_n - m)^2$$

Boundary Crossing Factor, $B(k,m)$. The function $B(k,m)$ is formulated in a very simple manner. Suppose there are N habitat areas or regions in the landscape in question. A constant is assigned to each habitat, e.g., β_n for the nth habitat. Expressed mathematically,

$$B(k,m) = \beta_n \left\{ (k,m) \text{ in region } n \right\} \tag{24}$$

where $n = 1, N$.

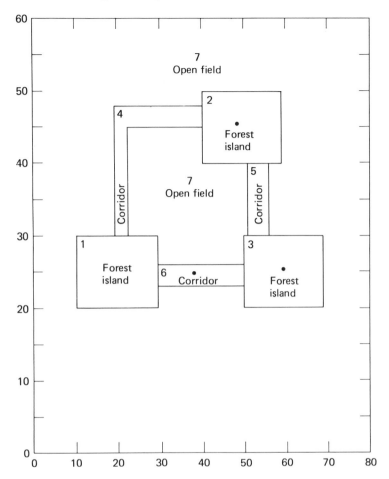

Fig. 10-4. Schematic diagram of forest islands (1, 2, 3) in a matrix that includes connections by corridors of reasonably similar habitat (4, 5, 6) such as fence rows or
drainageways, and separated and surrounded by open field (7).

Probability of Deposition of Seeds. Assume there are N habitat types in the landscape. A constant, s_n, is assigned to the nth habitat, to represent the probability of the
animal depositing the seed at a particular grid point in habitat n, to which the animal
has moved. For a given point (k,m), the probability of deposition is

$$S(k,m) = s_n \left\{ \ (k,m) \text{ in habitat } n \ \right\} \tag{25}$$

where $n = 1, N$.

Model Simulation

How accurately can the model simulate the movements and seed dispersal patterns of real animals? Earlier experience (DeAngelis et al. 1977) indicates that suitable choices of parameter values result in plausible simulations of important general types of animal movement and seed dispersal patterns. Here we parameterize the model with limited data from dissected landscapes in Europe to simulate the general dispersal of acorns by jays. The simulations raise hypotheses which need to be tested in man-dominated landscapes of eastern North America.

In the agricultural landscapes of Michigan, Linduska (1949) noted that red and fox squirrels (*Tamiasciurus hudsonicus loquax, Sciurus niger rufiventer,* respectively) commonly traveled to a number of different forest islands while gray and flying squirrels (*Sciurus carolinensis* and *Glaucomys volans volans*) were mostly restricted to single woodlots and border vegetation. But it is unlikely that squirrels carry acorns between islands. A large number of back and forth trips to scatter-hoard would contribute to high squirrel predation. It is more likely that squirrels scatter-hoard within respective source islands and/or nearby corridors for later retrieval. Rather, the blue jay (*Cyanocitta cristata*) is probably the most frequent long-distance (i.e., interisland) disperser of acorns in highly dissected landscapes. Avian scatter-hoarders such as jays and nutcrackers (Vander Wall and Balda 1977) seek out suitable caching sites, so that seeds may be transported long distances without any intermediate deposition.

General observations of the scatter-hoarding behavior of blue jays have been made, but to our knowledge, no detailed quantitative studies of their dispersal patterns in dissected landscapes of eastern North America have been conducted. However, dispersal information for the common jay (*Garrulus glandarius*) of Europe is available. Schuster (1950), Chettleburgh (1952) and Bossema (1979) working in Germany, England, and the Netherlands, respectively, observed the collection of acorns from source trees and their transport to burial sites up to 5 km from the source. Chettleburgh (1952) calculated that approximately 200,000 acorns were dispersed during one season by a flock of 35-40 jays. We observed from Chettleburgh's land use map, where the positions of burial sites were marked, that jays preferred to cache acorns in open scrub or open woodland conditions even though dense woods were located at equivalent or lesser distances. Apparently, a thin litter layer is required, but nearly all burial sites were in open conditions.

The animal dispersal model was parameterized to reflect the general findings in addition to what is generally known about jay movement. Flight between isolates is characterized by nearly straight-line movement, and boundaries between habitat types should have little hindering effect (as they would for ground-dwelling mammals such as squirrels). Therefore, to simulate the movement of a jay, the inertia of forward motion was set at a very large value and the boundary crossing factors, β_i, were made approximately equal to neutralize their effect. Birds flying away from source islands are strongly attracted to the wooded sites for seed burial (in contrast to agricultural fields), thus, $\rho_{a,q}$ was initially set equally high for the islands (by a factor of 20) and low for surrounding landscape. Initially, attractiveness of all isolates was equal, except for the seed source which was assumed to be a mature forest of large oaks and hence unattractive as a burial site. Because a litter layer is apparently required, the probability of seed deposition, S_i, in the forest was set high relative to the agricultural matrix.

With the above assumptions, a simulated deposition pattern of seeds is shown in Fig. 10-5. Island pattern and interisland distance (750 m) were generalized but are not atypical of landscapes in the eastern United States. Source trees were assumed to be located only in the easternmost island, where fencerows connected it with its nearest neighbors. Most seeds were dispersed to the northern and southern islands. Differences in numbers between the two islands are due to stochastic variability. The western island was equally attractive but had far fewer seeds dispersed to it. This dispersal pattern seems reasonable as there is an energetic advantage in flying to the closest patches, given that all are of equal attractiveness. A small fraction of the seeds was deposited in the agricultural matrix, corridors and the source island because small but finite deposition probabilities were assigned to each of these cover types.

A companion simulation examined the effect of habitat alteration of the western-most island which made it relatively more attractive to dispersing jays. A manipulation such as a clear-cut or heavy logging might create temporary open woodland conditions similar to those used as burial sites by European jays. A fivefold increase in the attractiveness of the westernmost island resulted in a large increase in the number of seeds deposited, relative to the pair of neighboring islands (Fig. 10-6). This second simulation is more likely to occur than the first because when dispersing over long distances, jays may focus on the highly attractive sites, thus improving the balance between the energetics of long-distance flights and the probability of recovering buried acorns.

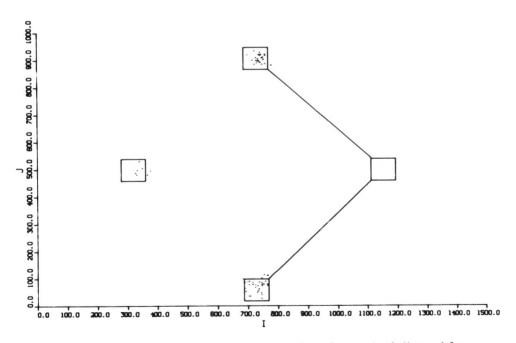

Fig. 10-5. Simulated seed deposition pattern resulting from animal dispersal from a seed source located in the eastern (right side) island. Forest islands are separated by 750 m, and the eastern island is connected to the northern and southern islands by fence rows (solid lines). Most seeds are deposited in the northern and southern forest patches.

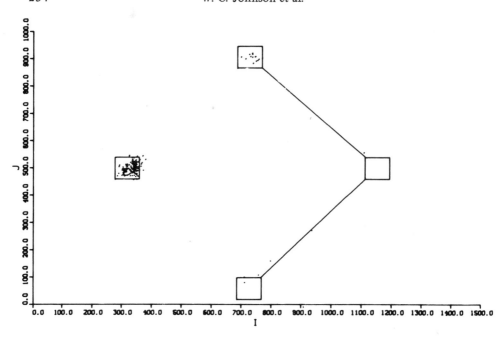

Fig. 10-6. Simulated seed dispersal pattern resulting from jay dispersal where the western (left side) forest island has been altered through cutting to make it more attractive as a caching site.

The model is best used in conjunction with field investigations of animal dispersal in specific landscapes. Initial field observations of dispersing jays could be used to crudely parameterize the dispersal model, much as was done in the preceding example using data from the literature. The model, in turn, could generate a dispersal pattern which could then be tested by more focused observation. This iterative process should culminate in a well-parameterized model for specific landscape conditions. A final test might be to alter the landscape pattern of the model (e.g., island configuration, inter-mix of land cover types) to correspond with a different field situation and then compare simulated dispersal maps with observations. The scientific value of the model should significantly exceed intuition when structurally complex landscapes are studied, such as those with a large number of islands and corridors widely varying in disperser attractiveness. Also, when all dispersers are considered, there are likely to be important interactive effects which could significantly affect dispersal patterns. In that case, the repeller term in the model (R_n) can be effectively used to simulate the interactive effects among dispersers.

Vegetation Dynamics Model

Background

Existing mathematical models exhibit varying assumptions concerning the importance of exogenous seed to the successional development of vegetation. To illustrate, JABOWA (Botkin et al. 1972) assumes that a seed source is available for each species, but only those species that can grow are added to the stand. A modified version of JABOWA for east Tennessee (FORET, Shugart and West 1977) contains the same assumptions, except for six old-field successional species which can be specified to have seed source limitations. These functional models thus assume that seed supply (amounts and species composition), at least for later successional stages, is not limiting. Markov process (Waggoner and Stephens 1970, Horn 1975) and differential equation (Johnson and Sharpe 1976) models parameterized from remeasurement data and applied to forest dynamics of large regions have tacitly assumed that a changing landscape pattern (i.e., as it relates to seed source proximity) does not alter the rates and direction of succession. Transition probabilities are constant, regardless of the spatial location of sample plots. Assumptions about the general unimportance of pattern may be satisfactory when these models are applied to landscapes that are heavily forested and where perturbation frequencies are low or moderate. But we expect this assumption in the above approaches to be less valid as the percentage of forest declines and as increasing perturbation frequency isolates seed sources of some species.

In an alternative modeling approach, Cohen (1970), Levins and Culver (1971), and Slatkin (1974), among others, estimate regional coexistence of species based on colonization potential, interspecific competition, and extinction rates for species occupying habitat islands (see review by Christiansen and Fenchel 1977). These fugitive-equilibria models have inherent assumptions about the nature of population interactions. Levins and Culver (1971) assume that colonization is proportional to the frequency of habitats already occupied. Furthermore, they assume that the species are independently distributed, i.e., the proportion of habitats jointly occupied is equal to the product of the proportions occupied by each species alone. Slatkin (1974) shows that these conditions are not biologically intuitive for all cases.

Existing fugitive-equilibria modeling approaches will not allow the examination of spatial and spatiotemporal heterogeneity initiated by the dispersal pattern itself (i.e., patches with poorly dispersed species may themselves exhibit a high degree of contagion). Colonization probabilities would need to depend on the distance of a habitat from the source of colonists. Incorporation of these necessary considerations into existing models, or development of new models, especially for a large number of species, is very complicated algebraically (Slatkin 1974, Levin 1977). However, DeAngelis et al. (1980) have introduced an approach that is computationally more simple for examining persistence and stability of seed-dispersed species in patchy environments.

Adapting FOREST

As an alternative to the above approaches, we have utilized the FOREST stand development model (Ek and Monserud 1974, Monserud 1975) because it explicitly considers seed production and local seed dispersal. Also, the model was parameterized for the forests of southern and central Wisconsin, many of which occur as isolates in a matrix of agricultural and urban land. Input is a set of real or generated tree locations and associated tree characteristics. Each tree is grown for a number of projection periods based on potential growth functions modified by competition measures developed from relative tree size, crowding, and shade tolerance. The competition measure, in effect, describes plant competition for moisture, nutrients, and light. Mortality is generated stochastically and depends on the competitive status of individual stems. Reproduction is introduced by the seed and sprout production of the overstory. Model simulations have been shown to accurately trace the development of stand and species attributes, judging from comparisons of observed versus predicted diameter distributions (Monserud 1975).

We have added an option to Monserud's (1975) version of FOREST such that amounts and species mixes of exogenous seed, assumed to be dispersed to the sample plot, can be specified. The endogenous seed supply routinely computed on the basis of stocking and stochastic variation in seed years has been retained. Thus, regeneration originates from seeds produced on the plot and those assumed to be dispersed to the plot from exogenous sources. Ranney (1978) has further modified FOREST to simulate the vegetation dynamics of forest island edges.

Briefly, reproduction and growth of the understory in the current version of FOREST are programmed as follows. Seed dispersed to the sample plot is the sum of endogenous and exogenous estimates. Endogenous supply is initially specified (seeds/ m^2) and can be held constant per year or allowed to vary stochastically based on the frequency of good and bad seed years. Exogenous seed is distributed uniformly across the sample plot whereas endogenous seed (a function of species basal area) falls within the respective seed shadows of seed trees.

Recruitment into the first height class equals the total seeds dispersed multiplied by percent viability, soundness, and germination. Percent values are actually chosen as random deviates from ranges reported in the literature. Sprouts are produced as a function of stump diameter and time since cutting. No estimate of a seed pool is made, i.e., seeds that do not germinate during a seed year as produced are not "held over" to germinate in successive years. The lack of a seed pool or seed storage component in the current version of FOREST is a conceptual inadequacy, but seeds of only a few species in the FOREST library retain significant viability for more than a few years (e.g., *Prunus serotina*).

Growth among four reproductive height classes is computed by competition index and a height growth model. Final understory structure is determined by applying species-specific and height class-specific survival rates between classes. Surviving outgrowth from the fourth understory height class becomes overstory ingrowth.

Model Simulation

The modified FOREST model was used to simulate the effects of order-of-magnitude decreases in seed supply on successional dynamics. The initial condition was a pure stand of *Prunus serotina* lacking *Acer saccharum* reproduction. Three separate simulations were made corresponding to three levels of exogenous *Acer* seed supply (low, 20,000 seeds/ha; medium, 200,000 seeds/ha; high, 2,000,000 seeds/ha). Each was replicated with three random starts. The same three random starts were used for each seed supply, thus the differences primarily reflect the effect of seed supply rather than stochastic variability. The three simulations can be viewed as forest plots occurring at three distances downwind to the northeast (adjacent, 350 m and 550 m) of a new *Acer* source of 3 million seeds (cf. Fig. 10-1). Annual exogenous supply of *Acer* seed was held constant while endogenous supply varied stochastically. No exogenous supply of *Prunus* was incorporated because of the already high amount of endogenous seed produced.

The simulations show considerable differences in the rates at which *Acer* invades the *Prunus* stand (Fig. 10-7). The high seed supply results in the most rapid invasion with 80% of the stand basal area being comprised of *Acer* after 120 years. At medium and low seed supplies, the 80% value is not reached until after 200 years. With low seed supplies, significant invasion of *Acer* is delayed beyond 100 years, at which time a local seed source develops and invasion accelerates. The simulation was performed in the absence of disturbance, therefore reproduction of *Prunus,* an opportunistic species, does not recolonize the stand.

The simulations lend support to the hypothesis that seed source proximity can significantly affect the vegetation dynamics of forest islands. Figure 10-8 shows that the three islands would reach the same end point or steady state (i.e., *Acer* dominance) but would require varying lengths of time to steady state. However, it is likely that the results would be far more complex if additional Wisconsin species were included. Other species more similar to *Acer* in niche characteristics but with greater dispersibility might occupy more of the space potentially available to *Acer* resulting in a different steady state. Thus, the composition at steady state would be a partial reflection of

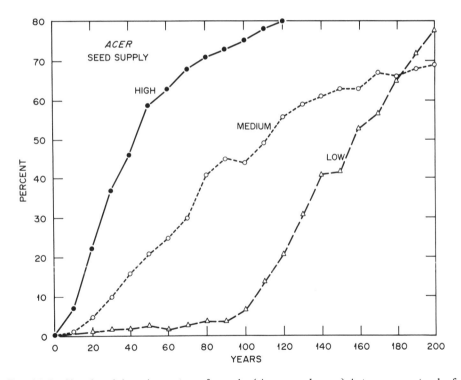

Fig. 10-7. Simulated invasion rates of maple (*Acer saccharum*) into a pure stand of black cherry (*Prunus serotina*) under three levels of exogenous maple seed supply (low, 20,000 seeds/ha; medium, 200,000 seeds/ha; high, 2,000,000 seeds/ha). Differences in seed supply generally correspond to three distances away from a source. Exogenous supply of cherry seeds was assumed to be zero. *y*-Axis is percent of stand basal area.

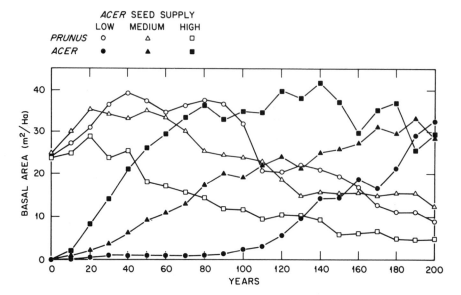

Fig. 10-8. Simulation results from FOREST showing invasion of maple (*Acer saccharum*) under high, medium, and low seed supply conditions, and subsequent decline of black cherry (*Prunus serotina*) over a 200-year period.

landscape pattern. This reasoning suggests that succession of isolated forest fragments is not convergent.

The question of whether succession is theoretically convergent has been raised by Horn (1976). Our preliminary simulations and other work on forest island landscapes suggests that the degree of convergence is, in part, a function of the spatial pattern of seed sources and the degree of their isolation. Thus, the question is perhaps not whether succession is or is not convergent, but under what conditions is it convergent. Succession is a spatially dependent process, and the likelihood of the same positions along an environmental gradient exhibiting divergent steady states, over and above that due to stochastic variation, would appear greater in highly fragmented than in continuously forested landscapes.

Conclusions

Animal and wind dispersal models were designed to be used in tandem with a vegetation simulator to gain an understanding of feedback relationships among disperser populations, landscape pattern and quality, and structural and compositional attributes of culturally fragmented ecosystems. The models presented are currently inadequate to characterize such complex feedback relationships in detail. However, initial simulations provide some indications to focus further research. For example, simulations of the wind dispersal model show that the juxtaposition of forest patches is important because of prevailing winds and that far less than 1% of the seeds of *Acer* are distributed much beyond several hundred meters from a single source tree in the absence of updrafts. Model output needs to be compared with field data acquired through the use of seed traps, and the importance of updrafts in extending the dispersal range of seeds needs to be evaluated. The animal dispersal model was designed to simulate the basic movement and deposition patterns of dispersers, but parameterization is more difficult than in the wind dispersal model. Simulations produced thus far reaffirm the model's ability to portray many kinds of dispersal events, but the real value of the model will emerge when used in conjunction with field studies which include observation, telemetry, seed traps, and radioactive tagging of seeds.

Simplified simulations of FOREST point to the significance of variable seed rain in the development of forest vegetation. Also, the model was relatively insensitive to a doubling or even tripling of seed supply, suggesting that it may be impractical to strive for high statistical accuracy and precision in seed sampling experiments. Wide confidence limits may be satisfactory. The model does, however, include several simplifying assumptions that limit the degree to which generalities can be made. A seed pool component needs to be added to improve the simulation of species with seeds of high longevity. Following this, it would be appropriate to examine the correspondence between seed source pattern and strength and the successional convergence of vegetation using a more complete mix of arboreal species in the forests of southeastern Wisconsin. Simulations could be tested by examining the composition of successional vegetation patches (e.g., old fields with comparable soils and past use) occurring at varying distances from major seed source islands.

The linking of seed dispersal and vegetation simulation models provides a rational basis for simulating forest dynamics in patchy landscapes. The outcome of the simulations is not prejudiced by assumptions that seed supply is adequate for "normal" succession to proceed. This approach to simulation is especially relevant in landscapes where seed sources have been isolated through deforestation by man or by extensive natural disturbance such as fire. It is believed that continued research in this direction can lead to the identification of island patterns, connectivities and distances that would assure adequate dispersal of most or all species and the maintenance of diversity. Presumably, these analyses would also yield coarse prescriptions on how landscape-level extinction of species could be avoided during the course of anthropogenic forest fragmentation.

Acknowledgments. The authors acknowledge the assistance of E. W. Stiles of Rutgers University and R. Kent Schreiber of the U.S. Fish and Wildlife Service during the development of the animal dispersal model. This research was supported by the Eastern Deciduous Forest Biome, USIBP, funded by the National Science Foundation under Interagency Agreement AG-199, DEB76-00761 with the U.S. Department of Energy under contract W-7405-eng-26 with Union Carbide Corporation. Contribution No. 346, Eastern Deciduous Forest Biome. Publication No. 1605, Environmental Sciences Division, Oak Ridge National Laboratory, Oak Ridge, Tennessee.

11. Optimization of Forest Island Spatial Patterns: Methodology for Analysis of Landscape Pattern

V. A. Rudis
Department of Forestry
University of Wisconsin-Madison

A. R. Ek
Department of Forestry
University of Wisconsin-Madison

Wildland habitats in the eastern United States have decreased in size over past centuries to the point where patches, or "islands," of similar habitat have become increasingly isolated from one another. Conservationists and land use planners have argued to save islands that harbor endangered flora or fauna, or that encompass critical habitat, as well as those of particular recreational value. Such areas that are preserved have been saved often for aesthetic or political reasons and have not always been of optimal benefit to the species they are meant to preserve.

In the future, it is likely that existing nature preserves may be reduced in size and/or isolated in space due to ever-increasing urbanizing influences as well as increased population demands for agricultural land. Truly "native" populations of organisms will decline through the loss of both habitat and biotic interaction with adjacent populations. Species and habitat diversity will be altered by biotic stresses inherent in man's activities or by genetic stagnation similar to the problems encountered with species propagation in zoos (Sullivan and Shaffer 1975). Further, preservation of specialized species habitats, e.g., the Curtis prairie on the University of Wisconsin Arboretum, will assume increased maintenance costs (and likely increased social value) as areas of similar habitat become more scarce. Consequently, the formulation of feasible and optimal strategies for natural areas and species preservation will likely grow in importance for the future.

In this chapter, the preservation of species in native forested landscape is of primary interest. This study emphasizes the forest land of southeastern Wisconsin, just north of Milwaukee. In recent decades, this area has developed into a discrete matrix of forest islands, composed largely of mesic maple-basswood communities, embedded in urban, suburban, and agricultural land uses. Eighty percent of these forest islands are now classed as northern hardwoods by cover type (Tyburski and Ek 1977), and are often connected by fencerows. Many of the forest islands have been disturbed through grazing and selective logging and are in various stages of succession. Some islands have been further reduced in size, and others obliterated since the first aerial survey (1937) of the region. The number of forest species in these islands varies with area, topography, degree of isolation from adjacent sources, and the amount of disturbance (Levenson 1976). In order to achieve optimal preservation of species, we need to define forest island conditions and their potential for influencing species composition through time. A necessary second step is then to impose these definitions as constraints on the objective of forest species preservation.

Objectives

Given the requirement for optimal species preservation, our methodology is developed in three parts:

(1) To conceptualize and define the constraints on this objective via a series of structural equations, collectively termed a systems model.

Beyond the theory from MacArthur and Wilson (1967), other aids to this modeling are spatial interaction concepts used in economic geography and urban population studies. The model is specifically designed to account for tree species diversity as a function of an island's characteristics, including location. Adaptation to preservation of other plant groups appears possible.

As a practical approach, variables used in the systems model represent those which are primarily available from remote-sensing data. This level of detail is specified to facilitate the adaptation of the procedures developed and their export to other regions. Field data on a subsample of islands collected by Levenson (1976) are available to estimate the coefficients of the structural equations. Remote-sensing data on a 6 × 12 mile (9.6 × 19.2 km) rectangular study area north of Milwaukee (Tyburski and Ek 1977), with complete ground checking, was used to examine the framework of the overall systems model.

(2) To demonstrate, in a simplified form, the utility of the systems model. This part deals with the optimization of the forest island matrix, i.e., to maximize the preservation of species while at the same time minimizing the number of islands needed to achieve that goal. A number of other options may be included. Two examples are (a) given that a fraction, say 50%, of the forest islands present today would likely disappear, which islands would be most important ecologically over the long term, and (b) given an established wildland habitat, which islands surrounding this preserve could be safely removed without adversely disturbing its species richness. The specific objectives to be optimized can be many and varied, once a valid framework for island location and species interaction has been established.

(3) To describe the forest island arrangement inherent in a given region by means of spatial pattern analyses. Pattern recognition statistics (Clark and Evans 1954, Liebetrau and Rothman 1977, Mead 1974, Pielou 1969) and spatial autocorrelation analysis

(Cliff and Ord 1973) can be used to characterize the spatial pattern present in a region. The resulting statistics provide a means of assessing the particular geography of the region and the correlation among variables in space and time. These may also indicate other variables which describe the region and which may be needed in the structural equations.

Because the procedures involve fitting models to a geographically restricted data set, the resulting equations may not be applicable to other regions. Spatial autocorrelation techniques provide a means of assessing variation in other regions, and permit characterization of significant differences among regions which can then be related to the structural equations. The end result is the development of a generalized systems model for forest island location and species interaction.

Background

The theory of island biogeography (MacArthur and Wilson 1967) is based upon observations that species richness (number of species) is lower on true oceanic islands than on the mainland. An equilibrium is established between the rate of immigration of new species to an island and the rate of extinction of existing species on that island. The rate of immigration decreases with distance from a source of colonizers. This pattern is much like the exponential decay of the abundance of seeds away from a seed tree. An island closer to a source will have more species than an island farther away. The rate of extinction decreases with area of the island. This is comparable to the negative exponential distribution of the death rates of a population for a decreasing resource. Area, it is assumed, is positively correlated with habitat complexity. A larger island will have a higher number of different habitats as well as larger populations of each species than a smaller one. The larger island will, therefore, support more species.

The theory has been found to be a satisfactory explanation for a wide variety of observed species distributions (Galli et al. 1976, Opler 1974, Sepkoski and Rex 1974, Simpson 1974, Vuilleumier 1970, Willis 1974, inter alios), and at least a few experimental observations (Rex 1975, Simberloff and Wilson 1969). Commonly, the relationship between the number of species (S) and the area of islands (A) has been expressed as

$$S = b_1 A^{b_2} \tag{1}$$

where b_1 and b_2 are constants. Indices other than area have been used to explain the variation in species among islands with sometimes greater statistical success. Some of these are a subjectively ranked index of ecological diversity (Opler 1974), foliage height diversity (MacArthur et al. 1966), elevation and distance-to-nearest-island (Mauriello and Roskoski 1974), and area and distance measures in past ages (Simpson 1974).

Based on island biogeography studies, the theory applied to mainland patches of native vegetation has suggested the preservation of the largest islands, those in close proximity to others, and those with corridors of similar habitat between them (Diamond 1975a, Terborgh 1974, Wilson and Willis 1975). However, mainland patches are spatially and temporally more interactive with the landscape. Consequently, management decisions regarding preservation are rarely as clear-cut as these simple principles would imply. In particular, knowledge regarding the numbers of islands to be preserved and species-equivalent spatial assemblages that vary with distance, area, successional stage, and surrounding landscape has been lacking (Sullivan and Shaffer 1975). It is hoped that future research on forest islands will clarify a number of these issues.

Discussion

Systems Model

For the southeastern Wisconsin study area, Levenson (1976) found that area alone did not account for significant variation in species numbers within 43 selected upland old growth forested islands. While other variables were thought to be at a minimum in choosing his islands, he conceded that the topographic variability, wide variation in disturbance, length of time the islands had been a particular size, and the degree of isolation were probably important determinants of tree and shrub species numbers. A generalized form for these and other characteristics used in the analysis for all mesic, northern hardwoods type forested islands within the study area may be stated symbolically as follows:

$$S_i = f(A_i, T_i, W_i, \tilde{t}, I_{ij}) \tag{2}$$

where S_i = species richness on island i; A_i = area of island i; T_i = topography, general soil conditions of island i; W_i = successional age and degree of disturbance on island i; $\tilde{t}$ = time-averaged changes in other variables affecting island i since its establishment; I_{ij} = interaction term relating the exchange of propagules, e.g., seeds, from other islands (j) with island i.

Area and Topography

The area of mainland island patches is an important variable in determining species richness. Levenson's (Chapter 3) data disagree with the predictive ability of the model [Eq. (1)] for upland tree and shrub species on islands greater than 2 ha. He concluded that area was important, but primarily for smaller-sized islands. A similar study on forest islands in New Jersey (Forman 1974) has indicated area to be of major importance in predicting tree species numbers. Further, area has been shown to be a significant predictor of other, more mobile species groups, particularly birds, in New Jersey (Galli et al. 1976) and in Wisconsin (Gustafson 1977).

A reanalysis of Levenson's data (Rudis, unpublished) suggests that topographic variability, in addition to area, may be important in explaining variance in tree species numbers. A plausible measure of topographic variation for this region is the Shannon-Wiener index of entropy or diversity (after Pielou 1969):

$$T_i = - \sum_k^{Q_i} p_k \ln p_k \tag{3}$$

where T_i = index of topographic diversity within a forested island i; p_k = proportion of area within a 3 m contour interval k, $k = 1,2,\ldots,Q_i$, relative to the total area of the forest (readily available from existing U.S. Geological Survey topographic maps). There are Q intervals in island i. A modification, such as a ranked weight, would be needed to account for islands with significantly greater-than-average density of contour changes (indicative of xeric conditions), or those having within-island pockets of lower elevation (heterogeneous moisture conditions). Nominally ranked soil conditions determined for this area (Miller and Niemann 1972) may be useful as well.

Consider now that topographic diversity and area are, relative to species richness on a given island, measures of habitat complexity. Area is correlated with topographic diversity. The probability of an increase in topographic diversity with an increase in area is expected to be large for the gently rolling terrain of the Milwaukee region. Thus, coupled with the species-area model [Eq. (1)], a reasonable expression for explaining species richness within an island would be:

$$S_i = b_1 (A_i T_i)^{b_2} + \epsilon_i \qquad (4)$$

where b_1 and b_2 are constants derived from least-squares procedures, and ϵ_i is the error term. In this context, T_i may be interpreted as a measure of the diversity (richness and equitability) of habitat, while A_i would be the territorial expanse over which the diversity extends.

To account for changes in species as a function of change in island size through time, we need to simulate such observations. This would permit calculation of the rate change per unit time and the "relaxation time" (Diamond 1972). The latter is the time needed for a given island to reach its new equilibrium species value. This has been done for birds (Diamond 1972, 1975b) on oceanic islands, but not for mainland islands or for tree species. It is likely, however that real changes in tree species composition cannot be calculated in the same manner. Problems arise due to the longer life expectancy of seeds and parent trees and variations in species composition resulting from previous selective logging and grazing in the study area (Levenson, Chapter 3).

Simulation modeling of forest stands is needed to determine species changes in mixed forest stands. For this region, adaptation of FOREST, a simulation model for northern hardwoods (Ek and Monserud 1974, Ek et al. 1977), is possible. For the present, with limitations on available historical data (earliest available aerial photos date only to 1937), we assume that the current species richness is some function of its area in previous years. A simple form to express this is as a weighted average of all observations through time, known as an exponential smoothing model (Brown 1963, Cliff et al. 1975). In the simplest case, such weights follow a geometric series. The exponentially weighted moving average for area A_i is given by

$$\widetilde{A}_i = \frac{(1-\lambda)^n A_{it} + (1-\lambda)^{n-1} A_{it-1} + (1-\lambda)^{n-2} A_{it-2} + \cdots}{(1-\lambda)^n + (1-\lambda)^{n-1} + (1-\lambda)^{n-2} + \cdots}$$

$$\frac{+ (1-\lambda)^{n-(n-1)} A_{it-(n-1)} + A_{it-n}}{+ (1-\lambda)^{n-(n-1)} + 1} \qquad (5)$$

where λ is constrained to lie between 0 and 1. Observations n years farther from the present time t are given greater importance. Though more precise estimates could be obtained through simulation, we might, for example, assume λ = 0.02. The value of the observations (or measurements) of area in 1977 would then be given approximately half as much weight as that of the area in 1937.

To weigh topography similarly across time would be difficult as the historical record is incomplete. Changes in topography through time are essentially a function of changes in forest island area. Consequently, the chance for error in using the current topographic diversity instead of its weighted average over time should be small.

Equation (4) may now be rewritten as:

$$S_{it} = b_1(\widetilde{A}_i T_i)^{b_2} + \epsilon_i \tag{6}$$

This form states that current species richness (at time t) is a function of the area, averaged over time, and of topographic diversity encompassed by the current area.

Successional Age and Degree of Disturbance

The successional age and the degree of disturbance of each forest island needs to be considered. Unfortunately, the characteristics are not readily discernible from available small scale (1:60,000) color aerial photographs. Forest stand height, density, and cover type can be quickly determined, however (see Tyburski and Ek 1977).

For the purposes of this model, height class is assumed to correspond to the successional age of the stand (1 = young forest, ≤ 8 m height; 2 = intermediate-aged forest, 8-16 m; 3 = mature forest, > 16 m). Density class is assumed to correspond to the degree of disturbance of the stand (1 = highly disturbed, 10-40% density; 2 = moderately disturbed, 41-70% density; 3 = little or no disturbance, 71-100% density). The density scale upon which this is based is that given by Avery (1969, p. 30).

Age and disturbance are ranked jointly as W_i = age-disturbance rank of island i, according to Table 11-1. Greater importance is given to stands which are older and less disturbed, i.e., those with greatest height and high density. This corresponds to the assumption that, on the average, older, undisturbed stands have a higher species richness than younger, or more disturbed stands.

This ranking scheme was derived from discussions presented by Helliwell (1973) and Smith et al. (1975). It is reasonable for the New Jersey Piedmont area, in view of succession studies there (Bard 1952). However, the suitability of the rankings for southeast Wisconsin is subject to criticism. Loucks (1970), in a study of *Pinus banksiana* stands in central Wisconsin, suggested that species richness declines after a peak at mid-age in the absence of fire disturbance. Rudis' observations on old growth stands in northern Wisconsin suggest that species richness, relative to disturbance and age of stands, is highly variable, especially if the disturbance has included selective logging.

Establishment of an appropriate ranking scheme for a given area requires both a knowledge of the successional pattern and the forest management practices prevalent in previous decades. Simulation studies would be required to determine the effect of age and kinds of disturbance on species composition for this region. However, for this chapter, we retain the rankings given in Table 11-1.

To consider the age-disturbance rank across the interval of measurable time, an exponentially weighted moving average for age disturbance is given by

Table 11-1. Age-Disturbance Rank (W_i) of Forest Islands, Based on Density and Height

Disturbance (density) class	Age (height) class		
	1	2	3
1	1/9	2/9	3/9
2	2/9	4/9	6/9
3	3/9	6/9	9/9

$$\tilde{W}_i = \frac{(1-\lambda)^n W_{it} + (1-\lambda)^{n-1} W_{it-1} + \ldots + (1-\lambda)^{n-(n-1)} W_{it-(n-1)} + W_{it-n}}{(1-\lambda)^n + (1-\lambda)^{n-1} + \ldots + (1-\lambda)^{n-(n-1)} + 1} \quad (7)$$

If we can assume that age-disturbance rank is independent of area and topographic diversity of a given island, then it is appropriate to add $\tilde{W}_i$ to the species prediction Eq. (6) as

$$S_{it} = b_1 (\tilde{A}_i T_i)^{b_2} + b_3 \tilde{W}_i + \epsilon_i \quad (8)$$

If the above assumption does not hold for the study region, i.e., that age-disturbance rank is correlated with area and topographic diversity, then another form of Eq. (8) is needed to avoid bias in the residuals. One possible form is

$$S_{it} = b_1 (\tilde{A}_i T_i)^{b_2} \tilde{W}_i^{b_3} + \epsilon_i \quad (8a)$$

Forest Island Interactions and Proposed Model

Finally, the degree of isolation of an island is considered. Interaction is the opposite of isolation. To measure the degree of interaction an analogy is drawn between the gravity interaction model and the concept of biotic interaction among forest islands. The gravity interaction model has been used successfully by economic geographers to study the interaction of human population centers with adjacent activity centers, such as shopping centers, factories, and towns (Wilson 1967, 1971, Griffith 1976).

In its simplest form, the gravity model applied to island biogeography theory is

$$I_{ij} = K \frac{P_i P_j}{d_{ij}^2} \quad (9)$$

This expression states that the amount of interaction I among two islands i and j is proportional to the size of their populations P, and inversely proportional to the square of the distance d between their population centers. The K term is a proportionality constant.

Transformation of the gravity interaction model into the biotic interaction conditions of tree species requires several assumptions and modifications. Interaction occurs essentially in the form of seed or propagule dissemination which decreases with distance from the source. The generalized distribution of propagule dissemination is hypothesized as a negative exponential distribution in the form $e^{-c_k d_{ij}}$, where d_{ij} is the distance between centers of islands i and j; c is constant for a particular species k.

The number of propagules produced by island j is some function of its area, where area is an estimate of the population of trees on island j. The number of propagules from island j that reach island i will decline with distance according to the negative exponential distribution discussed above. Propagules traveling from island j will have a better chance of reaching island i if island i is large (Whitehead and Jones 1969). Consequently, the number of propagules falling on island i will be some function of its own area and the area and distance of its neighbors, j.

Species are assumed to follow a similar pattern but topographic diversity and age-disturbance weights are added to account for differences in species richness of propagules. For convenience, the exponential decline of propagules is given as an average for all tree species in the region. Thus, the modified gravity interaction model for tree species for island i is

$$\sum_{\substack{j \\ j \neq i}}^{n} I_{ij} = I_{i\cdot} = \sum_{\substack{j \\ j \neq i}}^{n} \frac{W_j}{W_i} A_i T_i A_j T_j e^{-\bar{c} d_{ij}} \tag{10}$$

where

$$\bar{c} = \frac{\sum_{k=1}^{m} c_k}{m}$$ = averaged coefficient of exponential decline for all species k; m species in the region, and n = number of islands in the region.

The term W_j/W_i may be interpreted as the propagule flow rate. It reflects the theory that an older neighboring island, j, is more important (has a higher interaction value) to a younger island, i, since it is expected that the major flow of propagules is from the older to the younger island (also from the least disturbed island to the most disturbed). In this case, the propagule flow rate is $\gg 1.0$, according to the currently devised ranking of age-disturbance. The reverse situation, where propagules generated in a younger forest island invade an older one, has a rate $\ll 1.0$.

Interactions are also time dependent, so we again utilize the exponentially weighted moving average such that

$$\tilde{I}_{i\cdot} = \frac{(1-\lambda)^n I_{i\cdot t} + (1-\lambda)^{n-1} I_{i\cdot t-1} + \ldots + (1-\lambda)^{n-(n-1)} I_{i\cdot t-(n-1)} + I_{i\cdot t-n}}{(1-\lambda)^n + (1-\lambda)^{n-1} + \ldots + (1-\lambda)^{n-(n-1)} + 1} \tag{11}$$

with W_j, W_i, A_i, A_j varying simultaneously over time, and T_i, T_j, and $\bar{c}$ remaining constant. One of the advantages of this procedure is that consideration can be given to islands not now present, but which were recorded in earlier aerial photographs.

Thus, the species prediction model is generalized as

$$S_{it} = b_1 (\tilde{A}_i T_i)^{b_2} + b_3 \tilde{W}_i + b_4 \tilde{I}_{i\cdot} + \epsilon_i \tag{12}$$

Note that this expression contains the major elements important to the determination of species within a given forest island: area, topography, successional age and degree of disturbance, interactions with other islands, and changes in these elements over time. Nonlinear or linear regression analyses could be used to determine the coefficients of the variables and to form the basis for an evaluation of the amount of variation in species numbers. Some terms discussed above would not be significant, and others would be added, depending on the outcome of the analyses. One would anticipate that the best fitting and most useful expression would be simpler than Eq. (12).

Applications

Utility of Proposed Model. One is tempted to "forecast" the future configuration of forest islands, but too little is known regarding the probabilities of forest island disappearance and the factors determining the changes which occur in the landscape. Perhaps if yearly aerial photo data existed over a span of several decades, one could determine the transition probabilities for landscape change by a Markov process similar to the procedures developed by Voelker (1976) and others (Voelker et al. 1974).

With a model such as Eq. (12), one can at least "project" the future species richness on each island. We must assume, however, that other factors are held constant. To do this, say for a 40-year period, the age-disturbance ranking is projected one step higher. For example, a young forest with moderately disturbed vegetation (low height, medium density) is projected to an intermediate-aged, low level disturbance forest (medium height, high density). A mature forest with little disturbance is projected to remain at that level. One needs also to project the independent variables another 40 years, and recompute the moving average. This is expressed (after Cliff et al. 1975) as

$$\widetilde{Z}_{it+1} = \frac{\sum\limits_{j=0}^{n+1} (1-\lambda)^{n+1-j} Z_{it+1-j}}{\sum\limits_{j=0}^{n+1} (1-\lambda)^{n+1-j}} \tag{13}$$

where $\widetilde{Z}_{it+1}$ represents $\widetilde{A}_i$, $\widetilde{W}_i$, and $\widetilde{I}_{i\cdot}$, each recomputed separately. In essence this amounts to multiplying the previous expressions for each variable by a constant $(1 - \lambda)^{n+1}$ and including the new terms A_{it+1}, W_{it+1}, and $I_{i\cdot t+1}$.

The species prediction equation for $t + 40$ is then

$$S_{it+1} = b_1(\widetilde{A}_i T_i)^{b_2} + b_3 \widetilde{W}_i + b_4 \widetilde{I}_{i\cdot} + a_{it} \tag{14}$$

where a_{it} is generated from an appropriate random number algorithm.

Optimization. Operations research techniques are powerful tools in the development of management alternatives and guidelines for decision making. These tools have been used extensively in the fields of forest economics, resources management, agriculture, and other applied sciences. Use of these techniques in the application of island biogeography theory is suggested here.

In optimization, an objective to attain for a region is given, and a set of equations (a systems model), which in this case represents the tree species community, is developed. The desired objective is subject to the constraints of the relationships among the objective, the region under study, and the variables of the systems model. Due to these limitations, full attainment of any objective is rarely possible, so we must settle for the "optimal" solution, i.e., the best possible attainment of an objective, if such a solution exists.

Three simplified examples of optimization are presented in a linear programming form for ease of understanding. However, dynamic programming approaches will eventually be needed, as linear programming is too restrictive for large-scale systems such as the Milwaukee study region. Once a workable algorithm is developed, dynamic programming is far easier to handle. Computations for the dynamic programming approach are about two orders of magnitude lower than with the linear program specifications required.

The objective in the first example (given below) is to maximize species richness of a region when it is not possible to determine fully the species composition of each island in the region. Suppose that only half of the forest islands now present will exist in the near future, say 40 years from now. The projected species prediction Eq. (14) is used. We wish to know which islands are most important, ecologically, for optimizing that objective.

Symbols used are defined as follows:

N_t = total number of islands in region, time t.

Y_{it+1} = species richness on island i, time $t+1$; number of islands in region, $i = 1, 2, ..., N_t$.

I_{iit+1} = $b_1(\widetilde{A}_iT_i) + b_3 \widetilde{W}_i + a_{it}$. This is defined as the species richness of island i at time $t+1$, in the absence of interactions with other islands, j. Here I_{iit+1} is represented as a function of area A_i, topography T_i, and age-disturbance rank $\widetilde{W}_i$, projected to time period $t+1$, and averaged across time, as in Eq. (13).

I_{ijt+1} = $b_4 \widetilde{W}_j/\widetilde{W}_i (A_iT_iA_jT_je^{-cd_{ij}})$. The term I_{ijt+1} is defined as the "interaction," the additional species richness of island i which results from the inclusion of island j in the region, up to the time period $t+1$. Note that I_{ijt+1} approaches zero as distance d_{ijt+1} from island j to island i approaches infinity.

$(k = j|I_{ijt+1} \geqslant 1.0)$ where k equals islands j such that $I_{ijt+1} \geqslant 1.0$. Excluded from consideration are islands j such that I_{ijt+1} contributes less than one species to island i.

$X_{ii} = \begin{cases} 1 \text{ if island } i \text{ is to be preserved,} \\ 0 \text{ otherwise} \end{cases}$

$X_{ij} = \begin{cases} 1 \text{ if islands } i \text{ and } j \text{ are to be preserved,} \\ 0 \text{ otherwise} \end{cases}$

$X_{ii}^{(h)} = \begin{cases} X_{ii} \text{ if island } i \text{ is of type } h, \\ 0 \text{ otherwise} \end{cases}$

$I_{ii}^{(h)} = \begin{cases} I_{iit+1} \text{ if island } i \text{ is of type } h, \\ 0 \text{ otherwise} \end{cases}$

$X_{ii}^{(h)}$ and $I_{ii}^{(h)}$ are given; X_{ii} and X_{ij} are to be determined.

h = age and habitat class, $h=1, 2, ..., H$. H is the total number of age and habitat classes considered to harbor uniquely different species.

$\max(SR^{(h)})$ = the maximum species richness obtainable for a forest stand of a given successional age and definable habitat class h.

d_h = an estimate of the number of species common to the islands of each age and habitat class h.

Example 1. Maximize species richness where the species composition of each island is not known. Assumption: each successional age and definable habitat class h is a distinctly different community type and, in general, contains one or more unique species.

Maximize

$$\sum_{i=1}^{N_t} Y_{it+1} - \sum_{h=1}^{H} d_h \tag{15}$$

subject to

$$I_{iit+1}X_{ii} + \sum_{\substack{j=k \\ j\neq i}}^{n_i} I_{ijt+1}X_{ij} - Y_{it+1} = O\nabla_i \; (k=j|I_{ijt+1} \geqslant 1.0) \tag{16}$$

$$n_i X_{ii} - \sum_{\substack{j=k \\ j\neq i}}^{n_i} X_{ij} \geqslant O\nabla_i \; (k=j|I_{ijt+1} \geqslant 1.0) \tag{17}$$

$$\sum_{i=1}^{N_t} X_{ii} \leqslant N_t/2, \tag{18}$$

and

$$\sum_{i=1}^{N_t} I_{ii}^{(h)} X_{ii}^{(h)} + d_h = \max(SR^{(h)})\nabla_h \tag{19}$$

where ∇ indicates separate expressions for each subscript value.

Equation (16) is the projected species prediction Eq. (14) written as a system constraint. X_{ii} and X_{ij} are the decision variables which determine the value of Y_{it+1}. In the case where all islands are considered, $Y_{it+1} = S_{it+1}$, from Eq. (14). Equation (17) is a system constraint which states that the term X_{ij} must have a value equal to one if both islands i and j equal one. Equation (18) is a regional constraint. There is to be a maximum of half the total number of islands at time $t+1$ as at time t. Equation (19) links the region with the objective. It states that for every age and habitat class h, there exists a maximum number of species $SR^{(h)}$ which forces the algorithm to accept at least one from each h. To accept more than $SR^{(h)}$ for each h, the objective in Eq. (15) is reduced by an amount d_h. Equivalently, Eq. (19) states that other islands i are included in the solution only if they increase the species number Y_{it+1} via interactions with other h's in the region.

Equation (15) is the objective. We wish to maximize the number of species at time $t+1$ for each island i. The objective to be maximized, Y_{it+1}, is reduced by the estimated number of species held in common by two or more islands of the same class h.

The solution expected from this model would contain the desired assignment matrix of X_{ii}'s indicating the importance value (1 if island i should be preserved; 0 otherwise). The value of the objective would contain, when optimized, somewhat more than $\sum\limits_{h}^{H} \max(SR^{(h)})$ species. Obviously, $\sum\limits_{h}^{H} \max(SR^{(h)})$ is a crude overestimation of the total number of species in the region. However, the solution does have at least one of every type, h, which maximizes the successional age and habitat diversity of the region. Most likely the number of islands will be less than $N_t/2$, as the total number of islands in the solution will be constrained by H. Thus the number, location, and type of islands needed to optimize species richness would be specified by the solution.

A weakness in the above algorithm is that the number of common species must be substituted for a crude estimate, d_h, to compensate for the nonadditivity of Y_{it+1}. This weakness can be rectified only if data are available for each island and species in common can be compared. With this information, we can proceed to the second example.

Example 2. Maximize species richness, with species composition known. Define

$$Z_{ij} = \begin{cases} 1 \text{ if } X_{ij} = 1, \\ 0 \text{ otherwise} \end{cases}$$

CS_{ijt} = number of species common to islands i and j at time t. One may reasonably assume $CS_{ijt} \sim CS_{ijt+1}$.

Maximize

$$\sum_{i=1}^{N_t} Y_{it+1} - \sum_{i=1}^{N_t} \sum_{j=i+1}^{N_t} CS_{ijt} Z_{ij} \tag{20}$$

subject to Eqs. (16)-(18) and

$$\sum_{\substack{j=k \\ j \neq i}}^{n_i} X_{ij} - \sum_{i=1}^{N_t} \sum_{j=i+1}^{N_t} Z_{ij} = 0 \tag{21}$$

The solution expected from Example 2 would have an objective value [Eq. (20)] equal to the total number of species in the region minus the number of species common to all forest islands. Equation (21) links X_{ij} with the objective, thus forcing Z_{ij} to equal one when both islands i and j are in the solution matrix.

Example 3. The third example minimizes the number of needed new islands i as an objective for the preservation of species in islands already established. The number and kinds of species on each of several nearby islands, S_{it}, are given as constants. The projected species number Y_{it+1} is a constraint, and is calculated from the projected species prediction Eq. (14).

Given three established preserves, minimize the number of adjacent neighbor island patches needed which will maintain the current species richness of the established preserves in the future. Species compositions are known.

Minimize

$$\sum_{i=1}^{N_t} X_{ii} \tag{22}$$

subject to Eqs. (16), (17), (21) and

$$\sum_{i=1}^{N_t} Y_{it+1} - \sum_{i=1}^{N_t} \sum_{j=i+1}^{N_t} CS_{ijt} Z_{ij} \leqslant TOT \tag{23}$$

$$Y_{1t+1} \geqslant S_{1t} \tag{24}$$

$$Y_{2t+1} \geqslant S_{2t} \tag{25}$$

and

$$Y_{3t+1} \geqslant S_{3t} \tag{26}$$

The solution will yield the minimum number of other islands needed to maintain the number of species now present on islands 1, 2, and 3. The term TOT in Eq. (23) refers to the total number of species in the region. As before, this expression compensates directly for the nonadditivity of Y_{it+1} by the addition of CS_{ij} (the number of species common to islands i and j).

Suggestions for Further Refinement. We may not only want to maximize species richness, but to maximize species diversity as well, either within an island, or within a given region. To handle within-island diversity, we would need to develop structural equations as previously discussed for the species diversity index (Pielou 1969). For the Milwaukee region, species richness data have been easier and less costly to obtain.

The subject of maximizing species diversity across a region of forest islands needs further research. Derivation of diversity-maximizing formulations for biological populations across time and across a region is likely feasible through dynamic programming approaches. In essence, the procedures involve the maximization of the interaction between islands $I_{ijt+1} X_{ij}$, subject to a number of additional constraints and assumptions which need to be determined. Diversity-maximizing formulations are presented in recent papers by Webber (1976) and Coelho and Wilson (1977) for human population distributions within a given urban center. Applications to multiple sources of dispersion of populations, as in the case of forest islands, should be explored.

Criteria other than strictly biological constraints and species preservation objectives are, of course, needed to provide realistic optima for the management of a region's species on forest islands. Philip (1975) discusses some of the more important economic concerns to be evaluated, particularly with regard to forest woodlot owners. There is also a large body of literature dealing with the application of socioeconomic constraints in forest regulation for recreational management and related forest uses (see, for example, Menchik 1972, Whitaker and McCuen 1976).

Helliwell (1973) has discussed a number of other useful, although subjective, sets of criteria to be used as objectives in habitat and species preservation. A forest island habitat can be rated according to its scarcity in the region, size or age of its trees, educational and scientific value, genetic reserve potential, proximity to human population centers, wildlife potential, species diversity, and number of rare species it contains. Rating each island is expensive, as considerable time is spent in the field. Future technological advances in remote sensing should prove useful in this regard.

Pattern Analyses

Generalizations of the methodology and the model presented require an investigation of the pattern inherent in the data used in verification. Adjustments can then be made in the structural equations to apply basically similar optimization algorithms to a new set of forest islands in other regions.

Markedly different forest cover types do occur in any given landscape. Unfortunately, the species information currently available and the spatial location and number of forest island patches other than the northern hardwoods type are clustered and too few to permit adequate statistical and biological inferences. Interactions among widely variant habitats, such as between mesic northern hardwoods stands and tamarack bogs have not been quantified. A possible solution would be to design weights, in addition to successional age and degree of disturbance, which reflect generalized proportional probabilities of tree species colonization, in the aggregate, in both community types. The weights could be substituted in place of the propagule flow rate [Eq. (10)].

Consideration of the somewhat more detailed and largely unknown dynamics of the continental landscape have been omitted in favor of a simplified modeling and optimizing scheme. The shape of an island, the effects of wind pattern, vertical barriers, seed predation, and aspect of an island create nonstandard (in a mathematical and statistical sense) patterns of seed dispersal. These factors in turn affect the interaction between islands. The intervening sea of exogenous environmental conditions, including potential seed sources such as fencerows in agricultural areas, and shade trees in residential areas are additional variables that need to be considered.

Methods collectively known as pattern analyses are being developed in the form of easily handled computer algorithms. These methods would be useful in assessing some of the spatial characteristics of species distributions specific to a region.

Paloheimo and Vukov (1976) have shown that the nearest neighbor and point-to-plant indices are the least biased and most consistent statistics for describing spatial pattern or distribution of populations. However, the nearest neighbor distance index (Clark and Evans 1954) and the point-to-plant distance index (Pielou 1959), necessitate simplifying assumptions and adjustments for islands to be considered as points. To test the validity of these simplifying procedures, statistics based on a grid of contiguous quadrats at several scales (Liebetrau and Rothman 1977, Mead 1974) could be employed.

A more flexible approach to the study of spatial pattern is that of spatial autocorrelation analysis, first described by Moran (1948) and later expanded by Cliff and Ord (1973). Spatial autocorrelation is concerned with the spatial relationship between variate values of samples, points or islands in a region. Unlike nearest neighbor and other similar indices, spatial autocorrelation is not merely summarizing the pattern formed by their locations. Spatial autocorrelation extends a third dimension to spatial pattern analysis. In general (Cliff et al. 1975), if high values of a characteristic on one island are associated with high values of that characteristic on nearby islands, then the set of islands exhibits positive spatial autocorrelation for that characteristic. If high values alternate with low values, then the spatial autocorrelation is negative.

The procedures in spatial autocorrelation are considered in two steps (Cliff and Ord 1973, Cliff et al. 1975). The first involves the determination of a weighted "join" between islands. In this study, two islands are joined if the distance between island centers is less than a given critical distance. Critical distance refers to the maximum distance a seed might travel, all other things being equal. To avoid mis-specification of a limiting critical distance, joins are determined with a range of distances (e.g., between 0.125 and 2.000 km). Additional weights could also be employed to test assumptions regarding the direction and time dependence of pattern relative to urbanization of the landscape.

In the second step, joins are counted within the study region for a specified critical distance. In nominal data, such as forest cover type, the mean and variance of joined island data are separately determined according to their category, color, or code. In ordinal data, such as the area of each island, the mean and variance are determined such that the deviations of joined islands' observations from their mean are used. Tests of significance are calculated by comparison with expected mean and variance under assumptions of randomness and normality.

Preliminary spatial autocorrelation analysis utilizing 1976 aerial photos and ground information (unpublished, see Tyburski and Ek 1977 for general description) has been conducted for the characteristics of area, height, density, cover type, and adjacent land uses of forest stands on the islands of the Milwaukee study region. The results are briefly stated:

1. Area is randomly allocated to each island, i.e., the area of an island is independent of the area of its neighbors.

2. Forest stands of tall trees and those of intermediate height and density tend to be clustered near one another. Those of low height are scattered at random across the landscape.

3. Northern hardwood stands (176 of 223 islands) are randomly distributed. Bottomland, or swamp hardwood stands (35 islands), oak stands (4 islands), and conifer stands, mostly planted Norway spruce (8 islands), are clustered together.

4. Adjacent land uses are determined as a percentage of total area within the immediate perimeter and within a 40 meter (2 chain) radius surrounding each forest stand. The categories are agriculture, residential buildings, farm buildings, roads, low scrub vegetation, marsh, forest, and surface water. As expected, adjacent land use patterns are autocorrelated. That is, percent land uses adjacent to one island are similar for nearby islands.

5. Fencerows, the length of rows of trees within a 0.8 km (40 chain) radius of each stand, are also quantified. These are also positively autocorrelated.

While these findings in themselves could be of significant geographical interest, they also are likely to be useful in suggesting causes for observed species distributions such as those noted by Levenson (Chapter 3).

Spatial autocorrelation analysis may also be used in the spatial and temporal interpretation of regression residuals in the initial development of the coefficients for the species prediction Eq. (12). For instance, suppose that the residuals are autocorrelated. With any luck, the pattern of one or more of the already quantified island characteristics will be correlated with the residuals. One could then add one or more of the characteristics as variables to the species prediction Eq. (12) via a stepwise regression procedure. The residuals, however, may be related to some variable not quantified, such as local soil conditions. One may then add a "dummy" variable to account for the spatial and/or temporal autocorrelation. The "dummy" variable could be equal to one in the areas where the residuals are clustered, zero otherwise, after the manner of Bannister (1976) for space-time components of urban population change.

The preceding has described a plan for the development of a systems model, its applications and directions for further refinement through pattern analyses, particularly spatial autocorrelation analysis. We believe it is a useful approach to aid in the derivation of management strategies for species preservation in forest island habitats.

12. Artificial Succession—A Feeding Strategy for the Megazoo

ARTHUR L. SULLIVAN
SCHOOL OF DESIGN
NORTH CAROLINA STATE UNIVERSITY

Human domination of the world's landscapes takes many forms, but perhaps the most noticeable is the change in vegetation. The creation of new successional patterns of plants, and consequently of animals, is a major result of human population expansion. These artificial assemblages may or may not be of long-term benefit, but the short-term demand for agricultural and silvicultural products signals that it is possible for us to convert a major portion of the flora and fauna of Earth to the purposes of feeding, clothing, and housing ourselves. The climax mix of useful plants and animals may be foreshadowed by those forms that have succeeded on our supermarket shelves.

It is perhaps a small group of people who question that an entirely domesticated planet would be an acceptable place in which to live. The reasons cited, recreation, scientific study, visual amenity, or preservation of genetic diversity, are all persuasive, though perhaps not so to people who are hungry or at war. The expansion of people into regions of preexisting flora and fauna causes extreme reactions even among the groups that care. Whatever the description of growth, whether as euphoric as Herman Kahn or as pessimistic as Dennis Meadows (1975), it is certain that growth does produce change over whole regions of the earth. It is the thesis of this chapter that geographic ecology can be extended to the urbanizing ecosystem so that the process can be fairly evaluated and so that rules for floral and faunal "redevelopment" can be established. Ecologists normally have great affection for systems in which mankind is not dominant, especially ones that they helped preserve. Lawmakers and policy planners often have great affection for the social systems they have helped build. Consequently, attitudes toward growth range all the way from "multiply and subdue" to "leave only footprints."

The Geography of Dominance

The extension of geographic ecology to human-dominated systems required some logical breakthroughs in an ecology dominated by types of species interaction. MacArthur and Wilson's (1967) *Theory of Island Biogeography* did this by realizing that size, shape, and distance were important variables in determining (at least statistically) what might survive where. Because the physical process of urbanization, i.e., a sprawl of pavement and buildings across the land, leaves unaltered remnant islands in this sea of people and their artifacts, the theory has been extended to forest habitat islands (Whitcomb 1977).

Not since the *International Symposium on Man's Role in Changing the Face of the Earth* in 1955 (Thomas 1956) has there been a multidisciplinary approach to the geography of man's dominance of the ecosphere. Classification of land use into the 9999 categories used by government agencies is independent of environmental variation (U.S. Bureau of Public Roads 1965). Most environmental inventories lump developed areas into a "disturbed" class and set about finely dividing "natural vegetation." The remnant islands of vegetation which concern us here normally appear on planners' maps as "open space" or "undeveloped."

In England, where there has been no North American style wilderness since the fifteenth century (Westhoff 1971), residents have devalued their categories of vegetated environments. There, "natural" means entirely vegetated, even if extensively managed for forestry or pasture.

The "megazoo" concept set out here is a symbol of our managed biosphere and a reminder that energy and physical space are finite and must therefore be subdivided by all living things. As in any ecological system, dominance of one form quite often means the exclusion (physically, chemically, or behaviorally) of others. Man is probably not yet dominant in the conterminous United States but probably is in England. A reasonable measure of dominance might be exclusion of large-bodied carnivores, which, as Connell (1975) has pointed out, may be important to the success of organisms lower on the food chain, i.e., that predation to some extent promotes diversity (O'Neill 1976).

The path of dominance, or civilization, has proceeded from fertile river floodplains to temperate forests to high altitude tropics to temperate grasslands. Recent technologies have permitted agricultural conversions of deserts and wetlands. Low altitude tropics and short growing season, high latitude regions have resisted domestication and are therefore strongholds of remaining wild ecosystems. Although the word "domestic" does not quite fit the world oceans, they, too, have been profoundly altered.

The effects of urbanization on native plant and animal population distributions have been well recorded. Some populations, like starlings, have expanded. Others, like the bald eagle, have diminished. Loss or gain in habitat is the oft cited reason, yet causation is complex, mostly because we do not really know how plants and animals sense their habitats and thus whether they occur deterministically according to some small set of variables, or stochastically according to dispersal distances and success ratios (Drury 1974).

The growing number of applications of MacArthur and Wilson's equilibrium theory of island biogeography suggests that physical dimensions of habitat limit the number of species that may be supported. If environmental diversity (beta diversity or patchiness) is held constant (i.e., all possible niches are maintained), then the equilibrium number of species will diminish as "island" size is diminished (Simberloff 1976b). The critical aspect here, since size is nearly always given by urban development, is variability or patchiness. If local conditions of patch generation like fire, disease, wind, etc., are maintained (Wiens 1976), then the habitat should remain suitable throughout and size alone should determine population success (Galli et al. 1976).

In the case of birds, the mix of vegetation structure is normally satisfactory if the canopy is closed and size is sufficient to produce the light levels normally associated with the forest floor. A rough rule of thumb for the minimum habitat island is three times the height of the tallest trees in E-W dimensions, two times in the N-S dimensions. This allows space for low sun angles to be shaded so that an interior patch of shade-tolerant species might exist. In much of the eastern deciduous forest, 20-m trees would suggest a unit 60 X 40 m or about 0.25 ha. Galli et al. (1976) suggest a larger minimum (0.8 ha). Coincidentally, this is about the residential density compatible with high crown closures.

The Ecology of Dominance

Effects of dominance on plant and animal geography include range size changes and within-range changes in density. For those islands of habitat remaining, size has been shown to limit the numbers of different species that will maintain successful local populations. These effects can be selective by acting at apparently different intensities as sensed by possible floral and faunal residents.

The predictability of effect of change on ecosystems is not as good as we once thought, however. Succession theory held that a directional and apparently purposeful progression of species populated a disturbed system (Odum 1969). Problems with the theory were anticipated by Raup (1964) and discussed by Drury and Nisbet (1973). Historically, over a sample of habitat islands, migrant birds with specialized food requirements are the first to respond to insularization by becoming locally extinct (Terborgh 1975). So, in general, we know that the creation of an island habitat will cause a decrease in the number of species supportable, and we can tentatively identify which members of a guild are likely candidates for local extinction.

There is some thought that short-term population data, as a base for ecosystem management, may be inconclusive (Platt and Denman 1975). If the dynamics of patch formation force species diversification on the local level, then it is possible that a downward trend in a particular population size is determined by nonlinear oscillations among parts of the environment which are not being observed. Cyclic variation in space and time may be routinely hidden in the variance of many regressions which purport to describe nature. The point of this cautionary note is to suggest that categories of threatened and endangered species depend on population data which may be secondary to other, more fundamental fluctuations in environmental parameters.

As Lovejoy (1976) points out, some species may be quite naturally on the way out. This is not to suggest that society simply shrug off extinctions of organisms, although few show concern for unattractive or pestiferous species, but rather to suggest that we study the context in which survival is possible. The size distribution and linkage of habitat islands seems an appropriate first step (Sullivan and Shaffer 1975). The second might well be spectral analyses of such islands to determine rates of change in the species ensemble. Finally, the linkage between spatial parameters and species population parameters will need to be established to afford prediction of where plants and animals will occur, which will occur, and when.

A Preliminary Survey of Urban Islands

While it is obvious that urban development affects vegetation and consumers, there is little work done on the ecology and geography of this kind of dominance.

The relative stability (Medwecka-Kornas 1977) and diversity (Owen and Owen 1975) of man-dominated landscapes compare quite favorably with natural systems. There is a cultural bias, however, especially among naturalists and outdoor recreationists, that connotes that nature can exist only in "natural areas" and that human dominance is a kind of pollution. Certainly, the sport hunting assault on wildlife has been interpreted as a cause of species population reduction in number and range. However, hunters pursue only about two dozen species, many of whose ranges do not overlap. For any local species list, the hunting pressure on a few species of larger animals would likely be statistically insignificant, however morally controversial. Hunting and pollution, however, *do* increase the effect of natural pressures, occasionally to the point of extinction.

The danger of thinking that nature can exist only in the absence of people lies in the land use planning which results. Ecological determinism asks the metaphysical question, "Where is nature best off?" Quickly espousing community parameters of diversity and stability, some environmental planners have suggested that diversity serve as the criterion for the open space or natural area designation. Somewhat ironically, the floodplain normally chosen as an example of species richness is rich as a result of disturbance, i.e., instability (Smith 1972).

In hopes of avoiding a priori designation of where nature ought to persist, I undertook a preliminary survey of Raleigh, North Carolina. Along a transect from downtown to the beltline bypass highway, I took photographs with a Widelux camera at several arbitrary sites to record the relationship of artifacts and vegetation. Then I turned the camera upward to record canopy density. Estimates of percent cover were made for three structural layers: herb (ground level to 70 cm), shrub (70 cm-8 m), and canopy (8 m and over).

In addition, I analyzed 30 years of Audubon Christmas counts to determine if any correspondence between urban development and presence of bird species could be found. The number of species observed increased from 1940 to 1970, probably a result of increased effort as the local clubs secured more and more members (North Carolina Audubon Society 1940-1970).

Even with inconclusive data, it seems certain that bird species diversity in the city of Raleigh has changed very little with 30 years of urbanization, during which the human population increased from around 50,000 to over 150,000 people.

The aerial photographs (Figs. 12-1, 12-2, and 12-3) illustrate the balance of vegetation and development. Unfortunately, "Central" and "Hillsborough" aerials are in winter. In the "Central" area, about 80% of the vegetation has been removed and the canopy is almost nowhere closed. Some specimen trees remain. Land in the undeveloped category is cleared of vegetation. Is this structural truncation analogous to physical shrinkage, i.e., if the structural volume is reduced by 80% would one expect a 25% reduction in species diversity? According to Darlington's (1965) approximation, (ten times the area yields twice the number of species), this would be expected.

In the "Hillsborough" neighborhoods there is more heterogeneity. Although stably developed in institutional and residential uses, there is an average of about 50% crown closure and greater than 50% herb coverage. Shrubs are still low because of the house density of between 10 and 15 per hectare. At this density, about half the property is taken up by roofs, walks, and drives. Shrubs, mostly ornamental and many exotic, serve as excellent nesting cover for birds. However, the production of vegetative food for consumers such as birds is diminished. Cauley (1974) observed (in a small sample) that 80% of urban households put out "food" (bread, donuts, birdseed) for the birds. The availability of food and nesting cover allows bird densities in such neighborhoods equal to or greater than in the wild. Of course, some species would not be represented (Geis 1974).

In the "West" neighborhoods (now 3-5 km from downtown), 0.2-0.4 ha residential properties permit 80% of the herb layer to remain. However, the ornamental lawns are mowed before setting seed. Insect consumers of grass are plentiful though, providing food sources for insectivorous birds. Except on abandoned farmland, the canopy approaches closure. True forest habitat islands begin to exist as the "undeveloped" category here means uncleared as well. At 0.2 ha densities, patches of the forest floor creep in where unmanaged patches of property coincide. At 0.4 ha, it is possible to leave half unmanaged. This, as mentioned earlier, might be a minimum size for true forest habitat in the Raleigh area.

A useful aspect of the aerial and Widelux photographs is the ability of untrained city personnel to provide information. Maps of available habitat will help establish policy for planting at various densities. For example, a hypothetical street tree program for the "West" neighborhoods might be cosmetic and therefore political, but perhaps ecologically and geographically redundant. For local amenity and for physiognomic redevelopment, it might be better advised to double up on the central neighborhood planting scheme. Another value of the photo-information base is the ability to generate maps of habitat types and to facilitate design of linking and stepping-stone devices between islands of known value. This kind of survey establishes a base inventory from which decisions about habitat preservation or redevelopment can be made.

Discussion

The city of Raleigh is just one exhibit in the megazoo. The resources which it can command are being continually subdivided by a growing human population. Many people lament the loss of wilderness and seek to preserve remants wherever possible. As these reserves become more valuable for parks or food production, however, a strategy for overall management will become necessary. The principles of island biogeography suggest that reserves can act as a distributive network and not simply as arks (MacClintock et al. 1977).

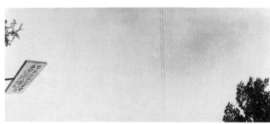

John Andresen (1976) and others are pioneering in giving scientific attention to urban vegetation, its measurement, management, and protection. Some work has been done on new designs for urban road development which consider habitat value (Kelcey 1975), while Doxiadis (1975) has considered the space needs of human settlement. There are many considerations of nature preserves (Helliwell 1976, Hooper 1970, Diamond and May 1976, Wilson and Willis 1975, Terborgh 1974, 1975). Sharpe (1975) has developed methods for mapping regional biological productivity. However, the design and maintenance of geographically and ecologically functional urban vegetation complexes is yet to be attempted. As more land is converted to agriculture, it will become the urban areas which will be best suited to preserve and maintain important plant and animal assemblages. The academic community has turned some attention to this opportunity in the symposia on *Restoration of Major Plant Communities in the United States* (Thorhaug 1977), and *The Recovery Process in Damaged Ecosystems* (Cairns 1980).

Management of the geography and phenology of farms and forests coupled with urban preservation schemes offers an artificial succession which gives humankind its share while maintaining maximum opportunities for everything else. Maps and schedules for type of crop planted, where, and when, can be developed to preserve the geography of patch formation native to various agriculture regions (Levin and Paine 1974). Road cuttings and harvesting patterns in forests can also be manipulated for effect on bird species diversity as implied in the work of B. L. Whitcomb et al. (1977).

The more developed countries like England lead us to believe that man and nature in a domestic condition can be a culturally satisfying environment even at eight times the present density of the United States. We do have the opportunity, however, of preselecting a settlement pattern that will retain optimal environmental values and of redeveloping areas of dysfunctional habitat. Analysis of the remaining habitat islands is the first step in establishing an artificial succession, or managed dynamic vegetation of wilderness, recreation areas, farms, forests, and cities which most efficiently allocates the primary productivity of the biosphere into an ecologically and culturally satisfying geography.

Fig. 12-1. Aerial view of "Central" neighborhoods and Widelux photos (horizontal and vertical) at three selected points. Estimates of vegetation cover (of a possible 100%) in each layer at each point are: Point 1: Herb 5-10; Shrub 0-5; Canopy 10-20. Point 2: Herb 20; Shrub 15; Canopy 5. Point 3: Herb 25; Shrub 15; Canopy 5. Area, by land use category, is as follows:

Land use	Hectares	Percent
Single family residential	83	25
Multiple family residential	18	6
Commercial	29.5	7
Industrial	6	2
Institutional	24	7
Open improved	8	2
Undeveloped	81	24
Rights-of-way	88	27

Fig. 12-2

Fig. 12-2. (*Facing page*) Aerial view of "Hillsborough" neighborhoods and Widelux photos (horizontal and vertical) at three selected points. Estimates of vegetation cover (of a possible 100%) in each layer at each point are: Point 4: Herb 50; Shrub 10; Canopy 75. Point 5: Herb 50; Shrub 25; Canopy 60. Point 6: Herb 75; Shrub 25; Canopy 25. Area, by land use category, is as follows:

Land use	Hectares	Percent
Single family residential	56	12
Multiple family residential	29	6
Commercial	17.5	5
Industrial	2	0.5
Institutional	190	41
Open improved	28	6
Undeveloped	32	7
Rights-of-way	80	18

Fig. 12-3. (*Overleaf*) Aerial view of "West" neighborhoods and Widelux photos (horizontal and vertical) at four selected points. Estimates of vegetation cover (of a possible 100%) in each layer at each point are: Point 7: Herb 85; Shrub 25; Canopy 60. Point 8: Herb 100; Shrub 100; Canopy 100. Point 9: Herb 80; Shrub 15; Canopy 10. Point 10: Herb 100; Shrub 5; Canopy 85. Area, by land use category, is as follows:

Land use	Hectares	Percent
Single family residential	574.5	15
Multiple family residential	102	3
Commercial	38.5	1
Industrial	38.5	1
Institutional	512	14
Open improved	423	12
Undeveloped	1494	41
Rights-of-way	365	10

Fig. 12-3

13. Summary and Conclusions

ROBERT L. BURGESS
ENVIRONMENTAL SCIENCES DIVISION
OAK RIDGE NATIONAL LABORATORY

DAVID M. SHARPE
DEPARTMENT OF GEOGRAPHY
SOUTHERN ILLINOIS UNIVERSITY

A vast experiment is under way. Its unplanned and unwitting design is changing the spatial and temporal structure of terrestrial ecosystems. We trace the beginnings of the experiment in the eastern United States to European settlement, and its continuation in other parts of the world to the spread of Western technology. This is arrogant—human impact on the biosphere has occurred over millenia. Nevertheless, current research assumes that biotic response to changing landscape pattern occurs within a time frame of years, or at most a century or two. In other words, recent and current land use changes have demonstrable impacts on terrestrial ecosystems. The condition of fragmented forest landscapes, the sensitivity of biotic responses, and assumptions concerning the time frame of these responses have fostered the research reported in this volume.

As Botkin (1980) points out, there is no a priori equilibrium state for regional ecosystems. Since the circumboreal Arctotertiary forest clothed the North American continent, functioning ecosystems have been subject to disruption, migration, perturbation, and evolution. The problems, in part, lie in our inability to relate to the long temporal and large spatial scales at which these processes operate (Smith 1975). The question then becomes one of identifying the "useful" or "desirable" or "valuable" ecosystem states with reference to requisites and conditions of the twentieth century and beyond.

We are learning that fragmented ecosystems still play a vital role in the ecology of a region. Not surprisingly, as different regions face different issues, the importance of forest ecosystem preservation varies from place to place. The size, structure, species composition, and interisland dynamics may have little relevance to an area faced with massive acidification of its freshwater lakes, or to a region whose food supply continues to depend on the clearing of tropical forest. As species response to such ecosystem disruption is more clearly defined, we begin to recognize the reasons behind differing extinction rates, and to appreciate the problems of selected species that may be sensitive to a rapidly changing habitat (MacMahon 1980).

It is no longer possible to look at "the way things should be" in regional landscapes. Ecologists, planners, engineers, and politicians must deal with things as they are. Identification of regional norms may consequently provide prescriptions for planners that are more than simply advisory. And regional landscapes then become a resource to manage rather than an ecology to maintain.

The interaction of landscape pattern and ecology has been the central theme of these studies. The scientific and practical issues that pertain to landscapes of fragmented natural communities and to their ecology are numerous, as suggested by the several contributions. The concerns range from the character (or pattern) of such landscapes, to their assemblages of flora and fauna, the dynamics that preserve them or lead to species extinction, and to strategies to manage these landscapes to achieve defined objectives.

The issues of landscape management are not solely ecological, but encompass social, ethical, economic, and demographic considerations as well. In any drive for preservation in the face of economic demands, value judgments must incorporate ecology, along with aesthetics, recreation, religion, and psychology. And obviously, ecologists have to participate in the complex process of identifying issues and then finding ways to infuse the results of multivariate analyses into environmental planning and resource decision making.

Landscape Patterns

The pattern of forest islands has been expressed in several ways. The sizes of forest patches have been of primary concern, both because of the power of the area/species richness relationship hypothesized from island biogeography, and because portions of the theory may be pertinent to habitat islands in a sea of agricultural and urban land uses. The distances separating forest islands also have been considered, as in the interaction index of Rudis and Ek and its inverse, the isolation coefficient formulated by Whitcomb and his colleagues.

Other elements of pattern have been discussed as well. Rudis and Ek reinterpret the Shannon-Wiener function to derive an index of topographic diversity based on topographic variability within respective forest islands. When multiplied by the area of each forest island, it yields an index of habitat diversity of the region. These authors also complement some of the points of Whitcomb et al. concerning the possible effects of a variety of forest island sizes in proximity to each other. Rudis and Ek extended Curtis' descriptions of forest fragmentation in Wisconsin (1956) using spatial analysis to show that in southeastern Wisconsin, small and large forest islands are interspersed, not clustered. Thus, small forest islands are likely to have larger neighbors which may possess greater biotic diversity and serve as sources of colonizing flora and fauna. However, in this region, the more mature communities are clustered, while younger stands are randomly distributed. Similarly, swamp hardwoods and some other forest types are clustered, while northern hardwoods are not. These spatial analyses provide a structural context for studying the region's biota and ecosystem functioning that is superior to the average, and largely aspatial, representations of landscapes heretofore used to characterize the Eastern Deciduous Forest Biome.

Species Richness of Forest Islands

Several authors have explored the central tenet of island biogeography theory that species richness of habitat patches is a direct function of their size. Levenson did not find this relationship for trees and shrubs in the interiors of 43 forest islands along a rural-urban gradient in southeastern Wisconsin, and Hoehne came to the same conclusion concerning the groundlayer. Scanlan, on the other hand, found that the flora of forest islands in Minnesota was more diverse in larger forest patches. Matthiae and Stearns, and Whitcomb et al. also found that mammalian and avian species richness was higher in larger forest islands than in small ones in Wisconsin and Maryland, respectively.

Lovejoy and Oren argued for the need to distinguish the minimum area needed to maintain self-reproducing populations in an ecosystem—its minimum critical size—from the area one would need to survey to assess its species richness, as shown by a species/area curve. They argue that a minimum critical size for an ecosystem is greater than suggested by species/area curves for three reasons. First, areas derived from species/area curves are based on the presence of one or more individuals of a species and do not necessarily assure a viable population. Additionally, species/area relationships are frequently derived by subsampling an extensive habitat, and are not necessarily transferable to fragmented habitat conditions. Finally, higher trophic levels are usually encountered only on the tail of the species/area curve, and may be excluded from smaller areas.

Levenson, in demonstrating this point for woodlots in southeastern Wisconsin, showed that maximum species richness was found in forest islands approximately 3 ha in size, in which both shade-tolerant and shade-intolerant species were present. At smaller sizes, shade intolerant species predominated, while at larger sizes, shade-tolerant species characteristic of mature northern hardwood forest were more important. Smaller forest islands are too exposed and thus too xeric for the major shade-tolerant species (sugar maple and beech) to compete with and eventually replace the array of shade intolerant species. Levenson concluded that 3 ha is the approximate minimum size required for a forest island to develop a mesic core that is able to sustain the mature northern hardwood forest type.

Ranney et al. draw a similar conclusion from analyzing the impact of forest island formation on the creation of an edge community, and the subsequent effect of the edge on the functioning of the forest island. In southeastern Wisconsin, edge is definable in terms of the continuum index, which incorporates both importance and adaptation values of the constituent species. Continuum index increases inward from the edge. Species with high adaptation values (shade-tolerant) significantly increase away from the edge, while those with low adaptation values (shade-intolerant species) decline. Importance of some species seemed to be independent of position vis-à-vis the edge. This analysis concluded that the edges on the east, north and south sides of forest islands extend 10-15 m into the forest island, and up to 30 m on the west side. Thus, the size of a forest island must be at least twice these dimensions in order to contain mesic species.

Furthermore, survival of the mesic community dominated by sugar maple depends on the course of regeneration. The seed rain into cores of small forest islands is dominated by the seed production of the edge species. Edges also function as filters for the interisland seed rain. Thus, the seed pools of small cores are dominated by the shade intolerant species, the environment is conducive to their survival, and they can successfully compete against the regeneration of mesic species. Forest islands larger than 2 ha are needed both to maintain the mesic environment and to enable the mesic species to develop an adequate seed pool for self-regeneration.

Ranney et al. believe that shade-intolerant species are attaining the unusually high positions of relative abundance predicted by Curtis (1956). This component is more abundant in the interiors of the forest islands examined than it would be if the stands were embedded in more extensive forest. The proximity of forest edge species to the core creates seed pools weighted toward shade-intolerant species for which small stand size and disturbance history create appropriate environmental conditions.

Much island biogeography research has assumed equilibrium conditions resulting from a balance between invasions and extinctions. However, the study of forest island landscapes is more akin to the experimental studies that have explored the impact of changing island size, shape, or isolation. The forest island landscape is ever-changing as land use patterns change, and individual forest patches are subjected to disturbance of varying types, intensities, and frequencies. These changing patterns have been amply documented in Latin America (Alvim 1977, D'Arcy 1977, Vovides and Gomez-Pompa 1977), but are only now beginning to concern North American researchers.

With disruption, the species composition of forest islands inevitably lags behind changing forest island patterns. Hoehne, for example, found that species richness of the groundlayer of forest islands in southeastern Wisconsin has declined significantly in the past 25 years, even though the size of the islands has remained stable during this period. The absence of a relationship between island size and overstory species richness in the same islands (Chapter 3) perhaps is attributable to the long relaxation times (the period required for a return to equilibrium) for this component of forest islands in contrast to the elapsed time since fragmentation. Avifauna and mammals appear to have short relaxation times, the groundlayer a moderate relaxation time, and the dominant tree species relatively long relaxation times. Ranney et al. provide evidence that changes in species composition resulting from reduction in island size are predictable, as Lovejoy and Oren have concluded. Their simulations indicate that reduction in forest island size below 4-5 ha leads to a reduction in the equilibrium continuum index, i.e., shade-tolerant species are lost in a predictable sequence.

The spatiotemporal pattern of land use change and creation of a forest island landscape thus has a strong impact on species structure and composition. Rudis and Ek incorporate this concept in a systems model of fragmented landscapes. Past landscape states, forest island sizes, shapes, and disturbance are given greater weight in the current species composition than are recent changes in landscape patterns. Interaction between forest islands is time dependent, and the sequence of spatial changes, and past disturbances within forest islands and their past age and species structures determine the current species composition of forest island landscapes. Johnson et al. also point out that forest succession is a spatially dependent process. The likelihood that forest islands occupying the same positions along an environmental gradient will have a predictable species composition, over and above that due to stochastic variation, is lower in highly fragmented landscapes than in landscapes in which the forest ecosystem is more contiguous.

Implications for Environmental Management

Our final theme concerns the implications of these developing concepts for landscape management, from the landscape as a laboratory to the landscape as a manageable resource. Early use of island biogeographic theory for environmental management called for larger clustered preserves (Diamond 1975a, Sullivan and Shaffer 1975). The research in this volume poses a more complex management issue, the management of regions with mosaics of land uses in order to preserve species genetic pools and intact ecosystems. The size, shape, and spacing of forest islands is but one element of that mosaic, recently addressed by Gilpin and Diamond (1980) and Higgs and Usher (1980). Ecologically effective management must encompass the mosaic, and must be based upon knowledge of species responses to the complexities of a man-dominated environment. Whitcomb et al. forcefully demonstrate this need and May provides a model to deal with it.

Rudis and Ek developed a conceptual framework for managing forest island landscapes. They suggest that operations research techniques can be used to optimize landscapes. This approach has the potential to achieve such plausible goals as maximizing species richness at some time in the future, or estimating the minimum number of satellite forest islands that would be needed around already established preserves in order to maintain their present species richness.

Finally, Sullivan underscores the tension between ecologists who favor pristine ecosystems and lawmakers and planners dedicated to the human systems that create man-dominated landscapes. This tension is expressed in research issues tackled, goals of landscape management, and in the paucity of interchange between ecologists and planners. Until recently, it resulted in research that ignored the human role in ecosystem functioning, and land use planning that proceeded in isolation from vital ecologic insight. Ranney and collaborators present both the problem and the hope.

> It should be realized that the faults of many regional plans with respect to large scale conservation efforts have resulted from their inability to directly connect broad scale planning goals with objectives that are meaningful to the small landowner and heed ongoing ecological phenomena. Perceived regional changes in forest composition from the perspective of naturally functioning but artificially maintained habitat units (the forest islands and their linkages) can connect human intent with ecological phenomena in landscapes already dominated by men.

Conclusions

The pattern of man-dominated landscapes undergoes rapid change. Removal of over 90% of the natural habitats in southcentral Wisconsin in less than a century (Curtis 1956) and similar clearing followed by large-scale secondary forest reversion along the eastern seaboard (Raup 1937, 1966, Bond and Spiller 1935) may be ecologic perturbations that are unprecedented in rate, intensity, and extent. Landscape pattern continues to change in these regions, and the pace of change in other regions, such as the Mississippi delta, the Georgia Piedmont, parts of western Europe, the British Isles, New Zealand, and much of the tropics, is increasing.

Some (perhaps all) components of the biota are in disequilibrium with the current landscape patterns. As a result, judgments about relationships between landscape pattern, species composition, and ecosystem functioning are difficult to draw. Research design needs to reflect this fact.

The strong interaction between elements of landscape mosaics, especially between habitat islands and the agro-urban matrix, is as fundamental to the dynamics of these landscapes as is the size and location of discrete habitats. The impact of size, shape, disturbance level, location—all the variables attributable to individual habitats—is mitigated by the surrounding components of the mosaic. Indeed, major unknowns are the time and distance decay rates of these interactions between any landscape element and those surrounding it.

"If we accept the basic tenet that [ecological] diversity is a legitimate concern for the future well-being of mankind, we must say so in terms that the public can comprehend and with facts upon which politicians can act" (Irwin 1977). All decisions are ultimately political, but choices are not necessarily between clear streams or cheaper garbage disposal. Perhaps the knowledge and perspectives of regional landscape ecology portrayed in the preceding chapters constitute a beginning toward a developing strategy to address these problems. It does no good to decry the loss. What we need are clear statements of the utility of the resource value. Just as the National Environmental Policy Act of 1969 was aimed at preserving options for the future, it must be remembered that research is option-creating, while management is often option-constraining. We need to analyze alternatives to continued forest fragmentation, distinguish the tradeoffs for management, and recognize that there is no real necessity for particular assemblages or patterns, only that these are artifacts with which modern society must deal. The quandary is still with us.

Acknowledgments. Research supported in part by the Eastern Deciduous Forest Biome, US-IBP, funded by the National Science Foundation under Interagency Agreement AG-199, BMS69-01147 A09 with the U.S. Department of Energy under contract W-7405-eng-26 with Union Carbide Corporation, and in part by NSF grant DEB 78-11338 to Southern Illinois University. Contribution No. 350, Eastern Deciduous Forest Biome, US-IBP. Publication No. 1650, Environmental Sciences Division, Oak Ridge National Laboratory.

References*

Able, K. P., and B. R. Noon. 1976. Avian community structure along elevational gradients in the northeastern United States. Oecologia 26:275-294. [8]

Alexander, R. R. 1969. Seedfall and establishment of Englemann spruce in clearcut openings: A case history. Research Paper RM-53. USDA Forest Service, Rocky Mountain Forest and Range Experiment Station, Fort Collins, Colorado. 8 pp. [10]

Allen, D. L. 1962. Our Wildlife Legacy. Funk and Wagnalls, New York. 422 pp. [6]

Allen, P. H., and K. B. Trousdell. 1961. Loblolly pine seed production in the Virginia-North Carolina coastal plain. J. For. 59:187-190. [10]

Alvim, P. deT. 1977. The balance between conservation and utilization in the humid tropics with special reference to Amazonian Brazil. pp. 347-351. In G. T. Prance and T. S. Elias (eds.), Extinction is forever. The status of threatened and endangered plants of the Americas. The New York Botanical Garden, Bronx, New York. 437 pp. [13]

Andresen, J. W. (ed.). 1976. Trees and forests for human settlements. Center for Urban Forestry Studies, University of Toronto, Toronto, Canada. 417 pp. [12]

Appleton Press. 1973. Appleton Area Centennial Issue. Appleton, Minnesota. [7]

Arrhenius, O. 1921. Species and area. J. Ecol. 9:95-99. [3]

Auclair, A. N. 1976. Ecological factors in the development of intensive-management ecosystems in the midwestern United States. Ecology 57:531-444. [1]

Auclair, A. N., and G. Cottam. 1971. Dynamics of black cherry (*Prunus serotina* Ehrh.) in southern Wisconsin oak forests. Ecol. Monogr. 41:153-177. [3,6,10]

Avery, T. E. 1969. Foresters' guide to aerial photo interpretation. USDA Forest Service Agric. Hdbk. No. 308. Washington, D.C. 40 pp. [11]

*Numbers in brackets indicate chapters where the reference is cited.

Bannister, G. 1976. Space-time components of urban population change. Econ. Geogr. 52:228-240. [11]

Bard, G. E. 1952. Secondary succession in the piedmont of New Jersey. Ecol. Monogr. 22:195-215. [11]

Barr, A. J., J. H. Goodnight, J. P. Sall, and J. T. Helwig. 1976. A user's guide to SAS 76. SAS Institute, Raleigh, North Carolina. 329 pp. [8]

Bent, A. C. 1946. Life histories of North American jays, crows, and titmice. U.S. National Museum Bull. 191. Smithsonian Institution Press, Washington, D.C. 495 pp. [8]

Bent, A. C. 1958. Life histories of North American blackbirds, orioles, tanagers, and allies. U.S. National Museum Bull. 211. Smithsonian Institution Press, Washington, D.C. 549 pp. [8]

Bent, A. C. 1968. Life histories of North America cardinals, grosbeaks, buntings, towhees, finches, sparrows, and allies. U.S. National Museum Bull. 237. Smithsonian Institution Press, Washington, D.C. 1889 pp. [8]

Benzie, J. W. 1959. Sugar maple and yellow birch seed dispersal from a fully stocked stand of mature northern hardwoods in the upper peninsula of Michigan. Technical Notes No. 561. USDA Forest Service, Lake States Forest Experiment Station, St. Paul, Minnesota. [10]

Bergen, J. D. 1975. Air movement in a forest clearing as indicated by smoke drift. Agric. Meteorol. 15:165-179. [10]

Bjorkbam, J. C. 1971. Production and germination of paperbirch seed and its dispersal into a forest opening. Research Paper NE-209. USDA Forest Service, Northeastern Forest Experiment Station, Upper Darby, Pennsylvania. 14 pp. [10]

Blair, W. F. 1977. Big biology. The US/IBP. Dowden, Hutchinson, and Ross, Stroudsberg, Pennsylvania. 261 pp. [Preface]

Blondel, J., C. Ferry, and B. Frochot. 1970. La method des indices ponctuels d'abondance (I.P.A.) ou des releves d' avifaune par "Stations d'ecoute." Alauda 38:55-71. [8]

Bond, R. R. 1957. Ecological distribution of breeding birds in the upland forests of southern Wisconsin. Ecol. Monogr. 27:351-384. [8]

Bond, W. E., and A. R. Spiller. 1935. Use of land for forests in the lower Piedmont region of Georgia. Occas. Paper 53. USDA Forest Service, Southern Forest Experiment Station, New Orleans, Louisiana. 47 pp. [1, 13]

Borchert, J. R. 1950. The climate of the central North American grassland. Ann. Assoc. Am. Geogr. 40:1-39. [1]

Bormann, F. H., and G. E. Likens. 1979. Catastrophic disturbances and the steady state in northern hardwood forests. Am. Sci. 67(6):660-669. [1]

Bossema, I. 1979. Jays and oaks: An eco-ethological study of a symbiosis. Behavior 70:1-117. [10]

Botkin, D. B. 1980. A grandfather clock down the staircase: Stability and disturbance in natural ecosystems. pp. 1-10. In R. H. Waring (ed.), Forests: Fresh perspectives from ecosystem analysis. Proc. 40th Annual Biology Colloquium. Oregon State University Press, Corvallis, Oregon. 199 pp. [13]

Botkin, D. B., J. F. Janak, and J. R. Wallis. 1972. Some ecological consequences of a computer model of forest growth. J. Ecol. 60:849-872. [10]

Boyce, S. G., and J. P. McClure. 1975. How to keep one-third of Georgia in pine. Research Paper SE-144. USDA Forest Service, Southeastern Forest Experiment Station, Asheville, North Carolina. 23 pp. [1]

Boyer, W. D. 1958. Longleaf pine seed dispersal in south Alabama. J. For. 56:265-268. [10]

Brackbill, H. 1977. Number of broods per season raised by some birds at Baltimore. Maryland Birdlife 33:113-115. [8]

Bray, J. R. 1956. Gap-phase replacement in a maple-basswood forest. Ecology 37:598-600. [3, 6]

Brewer, R., and L. Swander. 1977. Life history factors affecting the intrinsic rate of natural increase of birds of the deciduous forest biome. Wilson Bull. 89:211-232. [8]

Brown, J. H. 1971. Mammals on mountaintops: Nonequilibrium insular biogeography. Am. Nat. 105:467-478. [3, 8]

Brown, J. H., and A. Kodric-Brown. 1977. Turnover rates in insular biogeography: Effect of immigration on extinction. Ecology 58:445-449. [8]

Brown, R. G. 1963. Smoothing, forecasting and prediction. Prentice-Hall, Englewood Cliffs, New Jersey. [11]

Bruner, M. C. 1977. Vegetation of forest island edges. M.S. Thesis, University of Wisconsin, Milwaukee. 60 pp. [3]

Brush, G. S., C. Lenk, and J. Smith. 1976. Vegetation map of Maryland: The existing natural forest. Maryland Dept. Nat. Resources Power Plant Siting Program, Annapolis, Maryland. [8]

Brush, G. S., C. Lenk, and J. Smith. 1980. The natural forests of Maryland: An explanation of the vegetation map of Maryland. Ecol. Monogr. 50(1):77-92. [8]

Bullock, S. H., and R. B. Primack. 1977. Comparative experimental study of seed dispersed on animals. Ecology 58:681-686. [6]

Burgess, R. L. 1978. The changing face of eastern North America. Frontiers 42(3):8-11. [1]

Butcher, G. S., W. A. Niering, W. J. Barry, and R. H. Goodwin. 1980. Equilibrium biogeography and the size of nature preserves: An avian case study. Oecologia (in press). [8]

Bystrak, D., 1980. Application of Miniroutes to bird population studies. Maryland Birdlife (in press). [8]

Cain, S. A., and G. M. Castro. 1959. Manual of vegetation analysis. Harper, New York. 325 pp. [2]

Cairns, J., Jr. (ed.). 1980. The recovery process in damaged ecosystems. Ann Arbor Science Publ., Ann Arbor, Michigan. 167 pp. [12]

Carlquist, S. 1974. Island biology. Columbia University Press, New York. 660 pp. [8]

Carpenter, J. R. 1935. Forest-edge birds and exposure of their habitats. Wilson Bull. 17:106-108. [6]

Cauley, D. L. 1974. Urban habitat requirements for four wildlife species, pp. 143-147. In J. H. Noyes and D. R. Progulske (eds.), Wildlife in an urbanizing environment. Cooperative Extension Service, Springfield, Massachusetts. [12]

Cauley, D. L., and J. R. Schinner. 1973. The Cincinnati raccoons. Nat. Hist. 82(9):58-60. [5]

Cayford, J. H. 1964. Red Pine seedfall in southeastern Manitoba. For. Chron. 40:78-85. [10]

Challis, D. 1977. Breeding bird census: Mature mixed hardwood forest. Am. Birds 31:43. [8]

Chamberlin, T. C. 1873. Atlas of Wisconsin Geological Survey. Madison, Wisconsin. [3]

Chamberlin, T. C. 1877. Native vegetation of eastern Wisconsin. Geol. Wis. 2:176-187. [3]

Chaney, W. R., and T. T. Kozlowski. 1969. Seasonal and diurnal expansion and contraction of *Pinus banksiana* and *Picea glauca* cones. New Phytol. 68:873-882. [10]

Chettleburgh, M. R. 1952. Observations on the collection and burial of acorns by jays in Hainault Forest. Brit. Birds 45:359-364. [10]

Christiansen, F. B., and T. M. Fenchel. 1977. Theories of populations in biological communities. Springer-Verlag, New York. 140 pp. [10]

Clark, P. J., and F. C. Evans. 1954. Distance of nearest neighbor as a measure of spatial relationships in populations. Ecology 35:445-453. [11]

Clements, F. E. 1916. Plant succession: An analysis of the development of vegetation. Publ. 242, Carnegie Institution of Washington, Washington, D.C. 512 pp. [6]

Cliff, A. D., P. Haggett, J. K. Ord, K. A. Bassett, and R. B. Davies. 1975. Elements of spatial structure. Cambridge University Press, London. 258 pp. [11]

Cliff, A. D., and J. K. Ord. 1973. Spatial autocorrelation. Pion, London. 178 pp. [11]

Cody, M. L. 1974. Competition and the structure of bird communities. Monographs in population biology No. 7. Princeton University Press, Princeton, New Jersey. 318 pp. [8]

Coelho, J. D., and A. G. Wilson. 1977. An equivalence theorem to integrate entropy-maximizing submodels within overall mathematical programming frameworks. Geogr. Anal. 9:160-173. [11]

Cohen, J. G. 1970. A Markov contingency table for replicated Lotka-Volterra systems near equilibrium. Am. Nat. 104:547-559. [10]

Connell, J. H. 1975. Some mechanisms producing structure in natural communities. pp. 460-491. In M. L. Cody and J. M. Diamond (eds.), Ecology and evolution of communities. Belknap Press of Harvard University, Cambridge, Massachusetts. 545 pp. [12]

— Connor, E. F., and E. D. McCoy. 1979. The statistics and biology of the species-area relationship. Am. Nat. 113:791-833. [7]

Connor, E. F., and D. S. Simberloff. 1979. The assembly of species communities: Chance or competition? Ecology 60:1132-1140. [7]

Cope, J. A. 1948. White ash management possibilities in the northeast. J. For. 46:744-749. [3]

Cornell, H. V., and J. O. Washburn. 1979. Evolution of the richness-area correlation for cynipid gall wasps on oak trees: A comparison of two geographic areas. Evolution 33:257-274. [8]

Cottam, G. 1949. The phytosociology of an oak woods in southwestern Wisconsin. Ecology 30:271-287. [7]

Cremer, K. W. 1971. Speeds of falling and dispersal of seed of *Pinus radiata* and *Pinus contorta*. Austr. For. Res. 5:29-32. [10]

Criswell, J. H., and J. R. Gauthey. 1980. Breeding bird census: Mature deciduous floodplain forest. Am. Birds 34:46. [8]

Crockett, J. V. 1971. Landscape gardening. Time-Life Books, New York. 160 pp. [6]

Cromartie, W. J., Jr. 1975. The effect of stand size and vegetational background on the colonization of cruciferous plants by herbivorous insects. J. Appl. Ecol. 12:517-533. [6]

Crooke, R., and J. D. Webster. 1974. Breeding bird census: Mixed deciduous forest. Am. Birds 28:1007. [8]

— Crowell, K. L. 1973. Experimental zoogeography: Introduction of mice to small islands. Am. Nat. 107:535-558. [3]

Culver, D. C. 1970a. Analysis of simple cave communities I: Caves as islands. Evolution 24:463-474. [3]

Culver, D. C. 1970b. Analysis of simple cave communities: Niche separation and species packing. Ecology 51:949-958. [8]

Curtis, J. T. 1956. The modification of mid-latitude grasslands and forests by man. pp. 721-736. In W. L. Thomas (ed.), Man's role in changing the face of the earth. University of Chicago Press, Chicago. 1193 pp. [1, 3, 6, 10, 13]

Curtis, J. T. 1959. The vegetation of Wisconsin—An ordination of plant communities. University of Wisconsin Press, Madison. 657 pp. [3, 4, 7]

Curtis, J. T., and G. Cottam. 1962. Plant ecology workbook. Burgess, Minneapolis, Minnesota. 193 pp. [4]

Darby, H. C. 1956. The clearing of the woodland in Europe. pp. 183-216. In W. L. Thomas (ed.), Man's role in changing the face of the earth. University of Chicago Press, Chicago. 1193 pp. [1]

D'Arcy, W. G. 1977. Endangered landscapes in Panama and Central America: The threat to plant species. pp. 89-104. In G. T. Prance and T. S. Elias (eds.), Extinction is forever. The status of threatened and endangered plants of the Americas. The New York Botanical Garden, Bronx, New York. 437 pp. [13]

Darlington, P. J. 1965. Biogeography of the southern end of the world. Harvard University Press, Cambridge, Massachusetts. 236 pp. [12]

Davis, A. M., and T. F. Glick. 1978. Urban ecosystems and island biogeography. Environ. Conserv. 5:299-304. [3]

Day, G. M. 1953. The Indian as an ecological factor in the northeastern forest. Ecology 34:329-346. [1]

DeAngelis, D. L., E. G. Stiles, W. C. Johnson, D. M. Sharpe, and R. K. Schreiber. 1977. A model for the dispersal of seeds by animals. EDFB/IBP-77/5. Oak Ridge National Laboratory, Oak Ridge, Tennessee. 50 pp. [6, 10]

DeAngelis, D. L., C. C. Travis, and W. M. Post. 1980. Persistence and stability of seed-dispersed species in a patchy environment. Theor. Pop. Biol. 16:107-125. [10]

DeWalle, D. R., and S. G. McGuire. 1973. Albedo variations of an oak forest in Pennsylvania. Agric Meteorol. 11:107-113. [6]

Detwyler, T. R. 1972. Vegetation of the city. pp. 229-260. In T. R. Detwyler and M. G. Marcus (eds.), Urbanization and environment. Duxbury Press, Belmont, California. [3]

Dhont, A. A. 1979. Summer dispersal and survival of juvenile great tits in southern Sweden. Oecologia 42:139-157. [8]

Diamond, J. M. 1969. Avifaunal equilibria and species turnover rates on the Channel Islands of California. Proc. Nat. Acad. Sci. USA 64:57-63. [3]

Diamond, J. M. 1970. Ecological consequences of island colonization by southwest Pacific birds. I. Types of niche shifts. Proc. Nat Acad. Sci. USA 67:529-536. [8]

Diamond, J. M. 1972. Biogeographic kinetics: Estimation of relaxation times for avifaunas of southwest Pacific islands. Proc. Nat. Acad. Sci. USA 69:3199-3203. [2, 8, 10, 11]

Diamond, J. M. 1973. Distributional ecology of New Guinea birds. Science 179:759-769. [2, 3]

Diamond, J. M. 1975a. The island dilemma: Lessons of modern biogeographic studies for the design of natural reserves. Biol. Conserv. 7:129-146. [2, 3, 8, 11]

Diamond, J. M. 1975b. Assembly of species communities. pp. 342-444. In M. L. Cody and J. M. Diamond (eds.), Ecology and evolution of communities. Harvard University Press, Cambridge, Massachusetts. [11]

Diamond, J. M. 1976a. Island biogeography and conervation: Strategy and limitations, a reply. Science 193:1027-1029. [2]

Diamond, J. M. 1976b. Relaxation and differential extinction on land-bridge islands: Applications to natural preserves. pp. 616-628. In H. J. Frith and J. H. Calaby (eds.), Proc. 16th Int. Ornithol. Congr. (1974). Australian Acad. Sci., Canberra. [8]

Diamond, J. M. 1978. Critical areas for maintaining viable populations of species. pp. 27-40. In M. Holdgate and M. J. Woodman (eds.), Breakdown and restoration of ecosystems. Plenum Press, New York. [2]

Diamond, J. M., and A. G. Marshall. 1977. Distributional ecology of New Hebridian birds: A species kaleidoscope. J. Anim. Ecol. 46:703-727. [7]

Diamond, J. M., and R. M. May. 1976. Island biogeography and the design of natural reserves. pp. 163-186. In R. M. May (ed.), Theoretical ecology: Principles and applications. Saunders, Philadelphia-Toronto. 317 pp. [2, 12]

Dill, H. W., and R. C. Otte. 1971. Urbanization of land in the northeastern United States. ERS 485, USDA Economic Research Service, Washington, D.C. 11 pp. [1]

Dobbs, R. C. 1976. White spruce seed dispersal in central British Columbia. For. Chron. 52:225-231. [10]

Dony, J. G. 1977. Species-area relationships in an area of intermediate size. J. Ecol. 65:475-484. [7]

Dowden, A. O. 1975. To pollinate an orchid. Audubon Mag. 77(5):40-47. [4]

Doxiadis, C. 1975. The ecological types of space that we need. Environ. Conserv. 2(1): 3-14. [12]

Dritschilo, W., H. Cornell, D. Nafus, and B. O'Conner. 1975. Insular biogeography: Of mice and mites. Science 190:467-469. [8]

Drury, W. H. 1974. Rare species. Biol. Conserv. 6(3):162-169. [12]

Drury, W. H., and I. C. T. Nisbet. 1973. Succession. J. Arnold Arboretum 54(3):331-368. [12]

Dyrness, C. T. 1973. Early stages of plant succession following logging and burning in the Western Cascades of Oregon. Ecology 54:57-69. [7]

Ek, A. R., R. S. Meldahl, and R. A. Monserud. 1977 (revised). Instructions for implementing FOREST Version II on UNIVAC, IBM and other systems. Department of Forestry Staff Paper Series #3. University of Wisconsin, Madison. 8 pp. plus 3 microfiche. [11]

Ek, A. R., and R. A. Monserud. 1974. FOREST: A computer model for simulating the growth and reproduction of mixed species forest stands. Research Report R-2635, School of Natural Resources, University of Wisconsin, Madison. 64 pp. [6, 10, 11]

Elfstrom, B. A. 1974. Tree species diversity and forest island size on the Piedmont of New Jersey. M.S. Thesis, Rutgers University, New Brunswick, New Jersey. 55 pp. [10]

Faeth, S. H., and T. C. Kane. 1978. Urban biogeography. City parks as islands for Diptera and Coleoptera. Oecologia 32:127-133. [8]

Falinski, J. B. 1976. Durability of forest relicts in the agricultural landscape in light of investigations on permanent plots. Phytocoenosis 5:213-224. [1]

Falinski, J. B. 1977. Research on vegetation and plant population dynamics conducted by Bialowieza Geobotanical Station of the Warsaw University in the Bialowieza primeval forest 1952-1977. Phytocoenosis 6(1/2):1-145. [1]

Fenneman, N. M. 1938. Physiography of the eastern United States. McGraw-Hill, New York. 714 pp. [3, 4]

Fernald, M. L. 1970. Gray's Manual of Botany. D. Van Nostrand, New York. [7]

Ferry, C. 1974.Comparison between breeding bird communities in an oak forest and a beech forest, censused by the IPA method. Acta Ornithol. 14:302-309. [8]

Fields, D. E., and D. M. Sharpe. 1980. SEDFAL: A model of dispersal of tree seeds by wind. EDFB/IBP-78/2. Oak Ridge National Laboratory, Oak Ridge, Tennessee. 110 pp. [10]

Fitzpatrick, J. W. 1980. Th wintering of North American tyrant flycatchers in the neotropics. In J. A. Keast and E. S. Morton (eds.), Migrant birds in the neotropics: Ecology, behavior, distribution, and conservation. Smithsonian Institution Press, Washington, D.C. [8]

Fonaroff, L. S. 1974. Urbanization, birds and ecological change in northwestern Trinidad. Biol. Conserv. 6:258-261. [8]

Forman, R. T. T. 1974. Ecosystem size: A stress on the forest ecosystem. Bull. Ecol. Soc. Am. 55(2):38. [11]

Forman, R. T. T. 1976. Effect of woodlot size on selected ecosystem components (unpublished manuscript). [6]

Forman, R. T. T., and B. A. Elfstrom. 1975. Forest structure comparison of Hutcheson Memorial Forest and eight old woods on the New Jersey Piedmont. William L. Hutcheson Memorial Bulletin 3:44-51. [3, 10]

Forman, R. T. T., A. E. Galli, and C. F. Leck. 1976. Forest size and avian diversity in New Jersey woodlots with some land use implications. Oecologia 26:1-8. [2, 3, 5, 8, 10]

Forty-Second Congress. March 3, 1873. An act to encourage the growth of timber on western prairies. U. S. Statutes at Large 17:605-606. [7]

Forty-Third Congress. March 13, 1874. An act to amend the act entitled 'an act to encourage the growth of timber on western prairies.' U.S. Statutes at Large 18:21-22. [7]

Fowells, H. A. 1965. Silvics of forest trees of the United States. Agric. Handbook No. 271, USDA Forest Service, Washington, D.C. 762 pp. [3]

Franklin, I. R. 1980. Evolutionary change in small populations. pp. 135-149. In M. E. Soule and B. A. Wilcox (eds.), Conservation biology: An evolutionary-ecological perspective. Sinauer Assoc., Sunderland, Massachusetts. [2]

Freeland, W. J. 1979. Primate social groups as biological islands. Ecology 60:719-728. [7]

Frei, K. R., and D. E. Fairbrothers. 1963. Floristic study of the William L. Hutcheson Memorial Forest, New Jersey. Bull. Torrey Bot. Club 90:338-355. [6]

Fretwell, S. D. 1972. Populations in a seasonal environment. Monographs in Population Biology No. 5. Princeton University Press, Princeton, New Jersey. 217 pp. [8]

Fritz, R. S. 1979. Consequences of insular population structure: Distribution and extinction of spruce grouse populations. Oecologia 42:57-65. [8]

Gadgil, M. 1971. Dispersal: Population consequences and evaluation. Ecology 52:253-261. [8]

Galli, A. E., C. F. Leck, and R. T. T. Forman. 1976. Avian distribution patterns in forest islands of different sizes in central New Jersey. Auk 93:356-364. [2, 3, 8, 10, 11, 12]

Gallusser, W. A. 1978. Der Wiederaufbau der Nordamerikanischen Zivilisationslandschaft durch Staatliche Massnahmen, am Beispiel von Wisconsin. Erdkunde 32:142-157. [1]

Gashwiler, J. S. 1969. Seedfall of three conifers in west-central Oregon. For. Sci. 15(3):290-295. [10]

Geiger, R. 1966. The climate near the ground. Harvard University Press, Cambridge, Massachusetts. 611 pp. [10]

Geis, A. D. 1974. Effects of urbanization and type of urban development on bird populations. pp. 97-105. In J. H. Noyes and D. R. Progulske (eds.), A symposium on wildlife in an urbanizing environment. University of Massachusetts, Amherst. [8, 12]

Ghiselin, J. 1977. Analyzing ecotones to predict biotic productivity. Environ. Manage. 1:235-238. [6]

Gilpin, M. E., and J. M. Diamond. 1980. Subdivision of nature reserves and the maintenance of species diversity. Nature 285:567-568. [13]

Gleason, H. A. 1952. The new Britton and Brown illustrated flora of the northeastern U.S. and adjacent Canada. Hafner, New York. 3 Vols. 1732 pp. [3, 4]

Godman, R. M. 1953. Seed dispersal in southeastern Alaska. Note 16. USDA Forest Service, Alaska Forest Research Center, Fairbanks, Alaska. 2 pp. [10]

Goeden, G. B. 1979. Biogeographic theory as a management tool. Environ. Conserv. 6:27-32. [3]

Goff, F. G., and D. C. West. 1975. Canopy-understory interaction effects on forest population structure. For. Sci. 21:98-108. [6]

Goff, F. G., and P. H. Zedler. 1968. Structural gradient analysis of upland forests in the western Great Lakes area. Ecol. Monogr. 38:65-86. [3]

Goff, J. A. 1965. Saturation pressure of water on the new Kelvin Scale. pp. 289-292. In A. Wexler (ed.), Humidity and moisture measurement and control in science and industry, Vol. II. Reinhold, New York, 4 Vols. [10]

Gomez-Pompa, A., C. Vazquez-Yanes, and S. Guevara. 1972. The tropical rain forest, a non-renewable resource. Science 177:762-765. [1, 3, 10]

Gottfried, B. M. 1979. Small mammal populations in wood lot islands. Am. Midl. Nat. 102:105-112. [3]

Graber, R. R., and J. W. Graber. 1963. A comparative study of bird populations in Illinois, 1906-1909 and 1956-1958. Ill. Nat. Hist. Surv. Bull. 28:377-528. [8]

Grant, P. R., and I. Abbott. 1980. Interspecific competition, island biogeography and null hypotheses. Evolution 34:332-341. [8]

Greller, A. M. 1975. Persisting natural vegetation in northern Queens County, New York, with proposals for its conservation. Environ. Conserv. 2:61-69. [3]

Griffith, D. A. 1976. Spatial structure and spatial interaction: a review. Environ. Plan. A 8:731-740. [11]

Guenther, K. W. 1951. An investigation of the tolerance of white ash reproduction. J. For. 49:576-577. [3]

Gustafson, D. K. 1977. Effects of biogeographical island size and quality on avian diversity. Bull. Ecol. Soc. Am. 58(2):31. [11]

Gysel, L. W. 1951. Borders and openings of beech-maple woodlands in southern Michigan. J. For. 49:13-19. [6]

Haase, Edward F. 1965. A comparison of dispersal methods in species of prairie and upland hardwood forest in southern Wisconsin. M.A. Thesis, University of Wisconsin, Milwaukee. 52 pp. [4]

Hall, G. 1964. Breeding-bird censuses—Why and how. Audubon Field Notes 18:413-416. [8]

Hamilton, T. H., R. H. Barth, Jr., and I. Rubinoff. 1964. The environmental control of insular variation in bird species abundance. Proc. Nat. Acad. Sci. USA 52:132-140. [8]

Hamilton, T. H., and I. Rubinoff. 1963. Isolation, endemism and multiplication of species in the Darwin finches. Evolution 17:388-403. [8]

Hamilton, T. H., and I. Rubinoff. 1964. On models predicting abundance of species and endemics for the Darwin finches in the Galapagos Archipelago. Evolution 18:339-342. [8]

Hamilton, T. H., and I. Rubinoff. 1967. On predicting insular variation in endemism and sympaty for the Darwin finches in the Galapagos archipelago. Am. Nat. 101:161-171. [8]

Hann, H. W. 1937. Life history of the Oven-bird in southern Michigan. Wilson Bull. 49:145-240. [8]

Hansen, H. P. 1933. The tamarack bogs of the driftless area of Wisconsin. Milwaukee Public Museum 7(2):231-304. [3]

Harper, J. L., J. N. Clatworthy, I. H. Naughton, and G. R. Sagar. 1961. The evolution and ecology of closely related species living in the same area. Evolution 15:209-227. [6]

Harper, J. L., P. H. Lovell, and K. G. Moore. 1970. The shapes and sizes of seeds. Annu. Rev. Ecol. Syst. 1:327-356. [6, 10]

Harris, A. S. 1969. Ripening and dispersal of a bumper western hemlock-Sitka spruce seed crop in southeast Alaska. Research Note PNW-105. USDA Forest Service, Pacific Northwest Forest Experiment Station, Portland, Oregon. 11 pp. [10]

Heatwole, H., and R. Levins. 1973. Biogeography of the Puerto Rican Bank: Species turnover on a small cay, Cayo Ahogado. Ecology 54:1042-1055.[3, 5]

Heinselman, M. L. 1973. Fire in the virgin forests of the Boundary Waters Canoe Area, Minnesota. Quat. Res. 3:329-382. [1]

Heinselman, M. L. (ed.). 1974. The original vegetation of Minnesota, by Marschner, F. J., 1930. USDA Forest Service, North Central Forest Experiment Station, St. Paul, Minnesota. [7]

Helliwell, D. R. 1973. Priorities and values in nature conservation. J. Environ. Manage. 1:85-127. [11]

Helliwell, D. R. 1976. The extent and location of nature conservation areas. Environ. Conserv. 3(4):255-258. [12]

Henny, C. J. 1972. An analysis of the population dynamics of selected avian species. Wildlife Res. Rept. 1. Bureau of Sport Fisheries and Wildlife, Washington, D.C. 99 pp. [8]

Henny, C. J., F. C. Schmid, E. M. Martin, and L. L. Hood. 1973. Territorial behavior, pesticides, and the population ecology of red-shouldered hawks in central Maryland, 1943-1971. Ecology 54:545-554. [8]

Hett, J. M., and O. L. Loucks. 1971. Sugar maple (Acer saccharum) seedling mortality. J. Ecol. 59:507-520. [6]

Higgs, A. J., and M. B. Usher. 1980. Should nature reserves be large or small? Nature 285:568-569. [13]

Holt, B. R. 1972. Effect of arrival time on recruitment, mortality, and reproduction in successional plant populations. Ecology 53:668-673. [6]

Hooper, M. D. 1970. Size and surroundings of nature reserves. pp. 555-561. I. E. Duffy and A. S. Watt (eds.), The scientific management of animal and plant communities for conservation. Blackwell, London. [12]

Horn, H. S. 1975. Forest succession. Sci. Am. 232:90-98. [10]

Horn, H. S. 1976. Succession. pp. 187-204. In R. M. May (ed.), Theoretical ecology, principles and applications. Saunders, Philadelphia-Toronto. 317 pp. [10]

Howe, H. F. 1977. Bird activity and seed dispersal of a tropical wet forest tree. Ecology 58:539-550. [6]

Hutchison, B. A., and D. R. Matt. 1976a. Beam enrichment of diffuse radiation in a deciduous forest. Agric. Meteorol. 17:93-110. [6]

Hutchison, B. A., and D. R. Matt. 1976b. Forest meteorology research within the Oak Ridge site, Eastern Deciduous Forest Biome. U.S. IBP ATDL Contribution No. 76/4. Atmospheric Turbulence and Diffusion Laboratory, National Oceanic and Atmospheric Administration, Oak Ridge, Tennessee. 57 pp. [6]

Hutchison, B. A., and D. R. Matt. 1977. The distribution of solar radiation within a deciduous forest. Ecol. Monogr. 47:185-207. [6]

International Bird Census Committee. 1970. An international standard for a mapping method in bird census work recommended by the International Bird Census Committee. Am. Birds 24:722-726. [8]

Irwin, H. S. 1977. Preface. pp. 1-2. In G. T. Prance and T. S. Elias (eds.), Extinction is forever. The status of threatened and endangered plants of the Americas. The New York Botanical Garden, Bronx, New York. 437 pp. [13]

Isaac, L. A. 1930. Seed flight in the Douglas fir region. J. For. 28:492-499. [10]

Jackson, H. H. T. 1961. Mammals of Wisconsin. University of Wisconsin Press, Madison. 504 pp. [5]

Jaenike, J. 1978. Effect of island area on *Drosophila* population densities. Oecologia 36:327-332. [8]

James, F. C. 1971. Ordinations of habitat relationships among breeding birds. Wilson Bull. 83:215-236. [8]

Janzen, D. H. 1968. Host plants as islands in evolutionary and contemporary time. Am. Nat. 102:592-595. [3]

Janzen, D. H. 1973. Host plants as islands. II. Competition in evolutionary and contemporary time. Am. Nat. 107:786-790. [8]

Jemison, G. M., and C. F. Korstian. 1944. Loblolly pine seed production and dispersal. J. For. 42:734-741. [10]

Johnson, M. P., L. G. Mason, and P. H. Raven. 1968. Ecological parameters and plant diversity. Am. Nat. 102:297-306. [3]

Johnson, M. P., and D. S. Simberloff. 1974. Environmental determinants of island species numbers in the British Isles. J. Biogeogr. 1:149-154. [7]

Johnson, N. K. 1975. Controls on number of birds species on montane islands in the Great Basin. Evolution 29:545-567. [8]

Johnson, W. C., R. K. Schreiber, and R. L. Burgess. 1979. Diversity of small mammals in a powerline right-of-way and adjacent forest in east Tennessee. Am. Midl. Nat. 101(1):231-235. [6]

Johnson, W. C., and D. M. Sharpe. 1976. An analysis of forest dynamics in the northern Georgia Piedmont. For. Sci. 22(3):307-322. [1, 10]

Johnston, D. W. 1974. Decline of DDT residues in migratory songbirds. Science 186:841-842. [8]

Johnston, D. W., and E. P. Odum. 1956. Breeding bird populations in relation to plant succession on the Piedmont of Georgia. Ecology 37:50-62. [8]

Jones, R. E. 1969. Breeding bird census: Urban woodlots. Am. Birds 23:726-728. [8]

Keast, A., and E. S. Morton (eds.). 1980. Migrant birds in the neotropics: Ecology, behavior, distribution, and conservation. Smithsonian Institution Press, Washington, D.C. 578 pp. [8]

Kelcey, J. G. 1975. Opportunities for wildlife habitats on road verges in a new city. Urban Ecology 1:271-284. [12]

Kelker, G. H. 1964. Appraisal of ideas advanced by Aldo Leopold thirty years ago. J. Wildl. Manage. 28:180-185. [6]

Kellman, M. 1974. Preliminary seed budgets for two plant communities in coastal British Columbia. J. Biogeogr. 1:123-133. [10]

Kendeigh, S. C. 1944. Measurement of bird populations. Ecol. Monogr. 14:67-106. [8]

Kendeigh, S. C. 1948. Breeding bird census: Oak-maple forest and edge. Audubon Field Notes 2:232-233. [8]

Kendeigh, S. C. 1981. Bird populations in east central Illinois: Fluctuations, variations, and development over a half-century. Ill. Biol. Monogr. (in press). [8]

Kendeigh, S. C., and J. M. Edgington. 1977. Breeding bird census: Oak-maple forest and edge. Am. Birds 31:43-44. [8]

Kennard, J. H. 1975. Longevity records of North American birds. Bird-Banding 46:55-73. [8]

Klimkiewicz, M. K., and J. K. Solem. 1978. The breeding bird atlas of Montgomery and Howard Counties, Maryland. Maryland Birdlife 34:3-39. [8]

Knight, H. A. 1974. Land-use changes which affected Georgia's forest land, 1961-1972. Research Note SE-189. USDA Forest Service, Southeastern Forest Experiment Station, Asheville, North Carolina. 4 pp. [1]

Kohlerman, L. 1950. Untersuchungen über die Windverbreitung der Früchte and Samen der mitteleuropische Waldbäume. Forstw. C. 69:606-624. [10]

Kolata, G. B. 1974. Theoretical ecology: Beginnings of predictive science. Science 183:400. [3]

Koopman, K. 1958. Land bridges and ecology in bat distribution on islands off the northern coast of South America. Evolution 12:429-439. [7]

Lack, D. 1942. Ecological features of the bird faunas of British small islands. J. Animal Ecol. 11:9-36. [8]

Leopold, A. 1933. Game management. Scribners, New York. 481 pp. [6]

Leston, D. 1957. Spread potential and the colonization of islands. Syst. Zool. 6:41-46. [7]

Levenson, J. B. 1976. Forested woodlots as biogeographic islands in an urban-agricultural matrix. Ph.D. Thesis, University of Wisconsin-Milwaukee. 101 pp. and appendices. [3, 6, 10, 11]

Levenson, J. B. 1980. The southern-mesic forest of southeastern Wisconsin: Species composition and community structure. Contrib. Biol. Geol., Milwaukee Co. Public Museum, Milwaukee, Wisconsin. [3]

Levin, S. A. 1977. Population dynamic models in heterogeneous environments. Annu. Rev. Ecol. Syst. 7:287-310. [10]

Levin, S. A., and R. T. Paine. 1974. Disturbance, patch formation, and community structure. Proc. Nat. Acad. Sci. USA 71(7):2744-2747. [6, 12]

Levins, R., and D. Culver. 1971. Regional coexistence of species and competition between rare species. Proc. Nat. Acad. Sci. USA 68:1246-1248. [10]

Levins, R., and H. Heatwole. 1973. Biogeography of the Puerto Rican Bank: Introduction of species onto Palominitos Island. Ecology 54:1056-1064. [3]

Liddle, M. J. 1975. A selective review of the ecological effects of human trampling on natural ecosystems. Biol. Conserv. 7:17-34. [8]

Liebetrau, A. M., and E. D. Rothman. 1977. A classification of spatial distributions based upon several cell sizes. Geogr. Anal. 9:14-28. [11]

Lindsey, A. A. 1955. Testing the line-strip method against full tallies in diverse forest types. Ecology 36:485-495. [3, 4, 6]

Lindsey, A. A. 1956. Sampling methods and community attributes in forest ecology. For. Sci. 2:287-296. [3, 4]

Linduska, J. 1949. Ecology and land-use relationships of small mammals on a Michigan farm. Ph.D. Thesis, Michigan State University, East Lansing. 272 pp. [10]

Lineham, J. T., R. E. Jones, and J. R. Longcore. 1967. Breeding-bird populations in Delaware's urban woodlots. Audubon Field Notes 21:641-646. [8]

Löhrl, H. 1959. Zur Frage des Zeitpunktes einer Prägung auf die Heimatregion beim Halsbandschnapper (Ficedula albicollis). J. Orn. 100:132-140. [8]

Loucks, O. L. 1970. Evolution of diversity, efficiency and community stability. Am. Zool. 10:17-25. [1, 3, 11]

Lovejoy, T. E. 1975. Bird diversity and abundance in Amazon forest communities. Living Bird 13:127-191. [2]

Lovejoy, T. E. 1976. We must decide which species will go forever. Smithsonian. Aug: 52-58. [12]

Lovejoy, T. E. 1979. Genetic aspects of dwindling populations: A review. pp. 275-279. In S. A. Temple (ed.), Endangered birds: Management techniques for preserving threatened species. University of Wisconsin Press, Madison. [2]

Lussenhop, J. 1977. Urban cemeteries as bird refuges. Condor 79:456-461. [8]

Lynch, J. F., and N. K. Johnson. 1974. Turnover and equilibria in insular avifaunas, with special reference to the California Channel Islands. Condor 76:370-384. [8]

Lynch, J. F., and R. F. Whitcomb. 1978. Effects of the insularization of the eastern deciduous forest on avifaunal diversity and turnover. pp. 461-489. In A. Marmelstein (ed.), Classification inventory and analysis of fish and wildlife habitat. U.S. Fish and Wildlife Service, Washington, D.C. [8]

MacArthur, R. H. 1959. On the breeding distribution pattern of North American migrant birds. Auk 76:318-325. [8]

MacArthur, R. H. 1964. Environmental factors affecting bird species diversity. Am. Nat. 98:387-397. [3]

Mac Arthur, R. H., J. M. Diamond, and J. R. Karr. 1972. Density compensation in island faunas. Ecology 53:330-342. [8]

MacArthur, R. H., J. W. MacArthur, and J. Preer. 1962. On bird species diversity. II. Prediction of bird census from habitat measurements. Am. Nat. 96:167-174. [3]

MacArthur, R. H., and E. R. Pianka. 1966. On optimal use of a patchy environment. Am. Nat. 100:603-609. [6]

MacArthur, R. H., H. Recher, and M. Cody. 1966. On the relation between habitat selection and species diversity. Am. Nat. 100:319-327. [11]

MacArthur, R. H., and E. O. Wilson. 1963. An equilibrium theory of insular biogeography. Evolution 17:373-387. [2, 7, 8, 10]

MacArthur, R. H., and E. O. Wilson. 1967. The theory of island biogeography. Princeton University Press, Princeton, New Jersey. 203 pp. [1, 2, 3, 5, 6, 7, 8, 10, 11, 12]

MacClintock, L., R. F. Whitcomb, and N. E. MacClintock. 1978. Breeding bird census: Mature tulip-tree-oak forest. Am. Birds 32:60. [8]

MacClintock, L., R. F. Whitcomb, and B. L. Whitcomb. 1977. Island biogeography and "habitat islands" of eastern forest. II. Evidence for the value of corridors and minimization of isolation in preservation of biotic diversity. Am. Birds 31:6-16. [8, 12]

MacClintock, N. E., R. F. Whitcomb, and L. MacClintock. 1978. Breeding bird census: Selectively logged mature tulip-oak forest. Am. Birds 32:61. [8]

MacKinney, A. L., and C. F. Korstian. 1938. Loblolly pine seed dispersal. J. For. 36:465-468. [10]

MacMahon, J. A. 1980. Ecosystems over time: Succession and other types of change. pp. 27-58. In R. H. Waring (ed.), Forests: Fresh perspectives from ecosystem analysis. Proc. 40th Annual Biology Colloquium, Oregon State University Press, Corvallis, Oregon. 199 pp. [13]

Mair, A. R. 1973. Dissemination of tree seed; Sitka spruce, western hemlock and Douglas fir. Scottish Forestry 27(4):308-314. [10]

March, J. 1976. Wildlife abundance in upland woodlots on a southeastern Wisconsin study area. Project No. W-141-R-11, Final Report, Study Nos. 113.2 and 113.3. Wisconsin Department of Natural Resources, Madison. 13 pp. and 10 pp., respectively. [5]

Matthiae, P. E. 1977. The function of mammals in terrestrial forest islands in southeastern Wisconsin. M.S. Thesis, University of Wisconsin-Milwaukee. 93 pp. [5]

Mauriello, D., and J. P. Roskoski. 1974. A reanalysis of Vuilleumier's data. Am. Nat. 108:711-714. [11]

Mawson, J. C., J. W. Thomas, and R. M. DeGraaf. 1976. Program HTVOL: The determination of tree crown volume by layers. USDA Forest Service, Northeastern Forest Experiment Station, Upper Darby, Pennsylvania. 9 pp. [3]

Mayfield, H. F. 1973. Kirtland's Warbler census, 1973. Am. Birds 27:950-952. [8]

Mayfield, H. F. 1977. Brown-headed cowbird: Agent of extermination? Am. Brids 31:107-113. [8]

McConathy, R. K., S. B. McLaughlin, D. E. Reichle, and B. E. Dinger. 1976. Leaf energy balance and transpirational relationships of tulip poplar (*Liriodendron tulipifera*). EDFB/IBP-76/6. Oak Ridge National Laboratory, Oak Ridge, Tennessee. 109 pp. [6]

McCutchen, C. W. 1977. The spinning rotation of ash and tulip tree samaras. Science 197:691-692. [10]

McDiarmid, R. W., R. R. Ricklefs, and M. S. Foster. 1977. Dispersal of *Stemmadenia donnel-smithii* (Apocynaceae) by birds. Biotropica 9:9-25. [6]

McEwen, J. K. 1971. Aerial dispersal of black spruce seed. For. Chron. 47:161-162. [10]

McIntosh, R. P. 1957. The York Woods, a case history of forest succession in southern Wisconsin. Ecology 38:29-37. [6]

Mead, R. 1974. A test for spatial pattern at several scales using data from a grid of contiguous quadrats. Biometrics 30:295-309. [11]

Meadows, D., and H. Kahn. 1975. Different views of growth. Forum 13 (1&2):18-22. [12]

Medwecka-Kornas, A. 1977. Ecological problems in the conservation of plant communities with special reference to Central Europe. Biol. Conserv. 4(1):27-33. [12]

Mellinger, E. O. 1969. Breeding bird census: Mountain ravine mixed forest. Audubon Field Notes 23:711. [8]

Menchik, M. D. 1972. Optional allocation of outdoor recreational use in the presence of ecological carrying capacity limitations and congestion effects. Regional Sci. Assoc. 30:77-96. [11]

Miller, A. H., and B. J. Niemann, Jr. 1972. An interstate corridor selection process: The application of computer technology to highway location dynamics—Phase I. Environmental Awareness Center, Dept. Landscape Architecture, University of Wisconsin, Madison. 161 pp. and appendices. [11]

Mitchell, R. 1977. Bruchid beetles and seed packing in Palo Verde. Ecology 58:644-651. [6]

Moen, A. N. 1974. Turbulence and the visualization of wind flow. Ecology 55:1420-1424. [10]

Monitor and News. 1878. Benson, Minnesota. [7]

Monitor and News. 1970. Centennial Issue. Benson, Minnesota. [7]

Monserud, R. 1975. Methodology for simulating Wisconsin northern hardwood stand dynamics. Ph.D. Thesis, University of Wisconsin-Madison. 156 pp. [10]

Monserud, R. A. 1976. Simulation of forest tree mortality. For. Sci. 22:438-444. [6]

Moore, N. W., and M. D. Hooper. 1975. On the number of bird species in British woods. Biol. Conserv. 8:239-250. [2, 8]

Moran, P. A. P. 1948. The interpretation of statistical maps. J. Royal Statis. Soc. Ser. B 10:243-251. [11]

Morse, D. H. 1977. The occupation of small islands by passerine birds. Condor 79: 399-412. [8]

Morse, D. H. 1980. Population limitation: Breeding or wintering ground? pp. 505-516. In A. Keast and E. S. Morton (eds.), Migrant birds in the neotropics: Ecology behavior, distribution, and conservation. Smithsonian Institution Press, Washington, D.C. [8]

Morton, E. S. 1978. Reintroducing recently extirpated birds into a tropical forest preserve. pp. 379-384. In S. A. Temple (ed.), Endangered birds: Management techniques for preserving threatened species. University of Wisconsin Press, Madison. [2]

Mühlenberg, M., D. Leipold, H. J. Mader, and B. Steinhauer. 1977a. Island ecology of arthropods. I. Diversity, niches, and resources on some Seychelles islands. Oecologia 29:117-134. [8]

Mühlenberg, M., D. Leipold, H. J. Mader, and B. Steinhauer. 1977b. Island ecology of arthropods. II. Niches and relative abundances of Seychelles ants (Formicidae) in different habitats. Oecologia 29:135-144. [8]

Murphy, R. E. 1931. Geography of the northwest pine barrens of Wisconsin. Trans. Wis. Acad. Sci. Arts Lett. 26:69-120. [7]

Niering, W. A. 1963. Terrestrial ecology of Kapingamarangi Atoll, Caroline Islands. Ecol. Monogr. 33:131-160. [3, 7]

Nilsson, S. G. 1978. Fragmented habitats, species richness and conservation practice. Ambio 7:26-27. [3]

Nolan, V., Jr. 1979. The ecology and behavior of the Prairie Warbler *Dendroica discolor*. Ornith. Monographs 26. American Ornithologists' Union, Allen Press, Lawrence, Kansas. 595 pp. [8]

North Carolina Audubon Society 1940-1970. North Carolina Chat., Raleigh, North Carolina. [12]

Odum, E. P. 1959. Fundamentals of Ecology. Saunders, Philadelphia, Pennsylvania. 546 pp. [6]

Odum, E. P. 1969. The strategy of ecosystem development. Science 164:262-270. [12]

Odum, E. P. 1971. Fundamentals of ecology. Saunders, Philadelphia, Pennsylvania. 574 pp. [4]

Oelke, N. 1966. 35 years of breeding bird census work in Europe. Audubon Field Notes 20:635-642. [8]

Olaczek, R., and R. Sowa. 1976. Dying out of native flora in urbanized areas on the example of the forest reservation "Polesie Konstantinowskie" in the city of Lodz. Phytocoenosis 5(3/4):291-299. [1]

O'Neill, R. V. 1976. Ecosystem persistence and heterotrophic regulation. Ecology 57 (6):1244-1253. [12]

Opler, P. A. 1974. Oaks as evolutionary islands for leaf-mining insects. Am. Sci. 62:67-73. [8, 11]

Owen, J., and D. Owen. 1975. Suburban gardens: England's most important natural resource? Environ. Conserv. 2(1):53-60. [12]

Oxley, D. J., M. B. Fenton, and G. R. Carmody. 1974. The effects of roads on populations of small mammals. J. Appl. Ecol. 11(1):51-59. [12]

Paloheimo, J. E., and A. M. Vukov. 1976. On measures of aggregation and indices of contagion. Math. Biosci. 30:69-97. [11]

Patrick, R. 1967. The effect of invasion rate, species pool, and size of area on the structure of the diatom community. Proc. Nat. Acad. Sci. USA 58:1335-1342. [2]

Patton, D. R. 1975. A diversity index for quantifying habitat edge. Wildl. Soc. Bull. 394:171-173. [6]

Philip, M. S. 1975. The use of simulation in the management of privately-owned forest in Britain. Forestry 48(2):123-138. [11]

Picton, H. D. 1979. The application of insular biogeographic theory to the conservation of large mammals in the northern Rocky Mountains. Biol. Conserv. 15:73-79. [3]

Pielou, E. C. 1959. The use of point-to-plant distances in the study of the pattern of plant populations. J. Ecol. 47:607-613. [11]

Pielou, E. C. 1966. Measurement of diversity in different types of biological collections. J. Theor. Biol. 13:131-144. [4]

Pielou, E. C. 1969. An introduction to mathematical ecology. Wiley-Interscience, New York. 286 pp. [11]

Platt, T., and K. L. Denman. 1975. Spectral analysis in ecology. Annu. Rev. Ecol. Syst. 6:189-210. [12]

Poore, M. E. D. 1955. The use of phytosociological methods in ecological investigations: II. Practical issues involved in an attempt to apply the Braun-Blanquet system. J. Ecol. 43:245-269. [2]

Poore, M. E. D. 1968. Studies in Malaysian rain forest: I. The forest on Triassic sediments in Jengka Forest Reserve. J. Ecol. 56:143-196. [2]

Power, D. M. 1972. Numbers of bird species on the California islands. Evolution 26:451-463. [8]

Preston, F. W. 1960. Time and space and the variation of species. Ecology 41:611-627. [3]

Preston, F. W. 1962. The canonical distribution of commonness and rarity: Part II. Ecology 43:410-432. [8, 10]

Preston, F. W., and R. T. Norris. 1947. Nesting heights of breeding birds. Ecology 28: 241-273. [8]

Rabenold, K. N. 1978. Foraging strategies, diversity, and seasonality in bird communities of Appalachian spruce-fir forests. Ecol. Monogr. 48:397-424. [8]

Rabenold, K. N. 1979. A reversed latitudinal diversity gradient in avian communities of eastern deciduous forests. Am. Nat. 114:275-286. [8]

Raleigh Planning Department. 1974. Raleigh Neighborhoods. City of Raleigh, North Carolina. 63 pp. [12]

Randall, A. G. 1974. Seed dispersal into two spruce-fir clearcuts in Eastern Maine. Res. Life Sci. 21(8):1-15. [10]

Ranjitsinh, M. K. 1979. Forest destruction in Asia and the South Pacific. Ambio 8(5): 192-201. [1]

Ranney, J. W. 1977. Forest island edges—their structure, development, and importance to regional forest ecosystem dynamics. EDFB/IBP-77/1. Oak Ridge National Laboratory, Oak Ridge, Tennessee. 38 pp. [6, 8]

Ranney, J. W. 1978. Edges of forest islands: Structure, composition, and importance to regional forest dynamics. Ph.D. Thesis, University of Tennessee-Knoxville. 193 pp. [10]

Ranney, J. W., and W. C. Johnson. 1977. Propagule dispersal among forest islands in southeastern South Dakota. Prairie Nat. 9:17-24. [6]

Raup, H. M. 1937. Recent changes of climate and vegetation in southern New England and adjacent New York. J. Arnold Arbor. 18:79-117. [1, 13]

Raup, H. M. 1964. Some problems in ecological theory and their relation to conservation. J. Ecol. 52:19-28. [12]

Raup, H. M. 1966. The view from John Sanderson's farm: A perspective for the use of the land. For. Hist. April:2-11. [1, 13]

Reichman, O. J., and D. Oberstein. 1977. Selection of seed distribution types by Dipodomys merriami and Perognathus amplus. Ecology 58:636-643. [6]

Reifsnyder, W. E. 1955. Wind profiles in a small isolated forest stand. For. Sci. 1(4): 289-297. [10]

Reifsnyder, W. E. 1965. Radiant energy in relation to forests. USDA Forest Service, Tech. Bull. No. 1344. Government Printing Office, Washington, D.C. 64 pp. [6]

Rex, M. A. 1975. Laboratory experiments on stepping-stone colonization. Bull. Ecol. Soc. Am. 56(2):43. [11]

Robbins, C. S. 1969. Suggestions on gathering and summarizing return data. Migr. Bird Pop. Sta. USFWS, Laurel, Maryland. (mimeo). 11 pp. [8]

Robbins, C. S. 1978a. Census techniques for forest birds. pp. 142-163. In Management of southern forests for nongame birds. General Technical Report SE-14. USDA Forest Service. Southeastern Forest Experiment Station, Asheville, North Carolina. 176 pp. [8]

Robbins, C. S. 1978b. Determining habitat requirements of nongame species. North Amer. Wildl. and Natural Resources Conf. Trans. 43:57-68. [8]

Robbins, C. S. 1979. Effect of forest fragmentation on bird populations. pp. 198-212. In Management of north central and northeastern forests for nongame birds. General Technical Report NC-51. USDA Forest Service, North Central Forest Experiment Station, St. Paul, Minnesota. 268 pp. [8]

Robbins, C. S., and W. T. Van Velzen. 1974. Progress report on the North American breeding bird survey. Acta Ornithologica 14:170-191. [8]

Rombakis, V. S. 1947. Über die Verbreitung von Pflanzensamen und Sporen durch turbulent Luftströmungen. Z. F. Meteorol. 1(11):359-363. [10]

Root, R. B. 1967. The niche exploitation pattern of the Blue-gray Gnatcatcher. Ecol. Monogr. 37:317-350. [8]

Roth, R. R. 1976. Spatial heterogeneity and bird species diversity. Ecology 57:773-782. [8]

Roy, D. F. 1960. Douglas-fir seed dispersal in northwestern California. Technical Paper No. 49. USDA Forest Service, Pacific Southwest Forest and Range Experiment Station, Berkeley, California. 22 pp. [10]

Rusterholz, K. A., and R. W. Howe, 1979. Species-area relations of birds on small islands in a Minnesota lake. Evolution 33:468-477. [8]

Ruth, R. H., and C. M. Bengsten. 1955. A 4-year record of sitka spruce and western hemlock seedfall. Research Paper No. 12. USDA Forest Service, Pacific Northwest Forest Experiment Station, Portland, Oregon. 13 pp. [10]

Salibury, E. J. 1942. The reproductive capacity of plants. G. Bell, London. 277 pp. [6]

Salisbury, F. B., and C. Ross. 1969. Plant Physiology. Wadsworth, Belmont, California. 747 pp. [6]

Scanlan, M. J. 1975a. The geography of forest plants across the prairie-forest border in western Minnesota. Ph.D. Thesis, University of Minnesota, St. Paul. 228 pp. [7]

Scanlan, M. J. 197b. The geography of shrub and herb species across the prairie-forest ecotone in western Minnesota. Bull. Ecol. Soc. Am. 56:53. [10]

Schafer, J. 1927. Four Wisconsin counties: Prairie and forest. State Hist. Soc. Wis., Madison. 429 pp. [3]

Schoener, A. 1974. Colonization curves for planar marine islands. Ecology 55:818-827. [7]

Schoener, T. W. 1968. Sizes of feeding territories among birds. Ecology 49:123-141. [8]

Schopmeyer, C. S. 1974. Seeds of woody plants in the United States. Agric. Handbook No. 450. USDA Forest Service, Washington, D.C. [10]

Schreiber, R. K., W. C. Johnson, J. D. Story, C. Wenzel, and J. T. Kitchings. 1976. Effects of powerline rights-of-way on small nongame mammal community structure. pp. 263-273. In Proc. 1st National Symposium on Environmental Concerns in Rights-of-Way Management. Mississippi State University, Mississippi State. 335 pp. [6]

Schuster, L. 1950. Über den sammeltrieb des Eichelhahers (Garrulus glandarius). Vogelwelt 62:239-240. [10]

Sepkoski, J. J., Jr., and M. A. Rex. 1974. Distribution of freshwater mussels: Coastal rivers as biogeographic islands. Syst. Zool. 23:165-188. [11]

SEWRPC (Southeastern Wisconsin Regional Planning Commission). 1963. The natural resources of southeastern Wisconsin. Planning Report No. 5. Waukesha, Wisconsin. 163 pp. [3]

SEWRPC (Southeastern Wisconsin Regional Planning Commission). 1965. Potential parks and related open spaces. Technical Report No. 1. Waukesha, Wisconsin. 21 pp. [3]

SEWRPC (Southeastern Wisconsin Regional Planning Commission). 1970. A comprehensive plan for the Milwaukee River watershed. Planning Report No. 13. Vol. 1. Waukesha, Wisconsin. 514 pp. [3]

SEWRPC (Southeastern Wisconsin Regional Planning Commission). 1971. Residential land subdivision in southeastern Wisconsin. Technical Report No. 9. Waukesha, Wisconsin. 85 pp. [3]

SEWPRC (Southeastern Wisconsin Regional Planning Commission). 1972. The population of southeastern Wisconsin. Technical Report No. 11. Waukesha, Wisconsin. 98 pp. [3]

Sharpe, D. M. 1975. Methods of assessing the primary productivity of regions. pp. 147-160. In H. Leith and R. H. Whittaker (eds.). Primary productivity of the biosphere. Springer-Verlag, New York. [12]

Shearer, R. C. 1959. Western larch seed dispersal over clear-cut blocks in northwestern Montana. Proc. Montana Acad. Sci. 19:130-134. [10]

Sheldon, J. C., and F. M. Burrows. 1973. The dispersal effectiveness of the achene-pappus units of selected Compositae in steady winds with convection. New Phytol. 72:665-675. [10]

Shinners, L. H. 1940. Vegetation of the Milwaukee region. B.A. Thesis, University of Wisconsin. 101 pp. [3]

Shreve, F., M. A. Chrysler, F. H. Blodgett, and F. W. Besley. 1910. The plant life of Maryland. Maryland Weather Service 3:1-533. [8]

Shugart, H. H., and D. C. West. 1977. Development of an Appalachian deciduous forest succession model and its application to assessment of the impact of the chestnut blight. J. Environ. Manage. 5:161-179. [6, 10]

Siggins, H. 1933. Distribution and rate of fall of conifer seeds. J. Agric. Res. 47:119-128. [10]

—— Simberloff, D. S. 1974. Equilibrium theory of island biogeography and ecology. Annu. Rev. Ecol. Syst. 5:161-182. [7, 8]

Simberloff, D. S. 1976a. Species turnover and equilibrium island biogeography. Science 194:572-578. [2]

—— Simberloff, D. S. 1976b. Experimental zoogeography of islands: Effects of island size. Ecology 57:629-648. [5, 12]

Simberloff, D. S., and L. G. Abele. 1976a. Island biogeography and conservation: Strategy and limitations. Science 193:1032. [2]

Simberloff, D. S., and L. G. Abele. 1976b. Island biogeography theory and conservation practice. Science 191:285-586. [2, 8]

Simberloff, D. S., and E. O. Wilson. 1969. Experimental zoogeography of islands: The colonization of empty islands. Ecology 50:278-289. [3, 8, 11]

Simberloff, D. S., and E. O. Wilson. 1970. Experimental zoogeography of islands: A two-year record of colonization. Ecology 51:934-937. [3, 7]

Simons, M. M., Jr., and R. H. Simons. 1972. Breeding bird census: Second growth hardwood forest. Am. Birds 26:958. [8]

Simpson, B. B. 1974. Glacial migration of plants: Island biogeographical evidence. Science 185:698-700. [11]

Siniff, D. B., and C. R. Jensen. 1969. A simulation model of animal movement patterns. Adv. Ecol. Res. 6:185-219. [10]

Slack, N. G., S. A. Nicholson, and A. R. Breisch. 1975. Vegetation of the Lake George Islands: Disturbance, diversity and management. Bull. Ecol. Soc. Am. 56:53. [3]

Slatkin, M. 1974. Competition and regional coexistence. Ecology 55:128-134. [10]

Smith, C. C. 1968. The adaptive nature of social organization in the genus of tree squirrels *Tamiasciurus*. Ecol. Monogr. 38:31-63. [10]

Smith, D. M. 1962. The practice of silviculture. Wiley, New York. 578 pp. [6]

Smith, D. W., R. Suffling, D. Stevens, and T. S. Dai. 1975. Plant community age as a measure of sensitivity of ecosystems to disturbance. J. Environ. Manage. 3:271-285. [11]

Smith, F. E. 1972. Spatial heterogeneity, stability and diversity in ecological systems. Trans. Conn. Acad. Arts. Sci. 44:309-335. [12]

Smith, F. E. 1975. Ecosystems and evolution. Bull. Ecol. Soc. Amer. 56(4):2-6. [13]

Snedecor, G. W. 1956. Statistical methods. 5th ed. Iowa State University Press, Ames. 534 pp. [8]

Snedecor, G. W., and W. G. Cochran. 1967. Statistical methods. Iowa State University Press, Ames. 593 pp. [6]

Snyder, N. F. R., and H. A. Snyder. 1975. Raptors in range habitat. pp. 190-209. In Proc. Symp. Management of Forest and Range Habitats for Nongame Birds. General Technical Report WO-1. USDA Forest Service, Washington, D.C. 343 pp. [8]

Sork, V. L., and D. H. Boucher. 1977. Dispersal of sweet pignut hickory in a year of low fruit production, and the influence of predation by a Curculionid beetle. Oecologia 28:289-299. [6]

Sousa, W. P. 1980. The responses of a community to disturbance: The importance of successional age and species' life histories. Oecologia 45:72-81. [8]

Sprugel, D. G. 1976. Dynamic structure of wave-regenerated *Abies balsamea* forests in the north-eastern United States. J. Ecol. 64(3):889-911. [1]

Spurr, S. H. 1956. Forest associations in the Harvard Forest. Ecol. Monogr. 26(3):245-262. [1]

States, J. B. 1976. Local adaptations in chipmunk (*Eutamias amoenus*) populations and evolutionary potential at species borders. Ecol. Monogr. 46:221-256. [6]

Stearns, F. 1965. Present and future status of forests along surface waters of the north-central states. pp. 5-22. In Wood duck management and research. Wildlife Management Institute, Washington, D.C. [3]

Stearns, S. C. 1977. The evolution of life history traits: A critique of the theory and a review of the data. Annu. Rev. Ecol. Syst. 8:145-171. [8]

Stenger, J. 1958. Food habits and available food of Ovenbirds in relation to territory size. Auk 75:335-346. [8]

Stenger, J., and J. B. Falls. 1959. The utilized territory of the Ovenbird. Wilson Bull. 71:125-140. [8]

Sternitzke, H. S. 1976. Eastern hardwood resources: Trends and prospects. For. Prod. J. 24(3):13-16. [1]

Stewart, R. E., and C. S. Robbins. 1947. Breeding bird census: Virgin central hardwood deciduous forest. Audubon Field Notes 1:211-212. [8]

Strong, D. R., Jr. 1979. Biogeographic dynamics of insect-host plant communities. Annu. Rev. Entomol. 24:89-119. [8]

Strong, D. R., J., L. A. Szyoka, and D. S. Simberloff. 1979. Tests of community-wide character displacement against null hypotheses. Evolution 33:897-913. [8]

Sturman, W. A. 1968. Description and analysis of breeding habitats of the Chickadees, *Parus atricapillus* and *Parus rufescens*. Ecology 49:418-431. [3]

Suhrweir, D. E. 1976. Tree species richness: Farm woodlots as biogeographic islands. M.S. Thesis, University of Toledo, Toledo, Ohio. 33 pp. [3]

Sullivan, A. L., and M. L. Shaffer. 1975. Biogeography of the megazoo. Science 189: 13-17. [3, 8, 11, 12]

Swain, A. M. 1980. Landscape patterns and forest history in the Boundary Waters Canoe Area, Minnesota: A pollen study from Hug Lake. Ecology 61:747-754. [1]

Swift, L. W., Jr., and K. R. Knoerr. 1973. Estimating solar radiation on mountain slopes. Agric. Meteorol. 12:329-336. [6]

Tamm, C. O. 1972. Survival and flowering of perennial herbs. III. The behavior of *Primula veris* on permanent plots. Oikos 23:159-166. [7]

Terborgh, J. 1973. Chance, habitat, and dispersal in the distribution of birds in the West Indies. Evolution 27:338-349. [8]

Terborgh, J. 1974. Preservation of natural diversity: The problem of extinction prone species. Bioscience 24:715-722. [2, 8, 11, 12]

Terborgh, J. 1975. Faunal equilibria and the design of wildlife preserves. pp. 369-380. In Golley, F. B., and E. Medina (eds.), Tropical ecological systems: Trends in terrestrial and aquatic research. Springer-Verlag, New York. [2, 8, 12]

Terborgh, J. 1976. Island biogeography and conservation: Strategy and limitations. Science 193:1029-1030. [2]

Terborgh, J. W., and J. Faaborg. 1973. Turnover and ecological release in the avifauna of Mona Island, Puerto Rico. Auk 90:759-779. [3]

Tester, J. R. 1963. Techniques for studying movements of vertebrates in the field. pp. 445-450. In V. Schultz and A. Klement (eds.), Radioecology. Reinhold, New York. [10]

Thomas, W. L. (ed.). 1956. Man's role in changing the face of the earth. University of Chicago Press, Chicago, Illinois. 1193 pp. [12]

Thompson, C. F., and V. Nolan. 1973. Population biology of the Yellow-breasted Chat (*Icteria virens*) in southern Indiana. Ecol. Monogr. 43:145-171. [8]

Thompson, J. N., and M. F. Willson. 1978. Disturbance and the dispersal of fleshy fruits. Science 200:1161-1163. [3]

Thompson, L. S. 1978. Species abundance and habitat relations of an insular montane avifauna. Condor 80:1-14. [8]

Thorhaug, A. 1977. Symposium on restoration of major plant communities in the United States. Environ. Conserv. 4(1):49-50. [12]

Thornbury, W. D. 1965. Regional geomorphology of the United States. Wiley, New York. 609 pp. [3]

Tolle, D. A. 1974. Breeding bird census: Mature oak-hickory woodlot. Am. Birds 28: 1008-1009. [8]

Tramer, E. J., and D. E. Suhrweir. 1975. Farm woodlots as biogeographic islands: Regulation of tree species richness. Bull. Ecol. Soc. Am. 56(2):53. [3, 10]

Transeau, E. N. 1935. The prairie peninsula. Ecology 16:423-437. [1]

Trimble, G. R., and E. H. Tryon. 1966. Crown enroachment into openings cut in Appalachian hardwood stands. J. For. 64:104-108. [6]

Tyburski, J. A., and A. R. Ek. 1977. Forest type mapping accuracy in an urban area using small scale color and color infrared photography. College of Agriculture and Life Sciences. For. Res. Notes No. 204. University of Wisconsin, Madison. 4 pp. [11]

USBC (United States Bureau of the Census). Census of Agriculture. 1964. Vol. 1, Area Reports. Part 14, Wisconsin. Section 2, County Data. U.S. Government Printing Office, Washington, D.C. [3]

USBC (United States Bureau of the Census). Census of Agriculture. 1969. Vol. 1, Area Reports. Part 14, Wisconsin. Section 2, County Data. U.S. Government Printing Office, Washington, D.C. [3]

USDA (United States Department of Agriculture). 1970. Soil survey of Ozaukee County, Wisconsin. U.S. Government Printing Office, Washington, D.C. 92 pp. [3]

USDA (United States Department of Agriculture). 1971. Soil survey of Milwaukee and Waukesha Counties, Wisconsin. U.S. Government Printing Office, Washington, D.C. 177 pp. [3]

United States Bureau of Public Roads. 1965. Standard land use coding manual. U.S. Government Printing Office, Washington, D.C. [12]

Usher, M. B. 1979. Changes in the species-area relations of higher plants on nature preserves. J. Appl. Ecol. 16:213-215. [3, 7]

Van Balgooy, M. M. J. 1969. A study on the diversity of island floras. Blumea 17:139-178. [8]

van der Pijl, L. 1972. Principles of dispersal in higher plants. Springer-Verlag, New York. 161 pp. [10]

Vander Wall, S. B. V., and R. P. Balda. 1977. Coadaptation of the Clark's nutcracker and the Pinon pine for efficient seed harvest and dispersal. Ecol. Monogr. 47(1):89-111. [6, 10]

Veldman, D. J. 1967. FORTRAN programming for the behavioral sciences. Holt, Rinehart and Winston, New York. 406 pp. [3]

Vestal, A. G., and M. F. Heermans. 1945. Size requirements for reference areas in mixed forest. Ecology 26:122-134. [3]

Voelker, A. H. 1976. A cell-based land-use model. ORNL/RUS-16. Oak Ridge National Laboratory, Oak Ridge, Tennessee. 27 pp. and appendices. [11]

Voelker, A. H., M. M. McCarthy, M. L. Newman, R. C. Durfee, S. L. Yaffee, and C. W. Craven. 1974. Spatial disaggregation: Prerequisite to regional simulation. pp. 525-531. In Simulation in perspective. Proceedings of 1974 Summer Computer Simulation Conference. AFIPS Press, Montvale, New Jersey. [11]

Vovides, A. P., and A. Gomez-Pompa. 1977. The problems of threatened and endangered plant species of Mexico. pp. 77-88. In G. T. Prance and T. S. Elias (eds.), Extinction is forever. The status of threatened and endangered plants of the Americas. The New York Botanical Garden, Bronx, New York. 437 pp. [13]

Vuilleumier, F. 1970. Insular biogeography in continental regions. I. The northern Andes of South America. Am. Nat. 104:373-388. [3, 8, 11]

Vuilleumier, F. 1973. Insular biogeography in continential regions. II. Cave faunas from Tesoin, southern Switzerland. Syst. Zool 22:64-76. [8]

Waggoner, P. E., and G. R. Stephens. 1970. Transition probabilities for a forest. Nature 225:1160-1161. [10]

Wales, B. A. 1967. Climate, microclimate and vegetation relationships on north and south forest boundaries in New Jersey. William L. Hutcheson Mem. Forest Bull. 2:1-60. [6]

Wales, B. A. 1972. Vegetation analysis of north and south edges in a mature oak-hickory forest. Ecol. Monogr. 42:451-471. [3, 6, 10]

Ward, J. H. 1963. Hierarchical grouping to optimize an objective function. Am. Statis. Assoc. J. 58:236-244. [3, 4]

Ward, R. T. 1956. The beech forests of Wisconsin—Changes in forest composition and the nature of the beech border. Ecology 37:407-419. [3, 6]

Ware, G. H. 1955. A phytosociological study of the lowland forests in southern Wisconsin. Ph.D. Thesis, University of Wisconsin, Madison. 115 pp. [3]

Watt, A. S. 1947. Pattern and process in the plant community. J. Ecol. 35:12-22. [1, 3]

Weaver, J. E., and Albertson, F. W. 1936. Effects of the great drought on the prairies of Iowa, Nebraska, and Kansas. Ecology 17:567-639. [7]

Webber, M. J. 1976. Elementary entropy maximizing probability distributions: Analysis and interpretation. Econ. Geogr. 52:218-227. [11]

Weber, H. E. 1975. (The reaction of woody species to exposure and hints for a site-adapted choice of species for shelterbelts). Natur und Landschaft. No. 7. University of Osnabruck, German Federal Republic. From Allgemein Forst Zeitschrift 31:924. [6]

Webster, J. D., and D. L. Adams. 1971. Breeding bird census: Old growth bottomland forest. Am. Birds 25:965-966. [8]

Wegner, J. F., and G. Merriam. 1979. Movements by birds and small mammals between a wood and adjoining farmland habitat. J. Appl. Ecol. 16:349-357. [5]

Westhoff, D. 1971. The dynamic structure of plant communities in relation to the objectives of conservation. pp. 3-15. In E. Duffy and A. S. Watt (eds.), The scientific management of animal and plant communities for conservation. Blackwell, London. [12]

Whitaker, G. A., and R. H. McCuen. 1976. A proposed methodology for assessing the quality of wildlife habitat. Ecol. Model. 2:251-272. [11]

Whitcomb, B. L., R. F. Whitcomb, and D. Bystrak. 1977. Island biogeography and "habitat islands" of eastern forests. III. Long-term turnover and effects of selective logging on the avifauna of forest fragments. Am. Birds 31(1):17-23. [8, 12]

Whitcomb, B. L., S. Whitcomb, and R. F. Whitcomb. 1975. Breeding bird census: Upland mixed forest with small creek. Am. Birds 29:1089-1091. [8]

Whitcomb, R. F. 1977. Island biogeography and "habitat islands" of eastern forest. Am. Birds 31:3-5. [3, 12]

Whitcomb, R. F., J. F. Lynch, P. A. Opler, and C. S. Robbins. 1976. Island biogeography and conservation: Strategy and limitations. Science 193:1030-1032. [2, 8]

Whitcomb, R. F., B. L. Whitcomb, and D. Bystrak. 1977. Breeding bird census: Mature tulip-tree-oak forest. Am. Birds 31:91-92. [8]

Whitehead, D. R., and C. E. Jones. 1969. Small islands and the equilibrium theory of insular biogeography. Evolution 23:171-179. [11]

Whitford, P. B. 1951. Estimation of the ages of forest stands in the prairie-forest border region. Ecology 32:143-147. [3]

Whitford, P. B., and P. J. Salamun. 1954. An upland forest survey of the Milwaukee area. Ecology 35:533-540. [3, 4, 6]

Whitford, P. B., and K. Whitford. 1972. An ecological history of Milwaukee County. Milw. Co. Hist. Soc. Historical Messenger 28(2):46-57. [3]

Wiens, H. J. 1962. Atoll Environment and Ecology. Yale University Press, New Haven, Connecticut. [3]

Wiens, J. A. 1976. Population responses to patchy environments. Annu. Rev. Ecol. Syst. 7:81-120. [12]

Wilcox, B. A. 1978. Supersaturated island faunas: A species-age relationship for lizards on post-Pleistocene land-bridge islands in the Gulf of California. Science 199:996-998. [2, 8]

Williams, A. B. 1936. The composition and dynamics of a beech-maple climax community. Ecol. Monogr. 6:317-408. [8]

Willis, E. O. 1974. Populations and local extinctions of birds on Barro Colorado Island, Panama. Ecol. Monogr. 44:153-169. [2, 8, 11]

Willmot, A. 1980. The woody species of hedges with special reference to age in Church Broughton Parish, Derbyshire. J. Ecol. 68(1):269-285. [8]

Wilson, A. G. 1967. A statistical theory of spatial distribution models. Transpn. Res. 1:253-269. [11]

Wilson, A. G. 1971. A family of spatial interaction models and associated developments. Environ. Plan. 3:1-32. [11]

Wilson, E. O., and D. S. Simberloff. 1969. Experimental zoogeography of islands: Defaunation and monitoring techniques. Ecology 50:267-278. [3, 8]

Wilson, E. O., and E. O. Willis. 1975. Applied biogeography. pp. 522-534. In M. L. Cody and J. M. Diamond (eds.), Ecology and evolution of communities. Belknap Press of Harvard University, Cambridge, Massachusetts. 545 pp. [2, 11, 12]

Woodward, J., and P. W. Woodward. 1974. Breeding bird census: Upland tulip-tree-oak forest. Am. Birds 28:1003-1004. [8]

Yahner, R. H. 1972. Breeding bird census: Mixed deciduous forest. Am. Birds 26:951-952. [8]

Yeaton, R. I., and M. L. Cody. 1974. Competitive release in island Song Sparrow populations. Theor. Pop. Biol. 5:42-58. [8]

Zach, R., and J. B. Falls. 1975. Response of the ovenbird (Aves: Parulidae) to an outbreak of the spruce budworm. Can. J. Zool. 53:1669-1672. [8]

Zar, J. H. 1968. Computer calculation of information. Theoretic measures of diversity. Trans. Ill. Acad. Sci. 61:217-219. [4]

Zasada, J. C., and L. A. Viereck. 1970. White spruce cone and seed production in interior Alaska, 1957-68. Research Note PNW-129. USDA Forest Service, Pacific Northwest Forest Experiment Station, Portland, Oregon. 11 pp. [10]

Zedler, P. H., and F. G. Goff. 1973. Size-association analysis of forest successional trends in Wisconsin. Ecol. Monogr. 43(1):79-94. [6]

Zeleny, L. 1976. The Bluebird: How you can help its fight for survival. Indiana University Press, Bloomington. 170 pp. [8]

Index

Ecological Studies